Teaching and Learning Mathematics Online

Teaching and Learning Mathematics Online

Edited by
James P. Howard, II and John F. Beyers

CRC Press
Taylor & Francis Group
Boca Raton London New York

CRC Press is an imprint of the
Taylor & Francis Group, an **informa** business

A CHAPMAN & HALL BOOK

First edition published 2020
by CRC Press
6000 Broken Sound Parkway NW, Suite 300, Boca Raton, FL 33487-2742
and by CRC Press
2 Park Square, Milton Park, Abingdon, Oxon, OX14 4RN

Library of Congress Cataloging-in-Publication Data

Names: Howard, James P., II (James Patrick), editor. | Beyers, John F.,
editor.
Title: Teaching and learning mathematics online / James P. Howard, II, John
F. Beyers.
Description: First edition. | Boca Raton : C&H/CRC Press, 2020. | Includes
bibliographical references and index.
Identifiers: LCCN 2019060115 | ISBN 9780815372363 (hardback) | ISBN
9781351245586 (ebook)
Subjects: LCSH: Mathematics–Computer-assisted instruction–Case studies. |
Mathematics–Study and teaching–Case studies.
Classification: LCC QA20.C65 T42 2020 | DDC 510.71/2–dc23
LC record available at https://lccn.loc.gov/2019060115

ISBN: 978-0-815-37236-3 (hbk)
ISBN: 978-0-815-37233-2 (pbk)
ISBN: 978-1-35124558-6 (ebk)

Typeset in Minion Pro
by Deanta Global Publishing Services, Chennai, India

We dedicate this text to the thousands of students
we have all taught over the years.

Contents

Preface

ONLINE EDUCATION HAS GROWN to be a major component of the higher education market in both the United States and worldwide. Students take entire degree programs online or selected courses online for a variety of reasons. In addition, many instructors have adopted online resources to support face-to-face instruction, either as a supplement or through "flipping-the-classroom," which pushes most instruction outside the traditional classroom. Teaching and learning online introduce a number of complications ranging from effective assessment to managing student engagement.

In mathematics and statistics courses, these complications emphasize already existing difficulties in working with students with math anxiety, the abstractness of the material, and notation.

Teaching and Learning Mathematics Online (TLMO) hopes to bridge these issues and present meaningful solutions for teaching and learning mathematics online. TLMO focuses on the problems seen by mathematics instructors working in the field and provides a set of standard practices which have demonstrated their use and viability to improve the quality of online mathematics instruction.

This book includes chapters that present interactive demonstration techniques, and addresses the divide between students with access to varying levels of technology, familiarity and comfort with mathematical content, and methods for developing reliable and rigorous assessment techniques across digital connections. We believe this will enhance our ability as educators to reach students and successfully instruct in a subject with well-known difficulties.

The objective of TLMO is to provide a set of practice standards for educators teaching mathematics and statistics in the online and hybrid environments. The text presents sound methods for delivering mathematical content to students grounded in both the research and practice of mathematics education. It also presents methods and tools for integrating online material to students in the traditional classroom.

We include chapters that focus on empirical research, theoretical frameworks, or detailed case studies showing what works in the classroom. The book is organized around four key themes.

First, we explore the various aspects of course design. In particular, we have several case studies meant to demonstrate aspects of course design and apply them to different course levels across mathematics and statistics. Second, we look at interaction in online and hybrid mathematics courses.

Third, we look at different technologies, above and beyond the learning management system, to make connections with students. Fourth, we provide two chapters exploring teaching the teacher using online learning. Finally, the book ends with a chapter summarizing the key threads of the text. As a whole, TLMO should provide a framework for developing, implementing, and teaching online mathematics and statistics courses to students across disciplines and skill levels.

The target audience for TLMO are collegiate educators and administrators, though high school and graduate-level instructors may benefit, too. These educators are familiar with mathematics and statistical instruction, but may be new to the online or hybrid classroom. These educators may also be interested in integrating online instruction in the traditional classroom.

Instructors will benefit from learning new techniques and approaches to delivering mathematical content. Their students will benefit from the new techniques presented as students are better able to assimilate and apply the material. Finally, we expect our readers will find better methods to work with students across skill levels.

In addition to practicing educators, TLMO should also appeal to mathematics education researchers and instructors of other natural sciences who must integrate mathematical and statistical content into online and hybrid courses. We hope this volume can be a handbook of methods and practices for teaching excellence across mathematical courses.

MATLAB® is a registered trademark of The MathWorks, Inc. For product information, please contact:
The MathWorks, Inc.
3 Apple Hill Drive
Natick, MA 01760-2098 USA
Tel: 508 647 7000
Fax: 508-647-7001
E-mail: info@mathworks.com
Web: www.mathworks.com

Acknowledgments

WE WOULD LIKE TO thank Mansi Kabra, Saf Khan, Callum Fraser, Rob Calver, and many others at CRC Press for making this book happen. We also appreciate the dedication and commitment of our contributors and authors around the world. Finally, we want to thank our families and colleagues for supporting us in this project.

James P. Howard, II and John F. Beyers

Editors

James P. Howard, II is a scientist at the Johns Hopkins Applied Physics Laboratory. Previously, he worked for the Board of Governors of the Federal Reserve System as an internal consultant on statistical computing. He has also been a consultant to numerous government agencies. Additionally, he has taught mathematics, statistics, and public affairs since 2010. He has a PhD in public policy from the University of Maryland Baltimore County.

John F. Beyers, PhD, is program chair and professor of mathematics and statistics at the University of Maryland Global Campus (formerly UMUC). Dr. Beyers leads a global department of 200+ faculty to serve the educational needs of over 20,000 non-traditional, underrepresented undergraduate students globally. He is responsible for new program development, curriculum planning, teaching effectiveness, and Learning Outcomes Assessment. Prior to his current position, Dr. Beyers was the Associate Director for the Center of Distance Education at Johns Hopkins University.

Dr. Beyers has an extensive academic background as a faculty member and leader of one of the largest online mathematics and statistics departments in the country. While earning a PhD in mathematics education from American University (his dissertation was the first to earn "pass with distinction" in over two decades), Dr. Beyers began his career in innovative education as Research Coordinator on the NCTM *Standards 2000* Project, which resulted in the national standards document *Principles and Standards for School Mathematics* (NCTM, 2000). He later worked with post-secondary students and faculty at Johns Hopkins University's Center for Distance Education on innovative distance education models to determine how difficult/easy it is for faculty to embed or integrate technologies into a course and in harmony with their pedagogic model.

Dr. Beyers has been recognized for his efforts as an innovative academic leader in higher education. In 2012, he received the UMUC Presidential Award and the University System of Maryland's Faculty Fellowship Award from the Chancellor's Office. In 2007, he received the Adelle F. Robertson National Educator of the Year Award (UPCEA) and in 2006, he received the Alexander Charters Mid-Atlantic Region Educator of the Year Award (UPCEA).

Dr. Beyers has published and presented extensively on the topics of course redesign, adaptive and accelerated learning models, distance learning, and mathematics education.

His current interest is focused on how adaptive learning can be used to deepen the quality of faculty–student engagement and enable development of higher-order thinking skills.

Dr. Beyers has an extensive network of academic colleagues and adaptive learning vendors to establish an *Adaptive Resources Community (ARC)*.

Contributors

Jim Albert
Bowling Green State University

Chiara Andrà
Università del Piemonte Orientale

Laurie Battle
Montana Technological University

Domenico Brunetto
Politecnico di Milano

Mine Çetinkaya-Rundel
University of Edinburgh

Larry Copes
*Institute for Studies in Educational
    Mathematics*

Jessica Deshler
West Virginia University

Allison Dorko
Oklahoma State University

Sarah Ferguson
Old Dominion University

Helen Forgasz
Monash University, Australia

Edgar J. Fuller
Florida International University

Stephan Ramon Garcia
Pomona College

K.G. Garegae
University of Botswana

Jennifer Hall
Monash University, Australia

Jingchen Hu
Vassar College

Harpreet Kaur
British Columbia Institute of Technology

S. Kesianye
University of Botswana

Shay Kidd
University of Montana Western

Igor' Kontorovich
The University of Auckland

Bhesh Raj Mainali
Rider University

Glenn F. Miller
Borough of Manhattan Community College

Steven J. Miller
Williams College

Atish J. Mitra
Montana Technological University

S. Mogotsi
University of Botswana

Reneé H. Moore
Emory University

Deborah Moore-Russo
University of Oklahoma

M.J. Motswiri
University of Botswana

A.A. Nkhwalume
University of Botswana

Kathleen H. Offenholley
Borough of Manhattan Community College

H. Smith Risser
Montana Technological University

Mina Sedaghatjou
Alfred University

Paul Seeburger
Monroe Community College

Zohreh Shahbazi
University of Toronto Scarborough

Charles E. Smith
North Carolina State University

Michael A. Tallman
Oklahoma State University

Eugenia Taranto
University of Turin

Rosaura Uscanga
Oklahoma State University

Monica VanDieren
Robert Morris University

Kimberly S. Weems
North Carolina Central University

Kelly A. Williams
Alfred University

E. Zimudzi
University of Botswana

Simone Zmood
Monash University, Australia

PART 1

Course Design

CHAPTER 1

Teaching Cross-Listed Mathematics Courses Online

Laurie Battle, Atish J. Mitra, and H. Smith Risser

CONTENTS

1.1 INTRODUCTION

This purpose of this chapter is to provide a detailed description of three cross-listed (combined upper-level undergraduate and graduate) classes that were taught online in the summer of 2017 at Montana Tech. All three courses were part of a funded project designed to prepare current high school teachers for teaching Dual Credit Mathematics Classes. Current state-level policies require all Dual Credit teachers within the state of Montana to have a Masters degree and 9 credits of graduate-level math. The rural nature and size of Montana are barriers for current teachers wishing to take face-to-face graduate courses.

As a result the funded project was designed to support teachers taking online graduate-level mathematics. *Mathematical Modeling* and *Groups and Geometry* were previously taught face-to-face to upper-level undergraduates. *Advanced Linear Algebra* was designed specifically for the project. However, some topics (vector spaces and numerical methods for matrix algebra) were previously taught by the instructor in other face-to-face courses within the undergraduate curriculum.

As mentioned above, all three classes had both undergraduate and graduate students enrolled. The students in the undergraduate section were typically current undergraduate students majoring or minoring in mathematics. The students in the graduate section were typically adults working in K-12 or higher education. The cross-listed nature of the courses, along with the online delivery of the materials, provided unique challenges. Each of the courses had to accommodate different expectations in learning objectives and different approaches in evaluation. For example in *Advanced Linear Algebra*, most of the graduate students needed substantial review of the prerequisite material in order to succeed. Similarly in *Groups and Geometry* some of the students did not have a previous background in group theory. This required the instructor to introduce the concept of a group in the special case of groups of transformations. In this chapter, we discuss how we dealt with challenges like assessment, technology use, ensuring rigor, and providing support to students at a distance. This paper would be relevant to any instructor wishing to teach graduate courses or dual-enrollment courses online, especially those including graduate students with nontraditional preparation.

1.2 THE COURSES

1.2.1 Groups and Geometry

The face-to-face version of *Groups and Geometry* was not cross-listed. It was intended for advanced undergraduates, typically seniors, who had already completed a course in abstract algebra. *Groups and Geometry* introduced students to the unity of abstract mathematics by demonstrating interconnections between geometry and algebra. Additionally, it guided the students into developing and communicating mathematical proofs in both oral and written forms.

Groups and Geometry was developed as a natural successor to a year-long sequence of abstract algebra. The instructor assumed that the student had at least a rudimentary background and understanding of algebraic structures, and guided the student through an exploration of various geometries and their algebraic connections. *Linear Algebra* was the formal prerequisite of the course. The course met twice a week for a total of three hours.

The face-to-face section started with a gentle review of the algebra of complex numbers and its connections to plane Euclidean geometry. This topic took about three weeks. By the end of this first topic the students would prove results in Euclidean geometry, starting from simpler propositions such as the concurrence of medians/angle bisectors of a Euclidean triangle, to more esoteric aspects such as the fact that the adjacent trisectors of the angles of an arbitrary triangle form an equilateral triangle. At this stage the proofs largely required only a level of comfort with manipulating algebraic expressions involving complex numbers, and understanding their geometric significance.

Once students had a good understanding of the application of complex numbers to Euclidean geometry, the concept of isometries of the complex plane was introduced with the aim of a complete classification of the isometries of the Euclidean plane. This part of the course lasted about seven to eight weeks and was the heart of the course. Here the level of the proofs became more conceptual than the algebraic manipulations in the first part of the course; students were able to see that non-trivial geometric structures can be reduced to study of algebraic structures. Here the group structure of the set of Euclidean isometries was introduced, and several non-trivial results, such as a study of the subgroups of the Euclidean isometry group, could be reached at this stage.

In the final part of the course the students saw how an algebraic structure such as a finitely generated group could be viewed as a metric space. This part lasted about two to three weeks and was a gentle introduction to geometric group theory.

The grade for the face-to-face version of the course was based on homework assignments and exams. The homework assignments typically were extensions of concepts discussed in class.

1.2.2 Advanced Linear Algebra

Advanced Linear Algebra introduced students to theoretical concepts from algebra using matrices. The first part of the course focused on building a theoretical understanding of algebra and vector spaces. Then the course showcased applications of linear algebra to other disciplines (e.g. the singular value decomposition applied to regression). Finally, students learned how numerical methods are applied to problems described within the course. The instructor had previously taught some of the topics in face-to-face *Linear Algebra* and *Numerical Computing* courses.

The face-to-face *Linear Algebra* class had a prerequisite of *Calculus 2* and included elementary row operations, vector spaces, and eigenvalues. The class was taught in a lecture format and met three hours per week. Students were expected to complete homework assignments from the textbook. These homework assignments included both computational work and short proofs. Students were also assessed through face-to-face examinations. The questions on the examinations were similar to those completed in the homework assignments.

The *Numerical Computing* class had a prerequisite of either *Linear Algebra* or *Differential Equations*. The curriculum included a unit on numerical linear algebra. Both direct and iterative methods for solving systems of equations were covered. The class was taught in a lecture format and met three hours per week. The homework assignments required students to use MATLAB® [1] to solve systems of linear equations. Homework assignments were completed in groups. Each group analyzed the errors and floating point operations for different types of matrices (e.g. ill-conditioned, banded). The homework assignments required students to compare results for the different methods. Students were also assessed using face-to-face examinations which required students to solve questions by hand and to analyze errors produced by numerical algorithms.

1.2.3 Mathematical Modeling

The *Mathematical Modeling* course taught students how to devise and analyze models in the form of difference equations as well as differential equations, including systems of both types of equations. Applications were taken from a variety of fields with an emphasis on population modeling. Students learned methods for finding analytical and numerical solutions, as well as geometric representations, including phase lines and phase planes. Spreadsheet software was the primary technology that students used to analyze difference equations numerically. Matrix calculators were used to assist with eigenvalue and eigenvector analysis of differential equations. The instructor had previously taught this as a face-to-face class, but the content was modified for the online version. The face-to-face class focused on differential equations only. Methods of analysis included finding equilibrium solutions, stability analysis, and finding both analytical and numerical solutions. MATLAB was the primary technology used. The content for the online course included difference equations in addition to a less in-depth selection of topics in differential equations. The primary technology used for the online course was spreadsheet software. The face-to-face version required *Differential Equations* as a pre-requisite.

The face-to-face version met three times per week for 50 minutes. This time was used primarily for lectures and exams. The types of assessment included homework, three quizzes, three exams, three group projects, and a final exam. The homework exercises came from the textbook and were collected weekly. Students worked in groups of two to three on the projects, and each group selected a project from a list of several options. Some of these projects came from the textbook, and some were written by the instructor. Each group prepared a written report and gave a presentation to the class. Students were asked to use MATLAB on homework and group projects, but they did not have access to MATLAB for quizzes and exams. As a result, the quizzes and exams tended to emphasize theory while the other assignments explored practical analysis in addition to theory. The face-to-face course was also cross-listed for both undergraduate and graduate levels. The graduate students attended the same lectures, but they were assigned additional homework problems, additional questions on quizzes/exams, and more in-depth projects.

1.3 ADAPTING THE COURSES TO ONLINE DELIVERY

Each of the three courses presented unique challenges for the instructor in conveying abstract material, using appropriate symbolic notation, and integrating appropriate activities using online delivery.

1.3.1 Groups and Geometry

The online version of *Groups and Geometry* had five students. The small size of the section meant that no TA was needed, and the instructor was able to provide individual attention to the progress of all the students. However, the design of the course allows larger class sizes without substantial change.

The students who took the face-to-face version of the course were usually well-prepared by the prerequisites; they all had taken a two-semester course in abstract algebra (many of them with the same instructor as *Groups and Geometry*) and were quite comfortable with the constructs of basic group theory. This was clearly not the case for the online course; only a couple of the students were undergraduate students at the university. In fact, many of the students took the course while living in other states across the country. In the online course it was assumed that some of the students would not know the definition of a group. Within the first few days of class the students had to learn the concept of an algebraic structure. To make this transition smooth, the instructor decided to introduce groups as groups of transformations instead of structures satisfying abstract axioms as in a usual abstract algebra class. This approach helped the students become accustomed to group theoretical concepts such as subgroups, center of a group, normality of subgroups, and group homomorphisms from the geometric point of view.

The second challenge was something that arises when adapting any traditional course of geometry to an online platform: pictures. This issue was more prominent in this course, where the main idea was to search for algebraic structures in familiar transformations of Euclidean geometry. Each of the recorded lectures contained many pictures and sketches. This did not pose any real problem, as the (asynchronous) lectures were prerecorded in a smart classroom (see Figure 1.1) where both simple free-hand sketches and more complicated Geogebra [2] constructed images were easy to insert. In the synchronous part of the course (e.g. regular online office hours), the online educational platform WizIQ [3] was very convenient for sketching pictures.

FIGURE 1.1   Picture of smart classroom.

Finally, as mentioned earlier, the face-to-face course was not cross-listed. As the online course was cross-listed, it was necessary to have extra material for the graduate students both in the synchronous and asynchronous parts. In the asynchronous part this was achieved by having separate pre-recorded lectures available for the graduate students (in addition to the common lectures for both undergraduate and graduate students). As an example, one of the special lecture sessions for graduate students had a complete description of the isometry group of the reals. For the synchronous part there were separate online office hours available for the graduate students where some sample problems were worked out on the extra material presented for graduate students. Again, the graduate students had separate office hours which used a sequence of problems solved by the students to explore the properties of various discrete subgroups of the isometry group of the Euclidean plane. This material and the related extra problem sets were not included in the material for the undergraduate students.

1.3.2 Advanced Linear Algebra

Five students enrolled in *Advanced Linear Algebra* the first time it was offered. Approximately half were undergraduates and half were graduate students. There was no teaching assistant for the course. The assignments and lectures were structured in a way to allow a larger enrollment without creating significantly more work for the instructor.

Whereas all of the graduate students did have a satisfactory prerequisite in matrix arithmetic, most had taken such a course more than ten years before attempting *Advanced Linear Algebra*. The graduate students needed some review of the prerequisite material. There were two ways in which this prerequisite review was provided. First, each section had a list of prerequisite skills for that section and links to open educational resources that could be used for review of these skills. The open educational resources included both videos and textbook materials. Second, students were able to ask questions on prerequisite material via a free online question-and-answer platform called Piazza [4]. This platform enabled students in the course to ask questions both of the instructor and other students in the course.

Neither of the face-to-face courses on which this class was based were cross- listed. While the cross-listing required the instructor to modify assignments for the graduate students, the cross-listing provided a unique advantage; the undergraduate students in the course had all recently completed the prerequisite. This enabled undergraduates to answer questions concerning prerequisite material that were posed by students in the graduate section.

Another issue faced in adapting the face-to-face course online was the availability of technology. The undergraduates in the *Numerical Computing* class all had access to MATLAB via the campus computer labs. In *Advanced Linear Algebra* many of the students did not have access to MATLAB. Instead of MATLAB, students used a cloud-based computational engine called SAGE [5]. Students in the course were provided with both manuals and video tutorials for SAGE. Many of the lectures included step-by-step instructions for using SAGE to perform computations.

Both the face-to-face *Linear Algebra* and *Numerical Computing* classes included formative assessments of student understanding. During class meetings students worked problems related to course content. These problems were not graded. Instead, the results were used to assess student understanding of the course material. In *Advanced Linear Algebra* the instructor included questions between sections of the videos. Students were asked to work these questions and to submit answers via the learning management system (LMS). The instructor was able to assess student understanding of the course material from answers to these questions.

1.3.3 Mathematical Modeling

During the project, approximately 20 students enrolled in *Mathematical Modeling*, the majority of whom were undergraduates. There was no teaching assistant for the course.

The instructor originally adapted this course to be cross-listed with the face-to-face version. The undergraduate and graduate students attended the same lectures, but there were some differences in assessment. In particular, the graduate students were assigned some additional problems in each of the homework assignments. These additional problems often required exploration of generalized abstract concepts. In addition, the quizzes and exams for the graduate students included some modified problems and some additional problems to assess student mastery of more advanced concepts. The projects assigned to the graduate students were completely different than those given to the undergraduates. The graduate students' projects require more advanced and more in-depth analysis. This same system was used for the cross-listed online course. In contrast to the other two courses, all of the graduate students in *Mathematical Modeling* had taken the prerequisite courses and had up-to-date mathematical backgrounds, so additional review materials were not necessary. Several modifications were made in adapting from a face-to-face to an online course, and technology is an important factor in these changes. The face-to-face course required students to use MATLAB for numerical solutions to differential equations. Since many of the off-campus students enrolled in the online version did not have access to MATLAB, the instructor chose to use spreadsheet software as the primary technology. However, this type of software did not include numerical approximation methods for differential equations, so the topics were modified to accommodate the new technology. The first half of the course dealt with modeling using difference equations, for which spreadsheet software is well-suited. The second half of the course dealt with modeling using differential equations, but the focus was shifted to a more theoretical rather than numerical approach. Little technology was used in the second half, although spreadsheet software was used to perform matrix operations. Using this type of software has the advantage that most students are familiar with at least some basics of spreadsheets. The instructor provided documents explaining some of these basics for students with less experience, and more advanced techniques were covered in the recorded lectures.

The online class was structured in four modules, and each module included a reading assignment, recorded lectures, one homework assignment, one quiz, and one project. It was taught during a five-week summer session, so each module was just over one week. The balance of work was distributed differently for the online version, including fewer hours of lectures and more time on independent work. To offer flexibility to the online students, most of the learning was asynchronous. All lectures were recorded in the Smart

Classroom™ for students to watch at their convenience. The only synchronous component was an optional discussion session offered once during each module, in which students had an opportunity to ask questions and the instructor reviewed important concepts. These sessions were recorded and posted for students who could not attend. The instructor also scheduled daily office hours as an additional opportunity for students to ask questions. The instructor posted a summary of work at the beginning of each module to make sure students were aware of the reading, videos, and assignments. This summary included suggestions for pacing their work over the module to ensure all assignments were submitted before the deadline.

The types of assessment were similar for the online and face-to-face courses, both including homework assignments, quizzes, and projects. The face-to-face course included three exams and a final exam. All of these exams were dropped for the online course, due to the difficulty in proctoring. Instead more emphasis was placed on the projects. This project-focused approach was well-suited for the applied nature of this course. For the online version the number of projects was higher, the projects were more in-depth, and the work was done independently rather than in groups. For both the online and face-to-face courses, students were allowed to consult each other, as well as other resources for homework and projects. For the face-to-face section, quizzes and exams were closed-book independent work administered during the class period. The quizzes were open-book for online students, but students were informed that they should not consult any people about the quizzes. The online course had a much smaller number of homework assignments (four in the online and 12 in the face-to-face), but each of these assignments was more extensive. This adaptation was due mainly to the course being offered in the summer rather than due to being offered online.

1.4 TECHNICAL ISSUES

1.4.1 Creating the Lectures

In a typical mathematics lecture, the material includes not only theorems, proofs, and their applications, but is typically supplemented by worked examples. In an ideal classroom lecture the instructors narrate the steps along with their writing on the board. This style of presentation can be challenging to adapt to online environments. All three classes discussed here used technology to create video-based lectures. These video-based lectures combined worked proofs and problems with an audio narration of the steps in the process.

For two of the courses (*Groups and Geometry*, and *Mathematical Modeling*), Montana Tech's Smart Classrooms were used (see Figure 1.1).

These smart classrooms allowed instructors to record their lectures, both the active whiteboard and the audio/video accompaniment. As the lectures are recorded without requiring technical support staff, instructors were able to record the lectures at their convenience. Once a recording was completed to the satisfaction of the instructor, the links for these pre-recorded lectures and lecture notes were made available for registered students on the LMS. This allowed the students in the course to attend the lectures asynchronously.

For the *Advanced Linear Algebra* course, pencasts were created using a platform called Explain Everything [6]. A pencast is a video that combines handwritten notes with audio.

The lecture slides for the pencast (created in LaTex) included important definitions, example problems, and diagrams. The PDF of the slides was then imported into the Explain Everything iPad app. The instructor could then write on the slides using an Apple Pencil and record audio narration. Unlike other pencast software, Explain Everything allowed the pencast for each slide to be recorded separately. After the pencasts for all of the slides were recorded, the entire set was exported as a single video. After the course had begun the instructor discovered that many of the graduate students needed more detailed review of prerequisite concepts during the lectures. Because of the features of Explain Everything the instructor could insert additional examples into the lectures without having to splice together additional video files.

1.4.2 Online Interactions with Students

In the face-to-face courses, the instructors primarily used the LMS to communicate information to students between class meetings and to share files with students. Student homework questions were typically addressed either during class meetings or during face-to-face office hours. With our online students we all had to find a way to address student misconceptions and questions at a distance. In online mathematics courses the heavily symbolic notation and necessity of pictures create a barrier to efficient and effective communication with students. We used several different tools to communicate with students in our online courses. The tools fell into two generic types: asynchronous and synchronous tools.

1.4.2.1 Asynchronous Tools

In all three courses, students would submit pictures or PDF scans of work on the assigned problems by email or through the LMS and ask for feedback. Many of the assigned practice problems required several pages of written work to complete. Using this method, it was possible for the instructor to locate arithmetic or process errors in a student's written work and respond. This type of feedback helped students to understand their mistakes and misconceptions. It also made it possible for a student to show the instructor partial work on a problem and to ask a question concerning how to proceed. One instructor used an iPad app to mark up the student's written work. The other two used Wacom tablets [7] to mark-up student work. Once the written work was marked, it was returned to the student by email or via the LMS.

The *Advanced Linear Algebra* course also used an additional method of asynchronous communication: an online text-based discussion tool called Piazza [8]. In Piazza, students asked questions not only of the instructor, but also of other students in the course. One of the primary benefits of Piazza over the discussion board in the LMS was the ability of students to post questions anonymously. This feature allowed students to ask questions without any risk of embarrassment. The second feature that made Piazza more desirable than the built-in discussion board in the LMS was the ability to use a LaTex-based equation editor to add symbols to the post. The LMS did not include the option of adding equations and symbols to discussion posts. During the course, a small number of students posed questions to the class. Primarily the students used Piazza to ask questions of the instructor. This was likely due to the small size of the course. In a larger course, Piazza might have been used for student-to-student interactions more frequently. Most of the questions were either on prerequisite material or on SAGE. Students in the course mostly used online office hours or email to ask questions concerning material.

1.4.2.2 Synchronous Tools: Office Hours

The instructors of *Groups and Geometry* and *Mathematical Modeling* used the online platform WizIQ [3] to offer virtual office hours. It was also used for the weekly discussion sessions for *Mathematical Modeling*, which were recorded and posted for students who missed the live session. This platform allowed students to log in remotely and engage in a face-to-face discussion with the instructor. WizIQ included an onscreen whiteboard which not only the instructor could use, but (with the instructor's permission) the students could use also. Wacom tablets [7] were used by the instructors for these two courses during online office hours. These tablets, when connected to the instructor's computers, worked as de-facto extended whiteboards which the instructors could use to communicate with the students.

In *Advanced Linear Algebra*, the synchronous interactions occurred via the videoconferencing tool Zoom [9]. During a session on Zoom, the instructor and students could share screens with one another or write on a whiteboard (see Figure 1.2).

The sessions could also be recorded and posted to the LMS for students who were not able to attend. Zoom was used primarily to host online office hours each week. During these online office hours, students could ask questions of the instructor in real-time. Student questions in the online office hours were similar to those asked in face-to-face office hours. Students asked the instructor to work additional examples, discuss graded work, and answer questions about assignments. In addition to the scheduled office hours, which were open to all students in the course, students were able to schedule one-on-one office hours with the instructor as well.

1.4.3 Assessment

One of the primary challenges faced in all three online courses was adapting assessments for online students. All three courses had previously used face- to-face exams as one way to

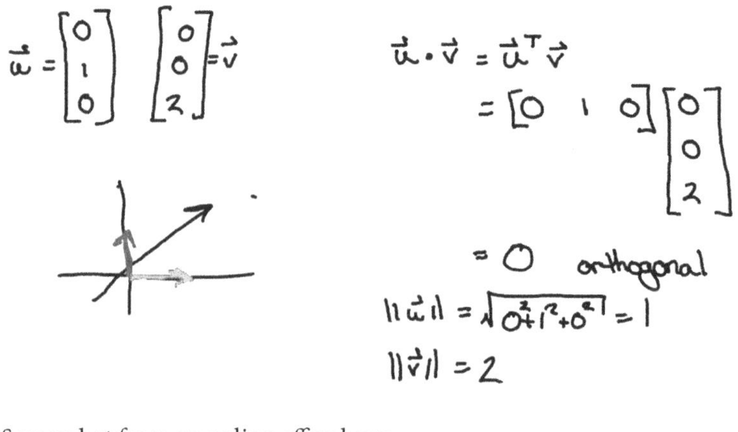

FIGURE 1.2 Screenshot from an online office hour.

assess student mastery of learning objectives. Proctoring exams for online students posed a significant challenge.

In *Groups and Geometry*, multiple students took the course from out of state, and two students from out of the country. The course grade was calculated using homework assignments and exams. To make the assessment fair, the exams (both midterm and final) had a substantial oral part. Each student had an individual assigned time to take a face-to-face oral exam over WizIQ [3] with the instructor, where the student could answer the instructor's questions using the whiteboard on WizIQ. This oral component was effective in assessing the student's understanding of the material.

All assignments for *Mathematical Modeling*, except the quizzes, were posted at the beginning of each module. Each quiz was posted at 8:00 am on the final day of each module and was due by 5:00 pm on that same day. For all assignments, students had the option to type or write solutions by hand. Most students chose to write by hand because of the time required to type equations. When students worked problems by hand, they could either scan or take a picture of their work. Students were required to upload their work to the LMS. The instructor used a Wacom to write comments on the student documents, which were posted back to the students through the LMS gradebook.

In *Advanced Linear Algebra*, all of the students completed weekly home- work assignments. The undergraduate students took a midterm and final examination. The instructor had planned to arrange proctors for any under- graduate students who were not able to physically travel to the university. However, all of the students enrolled in the course that summer were able to come to the campus to take the exams. The graduate students were assigned weekly projects. The graduate students chose one of the projects for the final exam. Each graduate student scheduled an individual online meeting with the instructor during the last week of the course. At this online meeting, the instructor conducted an oral examination of the student. This oral examination counted as the final exam in the course. The oral component was effective in assessing the understanding of the graduate students in the course.

1.4.4 Ensuring Rigor

One of the challenges faced in adapting all of the courses to the online environment was ensuring that the rigor of the original course was not diluted. All three face-to-face courses had closed-book proctored assessments. In an online environment, students have access to material that would not be allowed in a face-to-face examination. While it was not possible to give the same types of assessments in the online environment, we wanted to ensure that the rigor of the course was preserved. In order to ensure rigor, we took several different approaches. One approach was to set a strict time limit for online assessments.

This strict time limit ensures that students cannot get too much help from outside sources. A strong mastery of the subject matter helps students to complete the assignment within the time limit. In some cases, we gave our online students more challenging and less routine problems than we would have given in a face-to-face examination or quiz. Two of us also gave oral examinations in order to ensure that the online students had a strong mastery of the course content. The *Modeling* online course placed a stronger emphasis on projects and less emphasis on quizzes and exams compared to the face-to- face course. By assigning more in-depth projects, rigor could be maintained while focusing on assignments that allow the use of outside resources.

1.5 CONCLUSIONS

This chapter discusses some of the unique challenges faced by the authors in teaching three online cross-listed courses. The three courses discussed were from a wide cross-section of mathematics. While there are obvious differences between the course structures, instruction techniques, and evaluation methods of each course, there are noticeable similarities also. The three instructors all faced similar issues while designing the courses. One such issue is the cross-listed nature of the courses, where the basic material had to be presented at a level suitable for the entire group, and at the same time more advanced material had to be presented to the graduate students. One other recurrent theme that all three instructors faced was the challenge of adapting material from a traditional classroom to online delivery, where the theoretical material presented in the lectures had to mesh seamlessly with the various software used. A third common issue in all three courses was the careful choice of assessment methods. New assessment methods were designed to be effective for online delivery, and additional assessment materials were created for the graduate students in each course. All three instructors also needed to determine an appropriate balance of synchronous and asynchronous learning, and to select effective platforms for both types of learning. These platforms were selected to facilitate online learning and to allow for some shared and some different materials for the undergraduate and graduate students.

What comes out from the common experience of teaching these three classes are some of the recurring issues instructors need to consider when designing a course that is both online and cross-listed.

1. How will the course content need to be adapted for online students? For all three of our courses, the online students often had different mathematical preparation, access to campus, and time constraints than the students typically served by the face-to-face course. Many of our online students were not taking courses full time at our university. Not all of the students had access to university computer labs or campus resources like face-to-face office hours. In addition, some of our online students were working full-time while taking the course. This meant that we had offer online help in the evenings as well as during the day. We had to consider how to meet the needs of students that were different from those served in our face-to-face courses.

2. What technology will be needed by the students and the instructor? In all three courses, the instructors had to find technology that would allow both the creation of asynchronous lectures and the synchronous interactions with students. We used a variety of technologies including Smart Classrooms, tablets, and software. We also used different types of technology to create lectures than we used to interact synchronously with students. In synchronous interactions, it was very valuable to have some sort of virtual whiteboard that both students and the instructor could write on. In

the asynchronous lectures, either a physical or virtual whiteboard would work. It was also necessary to determine what technology students would be able to access from off campus. The instructors for all of the courses relied on freely available software like Geogebra and SAGE or common office software instead of the specialized software used in the face-to-face course.

3. How will students be assessed at a distance? In all three courses, instructors modified some or all of the assessment methods for the course. Typical proctored exams were often impractical. Instead all of the instructors used other means like projects or online oral examinations to measure student learning. The instructors also had to consider what types of assessments would be appropriate for graduate students versus undergraduate students.

4. What support will students need? In all three courses, the instructors had synchronous online meetings for students to ask questions and get help. As the distance students did not have access to on-campus support like tutoring services, the instructor for each course had to provide regular and extensive individual assistance on course material. Even though the lectures were provided asynchronously, all of the instructors provided opportunities for students to ask questions synchronously. The students appreciated these online office hours and help sessions immensely.

In conclusion, online cross-listed mathematics courses can be quite successful in today's learning environments as such courses offer students greater flexibility and allow the course to be offered to a substantially larger audience—as compared to traditional face-to-face courses. However to make such a course successful involves considerable effort and involvement of the instructor in terms of course design, preparation of online lectures, use of appropriate and conveniently available software, use of suitable assessment techniques, and willingness to be available for extensive (and often individual) virtual office hours.

BIBLIOGRAPHY

1. www.mathworks.com
2. www.geogebra.org
3. www.wiziq.com
4. piazza.com
5. www.sagemath.org
6. www.explaineverything.com
7. www.wacom.com
8. www.piazza.com
9. www.zoom.us

CHAPTER 2

What Do We Know about Student Learning from Online Mathematics Homework?

Allison Dorko

CONTENTS

*I*N THIS CHAPTER *I summarize the literature about university students' learning from online mathematics homework and use it to inform a list of considerations for instructors utilizing online mathematics homework systems in their courses. The chapter is aimed both at researchers and practitioners. Sections 2.1–2.4 are a literature review that will probably be primarily of interest to researchers; Section 2.5 is a list of research-based practices aimed at instructors looking to use online homework platforms more efficaciously.*

Educators largely agree that homework plays an important role in students' learning of mathematics. Further, research findings indicate that university calculus I students spend more time doing homework than they do in class (Ellis, Hanson, Nuñez, & Rasmussen, 2015; Krause & Putnam, 2016). As such, homework accounts for the majority of students' interaction with mathematics content. It follows that the choices instructors make regarding homework can greatly affect student learning. These choices include selecting problems, deciding which problems (if any) will be graded, determining the type of feedback provided in grading, selecting due dates, and so on. The increasing availability of online homework platforms presents instructors with another important decision: whether to assign online homework, traditional paper-and-pencil homework, or a combination of the two. Once an instructor has chosen to use an online platform for all or part of the course homework, there are more decisions to make within the online homework platform. How many attempts per problem will students have? Must students submit an entire question at once, or can they submit each part of a multi-part question individually? Should "see similar example"-type features be enabled?

The body of education research about online mathematics homework can help instructors make evidence-based decisions about homework in their courses. To that end, in this chapter I summarize the literature about student learning from online mathematics homework and provide a list of implications for instruction. For the purposes of this chapter, *online homework* refers to homework students access online and submit answers to online. In online homework (for the purposes of this chapter), students receive feedback immediately after submission. "After submission" may differ by platform and instructor choice; students might have to submit a whole assignment at once,* or might be able to submit each question (or part of a question) individually. Examples of online homework platforms include WeBWorK, WebAssign, and MyMathLab. *Paper-and-pencil homework* refers to homework students submit on paper; typically this homework would be a subset of problems from a textbook.

In Sections 2.1–2.4 of the chapter, I review the literature about online homework and student achievement, students' perceptions of online homework, and what we know about how students engage with online homework systems. These emerged as the main themes in the set of studies reviewed.[†,‡] In Section 2.5, I list evidence-based practices for instructors to consider when utilizing online mathematics homework systems in their courses. In the final section, I provide concluding remarks and suggestions for future research.

2.1 ONLINE HOMEWORK AND STUDENT ACHIEVEMENT

A key question about online homework is how it compares to paper-and-pencil homework in terms of student achievement. Researchers have focused on such comparisons because while online homework benefits instructors (e.g., decreased grading time), it might not be worthwhile if it negatively impacts students' grades. Hence much research has focused on comparing online homework to paper-and-pencil homework. Researchers generally agree that online homework is at least as effective as paper-and-pencil homework in terms of student achievement (e.g., Babaali & Gonzalez, 2015; Burch & Kuo, 2010; Hauk et al. 2015; Halcrow & Dunnigan, 2012; Hauk & Segalla, 2005; Hirsch & Weibel, 2003; LaRose, 2010; Lunsford & Pendergrass, 2016; Lenz, 2010; Weibel & Hirsch, 2002; Zerr, 2007). Researchers

---

* Some authors use the phrase "online quizzes" to describe out-of-class, online assignments to which students submit all answers at once, receive immediate feedback, and can re-take multiple times over a period of several days (e.g., Suzuki, 2003). This definition of "online quiz" fits my definition of "online homework," and I use the phrase "online homework" throughout the chapter.

† The body of literature reviewed includes educational research about online mathematics homework at the undergraduate level published since 1996. I chose 1996 as the cutoff date because WeBWorK, one of the first online homework systems, was introduced in this year. In all research reviewed for the chapter, students attended class in person. I did not find any studies about online homework for mathematics courses that are entirely online; this could be an area for future research.

‡ Several studies reviewed contained findings about how online homework changes teaching. These findings fall outside of the scope of this chapter. Readers are encouraged to see Hauk and Segalla (2005) and LaRose (2010) for more information.

have investigated relationships between homework and student success by considering exam scores, course grades, homework completion rates, and homework grades. I synthesize these findings below.*

2.1.1 Exam Scores

Many studies have compared students' exam scores across sections with different homework formats. By *format* I mean whether homework is done online, with paper and pencil, or a combination of the two. *Exam scores* refers to preliminary exams, final exams, or pre-/post-test scores, depending on the study.

In all but one of the studies reviewed for this chapter that used exam scores as a metric, researchers found exam scores for students with online homework were slightly higher than or the same as the scores for students with paper-and-pencil homework (Hauk, Powers, & Segalla, 2015; Hauk & Segalla, 2005; Halcrow & Dunnigan, 2012; Hirsch & Weibel, 2003; Lenz, 2010; Mathai & Olsen, 2013; Weibel & Hirsch, 2002; Zerr, 2007). Only Lenz (2010) found the opposite result. Lenz (2010) found that students with paper-and-pencil homework had a higher mean exam score than students with both paper-and-pencil and online homework.

Hirsch and Weibel (2003) found that students with online homework and paper-and-pencil homework scored significantly higher on the final exam than students with just paper-and-pencil homework ($p < 0.05$). The researchers divided calculus students into a control group (n = 296) and an experimental group (n = 715). Both groups were assigned the same number of homework problems. The control group was assigned paper-and-pencil homework. The experimental group was assigned roughly 11 problems per week on WeBWorK and the remaining problems as paper-and-pencil homework. Students in the experimental group scored an average of 4% higher on the final exam than students in the control group. While this result indicates a small increase in exam scores between groups, it is statistically significant at the $\alpha=0.05$ level when the researchers controlled for placement score (a measure of students' incoming skill level).

Other studies have found statistically significant differences between homework formats for some, but not all, course exams (Halcrow & Dunnigan, 2012; LaRose, 2010). These findings support the conclusion that online homework has either no effect or a slight positive effect on students' exam scores (as compared to other formats). Halcrow and Dunnigan (2012) compared exam grades for four calculus I sections. There were two instructors, each who taught one section with online homework and one section with paper-and-pencil homework. One instructor's online section scored significantly higher than the paper-and-pencil section on two of three preliminary (non-final) exams and the final exam ($p < 0.10$). Statistical tests could not be performed on one exam due to skewness in the data. The other instructor's online section scored significantly higher than the paper-and-pencil group on one of four exams ($p < 0.10$) and had no significant difference on the final exam

---

* When I say a p-value is significant, not significant, marginally significant, and so on, I follow the author's interpretations of that statement. Because some authors set = 0.05 and others set = 0.10, I have provided p-values so readers can better compare across statistical findings.

($p > 0.10$). In a different study with similar findings, LaRose (2010) compared scores for two preliminary exams and one final exam for three groups of calculus II students. LaRose's (2010) study focused on the effects of two variables: homework format and whether or not homework counted toward the course grade. One group (n = 225) completed paper-and-pencil homework that did not count toward the course grade. The second group (n = 222) completed online homework that did not count toward the course grade. The third group (n = 218) completed online homework that counted for 5% of the course grade. On one exam, the group with online homework that did not count toward the course grade scored significantly higher than the paper-and-pencil group ($p < 0.01$). On the same exam, the group with online homework that counted toward the course grade scored marginally significantly higher than the paper-and-pencil group ($p = 0.06$). In all other two-way comparisons, there were no significant differences between groups.

LaRose (2010) also investigated if homework format impacted students' integration skills. One way he measured this was via student performance (pass rates, number of attempts to pass) on an "gateway" exam consisting of seven integrals. Students could take and re-take the exam during a two-week period; not passing resulted in their course grade decreasing by a third to a full letter grade per failed test. LaRose (2010) found no statistically significant difference in the pass rates between groups with different homework formats. However, when grouping students with online homework (those with the homework counting toward the course grade and not counting toward the course grade) into a single group, students with online homework took significantly fewer attempts to pass the gateway test than students with paper-and-pencil homework.* Because the gateway test was composed of entirely procedural skill problems, LaRose (2010) concluded the online homework system may improve students' procedural competency more than paper-and-pencil homework does, though he cautions, "this conclusion is confounded by our use of the on-line homework system to administer the skills test" (LaRose, p. 678).

Several studies found no significant difference in exam scores based on homework format (Hauk et al., 2015; Hauk & Segalla, 2005; Lenz, 2010, Mathai & Olsen, 2013; Zerr, 2007). Hauk et al. (2015) gave a pre- and post-test to college algebra students, some of whom were enrolled in sections with WeBWorK and others who were enrolled in sections with paper-and-pencil homework. Controlling for pre-test score, a significant predictor variable of post-test score, Hauk et al. (2015) found no significant difference between post-test scores based on homework format. Zerr (2007) compared exam scores for calculus I students with online, graded homework to exam scores for calculus I students with paper-and-pencil homework. The latter group had access to the online homework, and the paper-and-pencil homework problems were "suggested" but not graded. Both sections were taught by the same instructor. Zerr (2007) found no significant difference in exam scores between the two groups ($p > 0.10$). Mathai and Olsen (2013) found no significant difference in exam scores between students with paper-and-pencil homework and students

---

* LaRose (2010) reported the mean number of attempts for the paper-and-pencil homework group was 4.01; the mean number of attempts for the group with online homework that did not count toward the course grade was 3.67; and the mean number of attempts for the group with online homework that counted toward the course grade was 3.58. LaRose did not report a mean number of attempts for the combined online group.

with online homework. However, they did find significant differences for sub-populations (see Section 2.1.3).

Finally, Lenz (2010) found a mix of significant and not significant results. However, in the significant result in Lenz' study, the students with paper-and-pencil homework scored significantly higher than the students with both types of homework. Lenz (2010) compared exam scores for students from seven sections of a finite mathematics course, some who had online homework (n = 84), some who had paper-and-pencil homework (n = 56), and some who had a combination of the two (n = 51). Lenz (2010) found the paper-and-pencil group had the highest mean exam score, but it was not significantly different from the mean exam score for the online homework group at the $\alpha=0.05$ level. There was also no significant difference between the online homework group and the group with both types of homework at the $\alpha=0.05$ level. However, the paper-and-pencil group scored significantly higher than the group with both types of homework at the $\alpha=0.05$ level. Lenz' (2010) qualitative findings suggested the students with both types of homework felt they had too much homework and consequently did less homework overall, even though students in all groups had the same number of problems regardless of format. Lenz (2010) found students often did either the online homework or the paper-and-pencil homework, but not both.

In the next section, I describe findings about homework format and course grades.

2.1.2 Course Grades

One study reviewed for the chapter included statistical tests involving course grade and homework format.* Using a covariate analysis and controlling for significant predictor variables (GPA, initial ALEKS score), Carpenter and Camp (2008) found students in an honors calculus I course with paper-and-pencil homework had statistically significantly higher course grades than students in the same course who had online homework. The authors followed the paper-and-pencil group through calculus II (where they again had paper-and-pencil homework) and calculus III (where they had online homework). There was no significant difference between calculus I course grades (with paper-and-pencil homework) and calculus III course grades (with online homework) or between calculus II course grades (with paper-and-pencil homework) and calculus III grades (with online homework). Using a regression analysis, Carpenter and Camp (2008) concluded that homework format was not a significant variable in predicting course grades.

In the next section, I explore findings about homework format and students' homework completion rates and homework grades.

2.1.3 Homework Completion Rates and Homework Grades

Researchers agree that students spend more time on homework and complete more homework when it is online (Gage, Pizer, & Roth, 2003; Hauk & Segalla, 2005; LaRose, 2010; Lenz, 2010). LaRose (2010) collected survey data about the time students spent on homework at three instances during the semester. Statistical tests on data from the first and third

---

* Lenz (2010) discusses computing a "modified course grade" based on students' exam grades. Because this grade included only exam grades, the results are included in Section 2.1.1.

surveys yielded statistically significant results, while tests on the second survey yielded a not-significant result. On the first and third surveys, LaRose (2010) found students with online homework that counted toward the course grade spent more time on homework than students with online homework that did not count as part of the course grade ($p = 0.03$). Similarly, students with online homework that counted toward the course grade spent more time on homework than students with paper-and-pencil homework that did not count toward the course grade ($p = 0.003$).

LaRose (2010) also collected survey data about the percentage of assigned homework students had completed within the three days prior to the survey. On all three surveys, students with online homework that counted toward their grade or paper-and-pencil homework that did not count toward the course grade completed a significantly higher percentage of homework than students with online homework that did not count toward their course grade ($p = 0.005$). Comparing the percentage of assignments that the students with online, counting-toward-course-grade homework completed versus the percentage of students with paper-and-pencil, not-counting-for-course-grade homework, LaRose (2010) found the online group completed marginally significantly more ($p = 0.06$) than the paper-and-pencil group on the first survey. There was no difference on the second survey and a significant difference ($p = 0.04$) on the third survey. Considering the statistical analysis of the data from the two survey questions (time spent, % of assignments completed), LaRose (2010) concluded, "[the results] suggest that the medium by which the homework is delivered matters less than whether it is graded" (p. 674).

Three other studies reviewed also found students with online homework completed more homework than students with paper-and-pencil homework. Hauk and Segalla (2005) found about 65% of college algebra students enrolled in paper-and-pencil homework sections turned in homework, while 78% of students enrolled in online homework sections completed the online homework. Similarly, Lenz (2010) found students with online homework completed a significantly higher percentage of assignments than students with paper-and-pencil homework ($p < 0.01$). Students with both types of homework attempted a higher percentage of assignments than students with paper-and-pencil assignments ($p < 0.01$; Lenz, 2010). While not a comparative study, in one of the first papers about WeBWorK (Gage, Pizer, & Roth, 2003), the researchers reported "we have been able to document a remarkable thoroughness toward full and accurate completion of homework: nearly all students using the system here at the University of Rochester completed virtually all of their homework sets until their answers are nearly 100% correct" (p. 5).

Gage, Pizer, and Roth (2003) are not the only researchers who have observed students working on homework until they reach perfect or near-perfect scores. Hirsch and Weibel (2003) found that the correlation coefficient between number of attempted WeBWorK problems and percentage of problems solved was $r = 0.944$, "suggesting that once students began a problem they persisted until they had solved it" (p. 14). Similarly, a study of 27 calculus students with online homework found 65% of assignments the students completed had a score of 90 or greater (Butler & Zerr, 2005; Zerr, 2007). Taken together, these studies suggest that students with online homework often persist until they have solved most of the problems correctly.

Like the findings that students seem to complete more homework when it is online, there is a consensus in the literature that students' homework grades tend to be higher with online homework. Lenz (2010) found in addition to attempting more homework assignments, students with online homework had higher homework grades than students with paper-and-pencil homework. She notes the higher grades likely follow from the higher amount of attempted assignments. Multiple attempts per problem also likely contribute to higher homework grades; for instance, Gage, Pizer, and Roth's (2003) students had unlimited tries on each problem. Similarly, Butler and Zerr (2005) and Zerr's (2005) findings discussed above indicate students had high homework averages. Butler and Zerr (2005) commented, "if students were not learning from their mistakes and retaking the assignments, it is likely that the grades would appear more normally distributed" (p. 56).

Mathai and Olsen (2013) compared the efficacy of online homework for students with differing levels of arithmetic and algebra skill. They found that students with high arithmetic skill and online homework had higher homework grades than students with high arithmetic skill and paper-and-pencil homework ($p = 0.06$). Similarly, students with high algebra skill and online homework had higher homework scores than students with high algebra skills and paper-and-pencil homework ($p = 0.01$). The researchers note students with online homework had unlimited attempts per problem, which may have contributed to their higher grades. Similarly, students with higher skill levels in the online group were more likely to complete their homework than students with higher skill level in the paper homework group, and so students in the online homework group may have had higher grades because they completed more homework overall. Mathai and Olsen (2013) found no difference in homework scores based on homework format for the low-skilled groups, and students with low skill level and online homework were less likely to complete their homework than students with low skill level and paper-and-pencil homework. These findings support Mathai and Olsen's (2013) claim that students with higher skill levels benefit more from online homework than students with low skill levels.

In the next section, I review findings about how students' homework grades relate to their exam grades.

2.1.4 Homework Grades and Exam Grades

Two studies investigated possible correlations between homework grades and exam grades. Hirsch and Weibel (2003) found a two-letter grade difference (on average, a D to B) between students who had done at least 80% of the WeBWorK and students who had done less than 50% of it. Zerr (2007) divided students with online homework into two groups, those who had fewer than 17 perfect homework scores (of 26 possible perfect scores) and those who had 17 or more perfect scores. Choosing 17 as a cutoff kept the group sizes roughly equal to allow for statistical comparisons. The exam scores for the students with 17 or more perfect scores were significantly higher than the exam scores for students with fewer than 17 perfect scores ($p < 0.01$). The two groups had similar Math ACT scores, making it unlikely that prior mathematical ability explained the difference in scores. Rather, both Zerr's (2007) and Hirsch and Weibel's (2003) findings suggest students who do more homework are likely to do better on exams than students who do less homework.

For the most part, the findings I have discussed so far are based on data from all students in the researchers' studies. In the next section, I describe findings about various subpopulations of students.

2.1.5 Findings about Online Homework for Various Sub-Populations

Some researchers have sought to investigate if the benefits of online homework differ for various sub-populations. Mathai and Olsen (2013) explored the relationships between students' arithmetic skill, algebra skills, homework format, and exam scores. Weibel and Hirsch (2002) and Zerr (2007) investigated relationships between homework completion rates and students' class standing. Hauk et al. (2015) sought to determine if there were differences in achievement among students of different ethnicities and genders.

Mathai and Olsen (2013) investigated whether the benefits that college algebra students received from online homework depended on their incoming arithmetic and algebra skills (as measured by a departmental placement exam). An analysis of variance indicated that online homework was more beneficial for the students with higher arithmetic skills, while students with lower arithmetic skills benefited more from paper-and-pencil homework. Mathai and Olsen (2013) found students with higher arithmetic skills and online homework (n = 15) scored significantly higher on the final exam than students with lower arithmetic skills and online homework (n = 26). In looking at the effects of students' incoming algebra skill, the researchers found no significant interaction effects between homework format and algebra skill level. Mathai and Olsen (2013) noted "even though the interaction effect was not significant, the students with low algebra skills did better with paper homework while students with better algebra skills performed better with online homework" (p. 678). Overall, the researchers found no statistically significant difference ($p = 0.28$) on the final exam score between a section of college algebra students with online homework (n = 48) and students with paper homework (n = 29). However, the finding that students with higher incoming algebra and arithmetic skills benefit more from online homework than students with lower incoming skill level is a good argument for including both types of homework in a course.

Other researchers have found that homework completion rate varies with population. First-year calculus students tend to attempt nearly every problem (Weibel & Hirsch, 2002) and have higher homework scores than students who have completed at least one semester (Zerr, 2007). Students repeating calculus attempt about a third of the problems (Wiebel & Hirsch, 2002).

Hauk et al. (2015) measured effects of homework format by statistically comparing students' pre- and post-test scores on a test of college algebra content. Controlling for previous mathematics achievement, their analysis of covariance found no significant difference in performance or achievement gain for students of different ethnicities or genders ($p > 0.10$). The authors concluded, "it seems that whatever may be culturally biased in the structure or process of college algebra, the use of web-based homework does not appear to significantly exacerbate or diminish it" (p. 74).

While statistics are useful in making decisions about homework systems, student perceptions matter as well. In the next section, I turn to literature about student perceptions of online homework.

2.2 STUDENTS' PERCEPTIONS OF ONLINE HOMEWORK

Researchers have found that student perceptions of pedagogy influence their attitudes about mathematics and their decisions to take future math courses (Ellis, Kelton, & Rasmussen, 2014; Sonnert, Sadler, Sadler, & Bressoud, 2014). Given students spend more time doing mathematics homework than they do in class (Ellis et al., 2015; Krause & Putnam, 2016), it is likely students' experiences with homework also influence their attitudes about mathematics and their decisions to take future math courses. Hence instructors should attend not only to statistics about achievement, but also to qualitative and quantitative data about what students find helpful for their learning. Research findings about student perceptions of online homework can help us make decisions about online homework systems so that these platforms contribute to positive course experiences for students.

In this section, I review literature about online homework and students' beliefs about mathematics, what students like about online homework, and what students dislike about online homework.

2.2.1 Online Homework and Students' Beliefs about Learning Mathematics and the Nature of Mathematics

Hauk and Segalla (2005) investigated if online homework influenced college algebra students' beliefs about mathematics. Overall, they found students who "evidenced expert-like views of mathematics learning as a complex and personal process of building conceptual understanding appeared to view WeBWorK as a tool that helped or hindered that process" while students who had a "novice-like view of mathematics learning as a disconnected collection of formulae and 'plug-and-chug' strategies appeared to view WeBWorK as either helping or hindering their procedural approach" (p. 16). That is, the online homework system did not seem to change students' views about the nature of mathematics.

Hauk and Segalla (2005) cited Spangler's (1992) summary of beliefs college students often hold about mathematics and mathematics learning:

> (1) mathematics is *computation* that does not involve reflection during task engagement; (2) mathematics must be done *quickly*, or, spending time is a more important task goal than sense-making; (3) mathematics problems have *one right answer* and no further action or evaluation is required once an answer is found; and (4) the *teacher is the agent* of mathematical learning, not the student (i.e., only intentional acts on the part of the teacher lead to learning, no intentionality on the part of the student is necessary.
>
> *(Spangler, 1992 as cited in Hauk & Segalla, 2005)*

Related to belief 1, Hauk and Segalla (2005) found that some students who viewed mathematics as more procedural felt they were not "doing math" in the online homework platform because they could input answers like x = (7 – 1)/3 and the computer would do the arithmetic. Related to belief 2, Hauk and Segalla's (2005) subjects commented "math homework shouldn't take so long" (p. 18). The opportunity to attempt problems multiple times contributed to these students' complaints that the homework took "too long." This finding has important implications for future research (see Section 2.5).

Related to belief 3, 11% of Hauk and Segalla's (2005) students expressed concern that the WeBWorK problems could have more than one right answer. The researchers noted these students may not have realized $x = (7 - 1)/3$, $x = 6/3$, and $x = 2$ were equivalent answers. Finally, related to belief 4, several studies have found that students express frustration with feedback that tells them only that an answer is incorrect, not why it is incorrect or hints about how to solve it (Ellis et al., 2015; Hauk & Segalla, 2005). Hauk and Segalla (2005) argue the desire for hints may stem from students' belief that the teacher is the agent of learning.

In the next section, I review literature about what students like about online homework.

2.2.2 What Students Like about Online Homework

Students tend to like doing homework online (as opposed to a paper-and-pencil format) and find it helpful for their learning (Burch & Kuo, 2010; Butler & Zerr, 2005; Ellis et al., 2015; Gage, Pizer, & Roth, 2003; Lenz, 2010; Lunsford & Pendergrass, 2016; Raines & Clark, 2013; Yushau & Khan, 2014, Zerr, 2007). Zerr (2007) found students with online homework were more likely to agree or strongly agree with the statements "the homework assignments in this class were helpful for learning the material" than students with paper-and-pencil homework ($p < 0.05$). Butler and Zerr (2005) found that 17 of 19 calculus students agreed with the statement "online homework was a worthwhile addition to this course." In the same study, 82% of high school students taking a college algebra course for college credit agreed with the same statement, with 50% of the students saying the online homework was the most helpful part of the course (Butler & Zerr, 2005). Lenz (2010) found similarly high percentages of students who liked the online homework format. She found 87.5% of students who had only online homework liked having online homework, and 69.2% of students who had both online and paper-and-pencil homework liked having online homework. Several studies have found students feel online homework helped them understand the course material better than paper-and-pencil homework (Hodge, Richardson, & York, 2009; Yushau & Khan, 2014) and that students feel online homework helps them understand material and prepare for exams (Raines & Clark, 2013; Yushau & Khan, 2014; Zerr, 2007).

Students express feeling more motivated to complete online homework than paper-and-pencil homework (Hodge, Richardson, & York, 2009). This higher motivation may explain findings that students tend to complete a larger percentage of online homework than paper-and-pencil homework (Hauk & Segalla, 2005; Gage, Pizer, & Roth, 2003; Lenz, 2010) and have higher homework grades on online homework (Butler & Zerr, 2005; Lenz, 2010). Findings mentioned above about students persisting until they had solved most questions correctly (Butler & Zerr, 2005; Gage, Pizer, & Roth, 2003; Weibel & Hirsch, 2002) are also evidence of high student engagement in online homework contexts. There is also a link between homework format, whether or not it is graded, and student perceptions. Specifically, students with online homework that counts toward their grade are more likely to make positive comments about online homework than students whose homework (online or paper-and-pencil) does not count toward their course grade (LaRose, 2010).

Several studies have linked student perceptions about online homework to other variables. Leong and Alexander (2014) found students with lower and average mathematics achievement were more positive about online homework than higher-achieving students. Hauk and Segalla (2005) found student attitudes about online homework were related to instructors' attitudes. They classified college algebra instructors' views about WeBWorK as "quite useful," "could be useful," and "not much use." Surveys from students whose instructors felt WeBWorK was not of much use made comments like "'useless' … 'hated it' … 'a colossal waste of time'" (Hauk & Segalla, 2005, p. 21). Students whose instructors felt WeBWorK could be useful, but might not be a replacement for paper-and-pencil homework, made comments to the effect that the online homework was helpful but they liked individual feedback from the professor. Finally, students whose instructors felt WeBWorK was quite useful generally commented that the online homework was helpful, and "made suggestions for how the interface might be improved. These students accepted WeBWorK as valuable and wanted to improve their efficacy in using it, a sign of a strong intention to engage with mathematics in the way supported by WeBWorK" (Hauk & Segalla, 2005, p. 21). These findings indicate that students will be most positive about online homework when instructors see it as a useful tool and communicate to students that they value the learning opportunities online homework offers. It also seems important that instructors take into consideration student input about the online homework systems.

Most students may prefer the online format (Hodge, Richardson, & York, 2009; Lenz, 2010; Raines & Clark, 2013), but they also seem to value the individual feedback they receive on paper-and-pencil homework. For example, Hodge, Richardson, and York (2009) found that about a third of students felt online homework increased their mathematical understanding more than paper-and-pencil homework, but on the whole students in their study and in other studies expressed a preference for having both types of homework (Hodge, Richardson, & York, 2009; Yushau & Khan, 2014). Similarly, only 57.69% of the students with both types of homework in Lenz' (2010) study indicated that they would prefer to take the course with only online homework. Lenz (2010) concluded, "it is possible that the students in the [combination online and paper-and-pencil homework] sections recognized the benefits of receiving individualized feedback on their traditional homework, making them more reluctant to choose a course with purely web-based homework" (p. 239).

Students like features such as hints, being shown the correct solution when they submit a wrong answer, or "see a similar example" features (Burch & Kuo, 2010; Halcrow & Dunnigan, 2012; Lenz, 2010; Leong & Alexander, 2014; Raines & Clark, 2013; Yushau & Khan, 2014; Zerr, 2007). They also like having multiple attempts per problem (Ellis et al., 2015; Halcrow & Dunnigan, 2012; Raines & Clark, 2013; Yushau & Khan, 2014; Zerr, 2007). Students like the immediate feedback online homework systems offer (Ellis et al., 2015; Gage, Pizer, & Roth, 2003; Leong & Alexander, 2014; Roth, Ivanchenko & Record, 2008). In the next section, I describe findings about what students dislike about online homework.

2.2.3 What Students Dislike about Online Homework

Many researchers have noted students' negative feedback about online homework centers around difficulties inputting answers into the programs (Ellis, Hanson, Nuñez, &

Rasmussen, 2015; Hauk & Segalla, 2005; Gage, Pizer, & Roth, 2003; Heenehan & Khorami, 2016; Leong & Alexander, 2014; Roth, Ivanchenko, & Record, 2008; Yushau & Khan, 2014). Heenehan and Khorami (2016) found students' attitudes about online homework were neutral at the start of the term and decreased over time. The decrease was correlated with an increase in technological issues related to inputting answers. The particular homework platform was inconsistent in the format and notation required for similar problems (e.g., "infinity" having to be written as INF, inf, infinity, or ∞ in different questions), and students became increasingly discontent as the mathematical content became more complex and required more notation (Heenehan & Khorami, 2016). Other researchers have found that students become frustrated with online homework platforms that do not have a margin for error (Leong & Alexander, 2014). For example, Leong and Alexander (2014) noted that students became particularly frustrated by a problem that required them to graph coordinate points and marked answers wrong when the points were off by a tiny amount.

Solutions to these difficulties may lie in choosing a system that is consistent in formatting; one that will accept multiple forms of correct notation for infinity, intervals, and so on; and one that has error tolerance built into some problems. Researchers have also found answer-preview buttons and alerts when a student is about to submit an answer they have already tried decrease student frustration with online homework platforms (Roth, Ivanchenko, & Record, 2008).

Aside from technological issues, students' negative comments about online homework tend to be related to the feedback they receive (or lack thereof). Students may like the immediate feedback online homework offers, but some express frustration that they receive only a red X when a problem is wrong, because it does not tell them why the answer is wrong (Ellis et al., 2015). Similarly, students note that a downside of online homework is it does not help them know what professors look for in terms of showing work, as paper-and-pencil homework would (Lenz, 2010).

Students also seem to want online homework to align well with the way the instructor teaches and poses problems. Some students in Lenz's (2010) study commented "web homework teaches differently than the teacher does, which makes it hard to follow" (Lenz, 2010, p. 240). Students may not feel online homework is useful if the problems are much easier than exam problems (LaRose, 2010).

Finally, while the findings above are largely logistical in nature, students also care about the content of a homework assignment. Ellis et al. (2015) found some students disliked the procedural nature of online homework and how it tended to involve "cranking out formulas" with few applications. Hauk and Segalla (2005) found that some students did not feel like they were really "doing math" in online homework because they could input answers that were not in their simplest forms. I return to the importance of content later in the chapter.

2.3 HOW DO STUDENTS ENGAGE WITH ONLINE HOMEWORK?

In addition to studying the effects of homework format on grades and student perceptions of online homework, researchers have investigated what it is that students *do* when they are doing online homework. The body of research about how students engage with online homework can help instructors make decisions about how many attempts students have per problem, whether or not to enable help features, and so on. While the literature about how students engage with online homework is not extensive, we know a little bit about how students reason in the context of online homework, how multiple attempts per problem influence that reasoning, and the resources students employ while doing online homework (including how they use "see similar example" features).

2.3.1 Multiple Attempts per Problem

Students tend to attempt online homework problems multiple times when given the opportunity (Butler & Zerr, 2005; Dorko, 2018; Dorko, accepted; Gage, Pizer, & Roth, 2003; Hauk & Segalla, 2005; Hirsch & Weibel, 2003; Suzuki, 2003; Weibel & Hirsch, 2002; Zerr, 2007). In Suzuki's (2003) online homework assignments, students had to submit all their answers at once, but could repeat the assignment as many times as they liked. Looking at the number of attempts for students who did not score a 100% on the first try, Suzuki (2003) found each quiz was attempted an average of 2.3 times. Findings described above about students obtaining high homework grades also indicate students attempting assignments/problems multiple times (Butler & Zerr, 2005; Gage, Pizer, & Roth, 2002; Suzuki, 2003; Zerr, 2007).

Students may attempt and re-attempt problems, but their reasoning is not always mathematical in nature (Dorko, 2018; accepted; under review; Hauk & Segalla, 2005). For example, sometimes students guess answers (Dorko, 2018; accepted; Hauk & Segalla, 2005). In a study of the nature of students' activity while doing online homework, Dorko (2018; accepted) found nine calculus II students' work on online homework problems about sequences could be classified as based on mathematical thinking, guessing, attending to didactic features (defined below), or a combination of those categories. Students in her study had three attempts per problem and could submit each part of a multi-part problem individually. There were no "help" features enabled. Dorko (2018; accepted) found most students' answer submissions were based on some mathematical thinking (e.g., computing terms, applying L'Hopital's Rule). However, students frequently coupled mathematical thinking with guessing or attending to didactic features. For example, consider the problem shown in Box 2.1,[*] which a study participant named Susie[†] completed.

BOX 2.1   One of Susie's homework problems (Dorko, 2018; accepted)

Use the appropriate limit laws and theorems to determine the limit of the sequence or show that it diverges. (If the quantity diverges, enter DIVERGES.)

$$b_n = n^{3/n}$$

$$\lim_{n \to \infty} b_n = \underline{\hspace{1cm}}$$

---

[*] The text of the problem is presented exactly as it appeared to students.

[†] In accordance with standard guidelines for ethical research, 'Susie' (and other student names later in the chapter) are gender-preserving pseudonyms.

Susie computed b_2, b_{10}, b_{20}, and b_{50}. She submitted the answer $\lim_{n\to\infty} b_n = 1$ and explained…

BOX 2.2 Susie's explanation of her thinking about a problem (Dorko, 2018, p. 55–56)

Susie: I think I was starting [inaudible] if it was diverging to 1 or 0.

Interviewer: Okay, so when you typed 1 here [in the answer box] -

Susie: I was, I was thinking it was 1, but it could have also been 0 if you went down further…

Interviewer: So was this, do you think this was one of those things where you like have three chances, so –

Susie: Yeah.

Interviewer: - try one and then –

Susie: If it's wrong try another one … cuz like just a lot of graphs converges [sic] to 0. But it was, like it was going to 1, like it had 1 point something, so I figured it was 1, but there was also the possibility that if you went further if it was going to converge to 0 or not.

Interviewer: So it sounds like you were paying attention to both number patterns like the 1 and the experience that things go to 0 often.

Susie: Yeah.

Dorko (2018; accepted) characterized Susie's reasoning in the problem as based on mathematical thinking (computing terms) and guessing.

Other researchers have also found that students sometimes guess on online homework problems. Hauk and Segalla (2005) found that, given unlimited attempts in WeBWorK, some students guessed up to 35 answers. Students may guess randomly, but Dorko (2018; accepted) found the guesses tended to be based on students' experience of common answers (e.g., Box 2.2) or based on didactic features. She defined a didactic feature as related to the didactic contract (Artigue, Haspekian, & Corblin-Lenfant, 2014; Brousseau, 1997), "a set of reciprocal obligations and mutual expectations [that is] the result of an often implicit negotiation" (Artigue et al., 2014, p. 53). For example, the didactic contract implies that students expect teachers to provide opportunities to learn and teachers expect students to engage in those opportunities. Rosalyn, one of Dorko's (2018) students, explained she guessed that a particular sequence diverged because she knew convergence and divergence were part of the content she was supposed to learn, and up until the problem shown in Box 2.3, all the sequences had converged. Rosalyn expected opportunities to learn both convergence and divergence ("the whole point is … you learn … a little bit of each"; Box 2.3).

BOX 2.3 Rosalyn's work on an online homework problem (Dorko, 2018, p. 57)

Rosalyn: Infinity to the infinity minus 1 would be, that doesn't make any sense. I'm going to enter diverges and hopefully that's it. [Submits 'DIVERGES', which is correct] Okay, it diverged. I don't know how to solve it, but I guessed. So – and probably one more of these diverges. Don't know which one though but most likely.

Interviewer: Why do you think probably one more diverges?

> *Rosalyn:* Usually like in WebAssign like the fact that there's four of them [problems that read 'If the quantity diverges, enter DIVERGES'] I want to say like probably at least another one diverges … because the whole point is you know you learn like, you know a little bit of each. So I feel like if they have one that diverges they most likely will have another one diverges.

In summary, providing students multiple attempts per question results in students both engaging in mathematical work and in guessing (Dorko, 2018; accepted; Hauk & Segalla, 2005). Guesses may be random but are also frequently related to students' experience with common answers or their expectations of an assignment (Dorko, 2018; accepted).

The number of attempts students have for a problem influences how students work on the assignment. For example, Dorko (2018; accepted) observed that students were willing to guess or make a guess coupled with mathematical thinking on their first attempt because they knew they had two more tries (e.g., Box 2.2). Similarly, students in Dorko's (2018; accepted) study could submit each part of a multi-part question individually. Dorko (2018; accepted) found that this structured student work. Students often submitted one part of a question and used the feedback of whether the answer was right or wrong as formative assessment. In contrast to guessing, this provides evidence of students taking advantage of the affordances of an online homework system.

2.3.2 Resources

We know a little bit about the resources students use to complete online homework. In a survey of calculus I students, Krause and Putnam (2016) found students commonly used class notes, "see similar example" features, online calculators, YouTube videos, and informational websites.

The vast majority of students use help features like "see a similar example" if such features are available to them (Dorko, under review; Krause & Putnamn, 2016; Raines & Clark, 2013). Raines and Clark (2013) surveyed college algebra students about their use of MyMathLab's "view a similar example" and "help me solve this" features and found 83.1% of students used these tutorial learning aids. Krause and Putnam (2016) found over 75% of students said they used a "see similar example feature" often or always while completing homework. Krause and Putnam (2016) describe,

> [we] observed students using this feature on many problems; this is often the first source that students turn to when they are stuck. Sometimes students look at the similar example immediately after they read the problem, other times they look at the example after they have attempted the problem and have submitted an incorrect answer, but students almost always turn to this resource first when they are not able to complete the problem on their own … informal conversations with the students match the observed behavior in that it appears as though the primary goal is to determine similarities in the structure of the problem to try to figure out how

the numbers in the example can be manipulated to yield the answer to the problem at hand.

(p. 273)

Dorko (under review) found similar themes in students' use of the "practice another version" (PAV) feature in WebAssign. She found students opened the PAV feature to see the steps for solving a problem, obtain a template to solve the problem, make sense of a solution method, troubleshoot if they had answered a problem incorrectly, check that their method was on the right track, obtain extra practice, and to maximize their score.

In terms of other resources, Krause and Putnam (2016) found students used online calculators both to troubleshoot errors and to circumvent a problem. When troubleshooting, students would use an online calculator after having submitted an incorrect answer to identify where they made a mistake. When circumventing a problem, students would type a question directly into the calculator before attempting it on their own. Krause and Putnam (2016) also found students looked at instructional websites and online forums when doing online homework. Krause and Putnam (2016) noted how different wording of web searches lead to different learning opportunities. For example, they described how one student searched "how do you find a second derivative?" and obtained information that allowed her to work through her own problem, while another student typed the exact wording of a problem into a search engine and found a page with a full solution. This student read the solution, then copy-and-pasted the answer into her homework platform. An implication of this finding is that instructors might tell students to search more generally (e.g., "how do you find a second derivative?") if they need help, rather than searching problems word-for-word.

2.4 THE IMPORTANCE OF HOMEWORK CONTENT

The findings from the literature suggest that online homework and paper-and-pencil homework have about the same effect on students' course grades. Both formats have benefits to students, but it is critical that concerns about format not usurp what is most important: homework content.

Ellis et al. (2015) found that one factor that distinguished more successful university calculus programs from less successful calculus programs was their homework systems. More successful programs were more likely to have an online component to their homework (sometimes with a written component), but they were also more likely to assign more novel or complex problems than less successful calculus programs. Students at universities with more successful programs were also asked to explain their thinking more often than students at universities with less successful programs. Instructors should keep in mind that online homework does not always provide opportunities for students to explain their thinking, and hence this opportunity might be important to include in written homework.

White and Mesa (2014) studied the cognitive orientation of calculus tasks across various types of coursework (homework, worksheets, quizzes, exams) for five different instructors at a two-year university. They classified each task as a simple procedure, a complex procedure, or a rich task, and found variation in the cognitive orientation percentages for homework by instructor. The variation provides evidence that even when instructors use the same textbook, students in different sections may have different opportunities to learn from homework. Relatedly, Dorko (2019) studied whether nine calculus II students achieved two instructors' attended learning outcomes from a set of homework problems about sequences. She found that students met instructors' goals of obtaining practice with operations (e.g., factorials), notation, and procedures. However, students largely missed goals more conceptual in nature, such as realizing there are multiple ways to define a given sequence.

One implication for instruction from Ellis et al. (2015), White and Mesa's (2014), and Dorko's (2019) findings is that instructors think about the cognitive orientation of the set of tasks they assign in homework in any one given assignment and over the entirety of the term. While White and Mesa's (2014) study did not link percentage of tasks at various cognitive orientations to student outcomes, Ellis et al. (2015) found successful calculus programs assign novel, complex tasks at a higher rate than less successful calculus programs. This finding suggests that instructors should aim to provide students opportunities to work on rich tasks. A review of literature about such tasks and their effects on student learning is outside of the scope of this chapter; interested readers should see Ellis et al. (2015) to learn more.

2.5 RESEARCH-BASED, EFFECTIVE PRACTICES FOR UTILIZING ONLINE HOMEWORK

An important goal of research is to inform practice. In this section, I synthesize the research findings into a list of considerations for instructors making decisions about the homework structures employed in their classes.

2.5.1 Instructors' Attitudes Influence Students' Attitudes

Hauk and Segalla (2005) found that when instructors were positive about online homework as a tool for learning, students were similarly positive. The opposite was also true: students whose instructors did not think online homework was useful tended to mirror that opinion. Similarly, Halcrow and Dunnigan (2012) found that when two instructors' online homework systems experienced technical issues, one instructor remained positive while the other became increasingly negative about the system. The students with online homework whose instructor retained a positive attitude scored significantly higher on most of the course exams than the students with paper-and-pencil homework (taught by the same instructor). There was no significant difference between groups for students whose instructor became increasingly negative in the face of technological issues. These findings suggest that instructors should be careful regarding their comments about any particular homework format or system, as it may influence students' attitudes and achievement.

2.5.2 Online Homework Has the Same Effects on Student Achievement, or Increases Achievement Slightly, as Compared to Paper-and-Pencil Homework

Researchers generally agree that online homework is at least as effective as paper-and-pencil homework in terms of student achievement (e.g., Babaali & Gonzalez, 2015; Burch

& Kuo, 2010; Hauk et al. 2015; Halcrow & Dunnigan, 2012; Hauk & Segalla, 2005; Hirsch & Weibel, 2003; LaRose, 2010; Lunsford & Pendergrass, 2016; Lenz, 2010; Weibel & Hirsch, 2002; Zerr, 2007). This means instructors can likely assign online homework for all or part of the homework in their courses without hurting students' grades. However, having all of a course's homework be online may be contraindicated (see Section 2.4).

2.5.3 Students Complete Online Homework at Higher Rates Than They Complete Paper-and-Pencil Homework

Research findings indicate that students complete online homework at higher rates than they complete paper-and-pencil homework (Gage, Pizer, & Roth, 2003; Hauk & Segalla, 2005; LaRose, 2010; Lenz, 2010). This may vary by population; one study found first-year calculus students are likely to attempt every problem, while students repeating calculus attempted less than a third of the problems (Wiebel & Hirsch, 2002). These results suggest that if it is important to an instructor that students do the homework, then including an online homework component is likely to result in a higher percentage of completed homework than would paper-and-pencil homework.

2.5.4 Students Like Having Both Online and Paper-and-Pencil Homework, and Aspects of Both May Be Important for Student Success

In a nation-wide study of college calculus programs, Ellis et al. (2015) found that institutions with more successful calculus programs (as measured by "students retain[ing] confidence, enjoyment, and interest in mathematics more than comparable universities involved in the [study]"; Ellis et al., 2015, p. 2) were significantly more likely to assign online homework than less successful calculus programs (p ≤0.001). This finding suggests that online homework may play a role in student success. Students seem to prefer online homework to paper-and-pencil homework, but like to have both types of homework in a course (Burch & Kuo, 2010; Butler & Zerr, 2005; Ellis et al., 2015; Lenz, 2010; Raines & Clark, 2013; Yushau & Khan, 2014; Zerr, 2007). Research findings suggest it may be important to assign both online and paper-and-pencil homework. In a literature review that included research about paper-and-pencil homework, Ellis et al. (2015) concluded, "there are aspects of both written and online homework that have been connected to student success" (p. 272). They use *success* to mean both grades and affective factors like students' confidence. Ellis et al. (2015) argue that content of homework assignments and the feedback students receive are critical factors in successful homework systems. Findings from Ellis et al. and others that students value the immediate feedback from online homework *and* the individualized feedback on paper-and-pencil assignments support the practice of assigning homework in both formats.

However, merely assigning paper-and-pencil homework is a necessary but not likely sufficient condition for supporting student success. Findings suggest that students reap the most benefit from paper-and-pencil homework when graded and returned with comments. I draw this conclusion from several threads in the literature. First, students are far more likely to complete homework when it is graded. Second, one of the features that students seem to value about paper-and-pencil homework is the individual feedback. Third, returning homework with helpful comments/feedback is a feature of successful calculus programs (Ellis et al., 2015).

These findings suggest that instructors have both paper-and-pencil and online homework components, and that the paper-and-pencil homework be both graded and returned with helpful feedback (that is, feedback beyond merely correct/incorrect). See also Section 2.5.

2.5.5 If Employing Both Online and Paper-and-Pencil Homework, Be Wary of the Total Workload

Given findings that students like online homework but also value the individualized feedback on paper-and-pencil homework, it might seem that the ideal solution would include both formats. However, Lenz' (2010) findings provide a cautionary tale. She found that the group of students with both types of homework had statistically significantly lower exam scores than students with just pencil-and-paper homework. (Contrastingly, Wiebel and Hirsch (2003) found students with both types of homework scored significantly higher on exams). Lenz (2010) noted of the section with both types of homework, "students seemed to feel that the homework burden was heavier when receiving assignments in two different formats even though the total number of problems was unchanged. There were some students who chose only to attempt assignments in one format and not the other, and some who attempted only one type of assignment per class meeting" (p. 244). This suggests that instructors should be wary of the total number of problems assigned. I interpret the comment "students attempted only one type of assignment per class meeting" as meaning both types of homework were due on the same day. It is possible that having the homework types due on different days could mitigate the problem of students doing only one type.

2.5.6 Be Intentional about Due Dates, Particularly If Employing Both Online and Written Homework Components

Due dates partially structure when students work on homework. For example, Butler, Pyzdrowski, Goodykoontz, and Walker (2008) found "many students wait to complete a first attempt of a homework quiz until either the evening before or the morning of an exam" (p. 134). Butler and Zerr (2005) assigned six online homework assignments, due at the end of the term. Students had three attempts per assignment. Students were told which assignments corresponded to which exam but at the exam times, many students had not done the homework. For example, exam 1 corresponded to homework assignments 1 and 2. The day of the exam, 76.5% of students had made at least one attempt on homework 1 and 50.6% of students had made at least one attempt on homework 2. Completion percentages for other assignments on the days of their corresponding exams fell in this range. Krause and Putnam (2016) found roughly 50% of students complete online homework a few days before the deadline. These findings suggest that regular due dates (as opposed to the end of the term) may increase the percentage of homework students complete.

Finally, Lenz (2010) found that when students had both written and online homework due on the same day, they tended to complete homework in one of the formats but not the other. This finding suggests that if instructors employ both online and paper-and-pencil homework, having each format due on a different day may increase the percentage of homework students complete.

2.5.7 Students Become Rapidly Frustrated with Online Homework Platforms That Make It Difficult to Format Answers, or Are Inconsistent in Formatting

Most of the students' negative comments about online homework systems relate to formatting answers (Ellis et al., 2015; Hauk & Segalla, 2005; Heenehan & Khorami, 2016; Leong & Alexander, 2014; Roth, Ivanchenko, & Record, 2008; Yushau & Khan, 2014). Implementing features that allay these difficulties increases students' satisfaction with online homework (Roth, Ivanchenko, & Record, 2008). Contrastingly, if formatting difficulties increase over the course of the term (say, due to increasingly sophisticated content), students' perception of the helpfulness of online homework declines (Heenehan & Khorami, 2016). Hence instructors should seek online homework platforms that make inputting answers as easy as possible. Students seem to appreciate supports like "preview my answer" and warnings that they are about to submit an answer they have already tried (Roth, Ivanchenko, & Record, 2008). It also seems important to students that a homework platform accept multiple versions of the same correct answer (e.g., "infinity" and "∞") and/or be consistent about how one answer should be inputted (e.g., always "infinity," not sometimes "infinity" and sometimes "inf"; Heenehan & Khorami, 2016). Finally, instructors can decrease student frustration with online homework by including some error tolerances when reasonable, such as graphing coordinate points (Heenehan & Khorami, 2016).

2.5.8 Students Do More Homework When It Counts toward Their Course Grade

Ellis et al. (2015) found that universities with successful calculus programs assigned homework more frequently and were more likely to collect homework than universities with less successful calculus programs. LaRose (2010) found that students with online homework that counted toward the course grade spent significantly more time on homework than (1) students with online homework that did not count toward the course grade and (2) students with paper-and-pencil homework that did not count toward the course grade. Students with online homework that counts toward their grade are more likely to make positive comments about online homework than students whose homework (online or paper-and-pencil) does not count toward their course grade (LaRose, 2010). Research findings indicate that including homework in the course grade may support student success (Ellis et al., 2015; LaRose, 2010).

These results suggest that if it is important to an instructor that students do the homework, then the homework should be part of the course grade.

2.5.9 Offering Unlimited Attempts per Problem Is Linked to Student Persistence in Repeating Assignments/Problems to Obtain High Scores, and It May Also Increase Students' Confidence in Their Ability to Do Math

Researchers have found evidence that when students have unlimited attempts to re-do problems or assignments, they often take advantage of it and work on the assignments until they have a high grade (Butler & Zerr, 2005; Gage, Pizer, & Roth, 2003; Hirsch & Weibel, 2003; Zerr, 2007). This may increase student confidence in their ability to do math (Halcrow & Dunnigan, 2012).

One reason instructors might care that students have higher homework scores is higher homework grades may boost overall grades and hence lower DFW rates. Higher grades as a result of higher homework scores are higher grades that students have earned, and might prevent an instructor from needing to scale grades at the end of a term.

2.5.10 Offering Unlimited/Multiple Attempts Is Linked Both to Students Employing Those Attempts to Make Sense of the Mathematics, and in Employing Those Attempts to Guess Answers

When students have more than one attempt per problem, sometimes they employ the multiple tries to work through the mathematics and sometimes they guess. Dorko (2018; accepted) found that, given three tries per problem, students engaged in both mathematical work and guessing as they submitted their answers. She noted guesses rarely were random. Rather, students often used their first attempt on a problem to test an idea they thought could be right, but were not entirely sure of. Students explained that they did this because they knew they had multiple tries, and could afford to use one on a guess. Similarly, four of Hauk and Segalla's (2005) students admitted to guessing "many times" (p. 18) if their first answer was incorrect. Reviewing answer submissions, Hauk and Segalla (2005) found some students submitted up to 35 guesses. Students may guess even when they have limited, multiple tries.

Further research is needed to determine if there is an optimal number of attempts per question that would minimize guessing behavior while allowing the benefits of multiple attempts. However, research findings support giving students multiple attempts. For example, Butler and Zerr's (2005) and Zerr's (2007) findings that a larger percent of students had scores 90 or greater on homework suggest that the students were employing the multiple attempts to engage with the mathematics. That is, it seems unlikely that students achieved so many high scores by guessing alone.

2.5.11 When Students Can Submit Each Part of a Question Individually, They Often Employ Feedback on One Part to Guide Their Work on the Other Parts of the Question

Dorko (2018; accepted) found that when students can submit each part of a question individually, they often do so. Students in her study had multiple attempts per problem. Students explained that they submitted answers to parts individually for several reasons. First, submitting each part of a question individually limits the number of attempts they "lose" if the answer is wrong. Second, they use the feedback for the answer to one part as formative assessment. That is, if an answer to the submission to part 1 of a multi-part question is correct, they know they are solving the problem correctly, and if the answer is wrong, they can try a different strategy. This finding supports allowing students to submit each part of a question individually.

2.5.12 Students Like "Help" Features. We Do Not Know Much about How They Employ These Tools

Zerr (2007) created an online homework platform that showed students a complete, correct solution if they inputted a wrong answer. The system then gave students a different (similar) problem. Students made exceedingly positive comments about having the full solutions. However, we do not know much about how students make use of full solutions. Krause and Putnam (2016) found students' primary goal when reading a "see similar example" feature was to figure out where they needed to substitute their own numbers. However, this finding is based on a small sample size (interviews and observations of four calculus students and survey data for an unreported number of calculus students). Further research is needed to determine which "help" features are most effective for student learning.

2.5.13 Content Is Incredibly, Incredibly Important

Regardless of whether an instructor employs an online homework platform, paper-and-pencil homework, or a combination of both, the content of homework problems is incredibly important. Homework systems are a distinguishing factor between more and less successful university calculus programs (Ellis et al., 2015). In a national study of college calculus programs, researchers found that more successful programs were more likely than less successful programs to (1) have an online component to their homework (sometimes with a written component), (2) assign more novel or complex problems than less successful calculus programs, and (3) ask students to explain their thinking on homework (Ellis et al., 2015). These findings imply that instructors should provide students opportunities to work on rich tasks, and opportunities to explain their thinking.

The task presented in Box 2.1 is an example of a problem from a textbook that was typed directly into an online homework platform by some online-homework-producing company. Readers might note the problem directs student to "show" something, yet the answer space does not allow this. Not all book problems are well-suited for an online format, and there is much research work that needs to be done about what sorts of problems *are* best for online environments such that students learn what the instructor intends they learn. This is a question of both content and format.

2.6 CLOSING REMARKS

In this chapter, I reviewed the literature about student learning from online homework and offered a list of considerations for instructors making decisions about the homework for their course. While the body of research about student achievement and student perceptions of online homework is extensive, there are still many open questions. For example, this review located only two studies about how online homework affects students' beliefs about mathematics and about themselves as doers of mathematics (Halcrow & Dunnigan, 2012; Hauk & Segalla, 2005). I wonder if having unlimited attempts per problem can be leveraged to help students shift from a "mathematics problems can be solved quickly" mindset to an understanding that mathematical problem solving is a cyclic process, often involving multiple attempts and reattempts (Carlson & Bloom, 2005; Schoenfeld, 1992; 2014).

Another area for research is what students learn from online homework. That is, many of the studies reviewed here measured learning using exam scores as a metric. A more fine-grained definition of learning could help us know more about what students gain from particular problems or assignments, and help us refine the content of online homework assignments. It also seems important to conduct research about the details of how students employ resources while doing homework and how that relates to their learning. I located only one study describing how students use various resources (Krause & Putnamn, 2016). When researching online homework, it would be useful to know if student learning and resource use differ with population (e.g., college algebra students versus upper-division students). That is, perhaps students who are more mathematically mature interact with online homework in different ways than students in developmental math classes. Future research about what and how students learn from online homework can help us improve student learning.

REFERENCES

Artigue, M., Haspekian, M., & Corblin-Lenfant, A. (2014). Introduction to the Theory of Didactical Situations (TDS). In A. Bikner-Ahsbahs, & S. Prediger (Eds.), *Networking of Theories as a Research Practice in Mathematics Education*, (pp. 47–65). Switzerland: Springer International Publishing.

Babaali, P., & Gonzalez, L. (2015). A quantitative analysis of the relationship between an online homework system and student achievement in pre-calculus. *International Journal of Mathematical Education in Science and Technology*, 46(5), 687–699.

Brousseau, G. (1997). *Theory of Didactical Situations in Mathematics*. Dordrecht, The Netherlands: Kluwer.

Burch, K. J., & Kuo, Y. (2010). Traditional vs. online homework in college algebra. *Mathematics and Computer Education*, 44(1), 53–63.

Butler, M., Pyzdrowski, L., Goodykoontz, A., & Walker, V. (2008). The effects of feedback on onlsine quizzes. *International Journal for Technology in Mathematics Education*, 15(4).

Butler, M., & Zerr, R. (2005). The use of online homework systems to enhance out-of-class student engagement. *The International Journal for Technology in Mathematics Education*, 12(2): 51–58.

Carlson, M. P., & Bloom, I. (2005). The cyclic nature of problem solving: An emergent multidimensional problem-solving framework. *Educational Studies in Mathematics*, 58, 45–75.

Carpenter, J., & Camp, B. (2008). Using a web-based homework system to improve accountability and mastery in calculus. In *2008 ASEE Annual Conference & Exposition*.

Dorko, A. (2018). Red x's and green checks: A preliminary study of student learning from online homework. In A. Weinberg, D. Moore-Russo, H. Soto, & M. Wawro (Eds.), *Proceedings of the 22nd Annual Conference on Research in Undergraduate Mathematics Education*. (pp. 172–179). Oklahoma City, OK.

Dorko, A. (2019). Professors' intentions and student learning in an online homework assignment. In A. Weinberg, D. Moore-Russo, H. Soto, & M. Wawro (Eds.), *Proceedings of the 22nd Annual Conference on Research in Undergraduate Mathematics Education*. (pp. 172–179). Oklahoma City, OK.

Dorko, A. (accepted). Red X's and green checks: A preliminary framework for how students engage with online homework. To appear in the *International Journal of Research in Undergraduate Mathematics Education*.

Dorko, A. (2020). How do students engage with 'Practice Another Version' in online homework? To appear in *Proceedings of the 23rd Annual Conference on Research in Undergraduate Mathematics Education*. Boston, MA.

Ellis, J., Hanson, K., Nuñez, G., & Rasmussen, C. (2015). Beyond plug and chug: An analysis of Calculus I homework. *International Journal of Research in Undergraduate Mathematics Education*, 1(2), 268–287.

Ellis, J., Kelton, M., & Rasmussen, C. (2014). Student perceptions of pedagogy and associated persistence in calculus. *ZDM – The International Journal on Mathematics Education, 46*(4), 661–673. doi:10.1007/s11858-014-0577.

Gage, M., Pizer, A., & Roth, V. (2003). WeBWorK: An internet-based system for generating and delivering homework problems. http://users.math.uoc.gr/~ictm2/Proceedings/pap189.pdf.

Halcrow, C., & Dunnigan, G. (2012). Online homework in calculus I: Friend or foe? *PRIMUS, 22*(8), 664–682.

Hauk, S., Powers, R. A., & Segalla, A. (2015). A comparison of web-based and paper-and-pencil homework on student performance in college algebra. *Primus, 25*(1), 61-79.

Hauk, S., Powers, R., & Segalla, A. (2004). A comparison of web-based and paper-and-pencil homework on student performance in college algebra. *PRIMUS, 25*(1), 61–79.

Hauk, S., & Segalla, A. (2005). Student perceptions of the web-based homework program WeBWorK in moderate enrollment college algebra classes. *The Journal of Computers in Mathematics and Science Teaching, 24*(3), 229.

Heenehan, M. E., & Khorami, M. (2016). Students' reactions to the homework assessment system WeBWorK. *Mathematics and Computer Education, 50*(1), 42.

Hirsch, L., & C. Weibel. 2003. Statistical evidence that web-based homework helps. *MAA Focus, 23*(2): 14.

Hodge, A., Richardson, J.C., & York, C.S. (2009). The impact of a Web-based homework tool in university algebra courses on student learning and strategies. *Journal of Online Learning and Teaching, 5*(4), 618–629.

Krause, A., & Putnam, R. (2016). Online calculus homework: The student experience. In T. Fukawa-Connelly, N. Infante, M. Wawro, & S. Brown (Eds.), *Proceedings of the 19th Annual Conference on Research in Undergraduate Mathematics Education*, (pp. 266–280). Pittsburgh, PA: West Virginia University.

LaRose, P. G. (2010). The impact of implementing web homework in second-semester calculus. *Primus, 20*(8), 664–683.

Lenz, L. (2010). The effect of a web-based homework system on student outcomes in a first-year mathematics course. *Journal of Computers in Mathematics and Science Teaching, 29*(3), 233–246.

Leong, K. E., & Alexander, N. (2014). College students attitude and mathematics achievement using web based homework. *Eurasia Journal of Mathematics, Science & Technology Education, 10*(6).

Lunsford, M. L., & Pendergrass, M. (2016). Making online homework work. *PRIMUS, 26*(6), 531–544.

Mathai, E., & Olsen, D. (2013). Studying the effectiveness of online homework for different skill levels in a college algebra course. *PRIMUS, 23*(8), 671–682.

Raines, J. M., & Clark, L. M. (2013). Analyzing the effectiveness of tutorial learning aids in a course management system. *Journal of Studies in Education, 3*(3), 120–136.

Roth, V., Ivanchenko V., & Record N.. 2008. Evaluating student response to WeBWorK, a web-based homework delivery and grading system. *Computers & Education, 50*, 1462–1482.

Schoenfeld, A. H. (2014). *Mathematical Problem Solving*. Elsevier.

Schoenfeld, A. H. (1992). Learning to think mathematically: Problem solving, metacognition, and sense making in mathematics. *Handbook of research on mathematics teaching and learning*, NCTM.

Sonnert, G., Sadler, P., Sadler, S., & Bressoud, D. (2014). The impact of instructor pedagogy on college calculus students' attitude toward mathematics. *International Journal of Mathematics Education for Science and Technology*. doi:10.1080/0020739X.2014.979898.

Spangler, D.A. (1992). Assessing students' beliefs about mathematics. *The Mathematics Educator, 3*(1), 19–23.

Suzuki, J. (2003). Using online quizzes: A report from the trenches. *FOCUS, 23*(9), 8–10.

Weibel, C., & L. Hirsch. 2002. WeBWorK effectiveness in Rutgers Calculus. preprint. 18 pages. http://www.math.rutgers.edu/~weibel/webwork.html.s

White, N., & Mesa, V. (2014). Describing cognitive orientation of calculus I tasks across different types of coursework. *ZDM, 46*(4), 675–690.

Yushau, B., & Khan, M. A. (2014). Student perceptions of online homework in preparatory year pre-calculus courses. *International Journal of Mathematics Trends and Technology, 8*(1), 12–17.

Zerr, R. (2007). A quantitative and qualitative analysis of the effectiveness of online homework in first-semester calculus. *Journal of Computers in Mathematics and Science Teaching, 26*(1), 55–73. Chesapeake, VA: Association for the Advancement of Computing in Education (AACE).

CHAPTER 3

Designing Mathematics Hybrid Classrooms in High School

The Case of Valeria

Chiara Andrà, Domenico Brunetto, and Igor' Kontorovich

CONTENTS

3.1 INTRODUCTION

In the recent decades, online teaching and learning are spreading among tertiary education: from online resources that students can obtain in a few clicks almost on every topic, to Massive Open Online Courses (MOOCs) that have become a standard in many universities worldwide. As a matter of fact, the majority of tertiary students around the world seems to engage in online learning of some sort and, as a consequence, some online mathematics forums have thousands of members and receive hundreds of posts every day (van de Sande, 2011). These new learning formats, such as MOOCs, promote self-directed learning, since the quality of students' engagement with and their understanding of the mathematical content in the MOOC videos depends on the extent to which they search for sources of knowledge in order to better understand and deal with the content of the video, when this is not immediately clear to them (Fredriksen, Hadjerrouit, Monaghan & Rensaa, 2017). Furthermore, to engage in learning formats like MOOC videos has been proven to be a crucial ability for the success of students in tertiary-level mathematics (Niegemann, Domagk, Hessel, Hein, Hupfer & Zobel, 2008).

Despite positive learning outcomes from MOOC videos and online resources in general, and despite the fact that our current and future students are part of an online generation by definition, students should not be assumed to be online learners from birth, at least not when mathematics and statistics are under discussion. Indeed, contemporary research recurrently shows that online environments are not as helpful as it could be assumed in overcoming classical challenges of higher mathematics education (e.g., see Tall, 1991 for abstractness, formality, and complexity; Moore, 1994 for proving; Bosch, Fonseca & Gascon, 2004 for new didactical contracts). Moreover, these environments entail new challenges of their own (Fredriksen et al., 2017), and tertiary students tend to resist online teaching formats. Our claim is that introducing online learning formats at secondary level would help both to smooth the transition from high school to university, lowering difficulties with mathematics, and to prepare the students to welcome these new formats once enrolled at university, getting used to them and appreciating the advantages. This claim is supported by numerous research findings, as we briefly recall in what follows. In their fundamental study, Clark and Lovric (2008) contend that at the basis of the leap between secondary school and university there is a shock: the students have to move from the procedural mathematics the students are used to at school, to the conceptual understanding that university mathematics entails. According to Hibert and Lefevre (1986), conceptual knowledge describes knowledge of the principles and relations between pieces of information in a certain domain, while procedural is the knowledge of the ways in which to solve problems quickly and efficiently. Some researchers (e.g., Gamer and Gamer, 2001) found that teacher-paced instruction favors the development of procedural knowledge, while student-directed instruction favors the development of conceptual knowledge. A teacher-paced lesson provides the students with a linear and organized exposition of knowledge, while a student-directed one engages students in groupwork activities, classroom discussion, and in the production of meanings that are inevitably other than final or authorized: they are personal and provisional, not universal and absolute. The use of video clips from university lectures in secondary mathematics classrooms has the potential of both acclimatizing the students to the language and the teaching style of university math professors and of promoting student-paced learning formats, since it may happen that the students watch the videos and try to get sense of the content without any guidance from the teacher, or in case parts of the videos are not clear for the students, search for other sources in order to make sense of the content, or propose new activities. All this entails a production of meanings, from the students, that is personal and that emerges from the mathematical activity in which each student is engaged. With Niegemann et al. (2008), we maintain that video-based mathematics lessons combine self-directed and externally regulated learning types of instructional formats: in fact, the math teacher plays a major role in introducing the videos to the class, and the focus of this chapter is on the role that teachers' beliefs, goals, and resources play in shaping the way(s) video clips are introduced in different mathematics classes.

The value of instructional videos for enhancing learning is recognized by many scholars, inside and outside mathematics education. Berk (2009), for example, notes that the variety of video formats, the ease with which the technology can facilitate their use in the classroom, the number of video techniques an instructor can use, and the research on multimedia learning that provides the theoretical and empirical support for their use as an effective teaching tool, prompt teachers to use video clips in their teaching. According to Berk (2009), findings from cognitive neuroscience support positive implications for educational practice, and the use of video clips in teaching is more appropriate for introductory courses, for introducing complex topics in any course, for lower-achieving students, and for visual/spatial learners. Furthermore, research findings on the effectiveness of videos embedded in multimedia classes are very encouraging: a large number of studies in teacher education, for example, have produced significant results favoring videos (e.g., Sherin, 2003). However, and despite the promising findings in education, few studies are dedicated to the use of math videos in classrooms, at secondary level. In order to contribute to this *problématique*, the present chapter focuses on the use of math videos in secondary school.

More specifically, the research project *Flip Math* at the Politecnico di Milano is aimed at preparing high school students to tertiary studies through turning their traditional face-to-face classrooms into hybrid learning environments. In the last six years, dozens of teachers in Italy engaged in the *Flip Math* project. The teachers engaged in the project driven by their interests in introducing online components into their mathematics lessons and in enhancing self-directed learning. The online components consisted of instructional videos that were taken from the first Italian mathematics MOOC initially developed by the Politecnico di Milano. The MOOC was developed for first-year engineering students who needed a recap of a variety of topics often captured under the term of "pre-calculus." The MOOC emphasized mathematical terminology, formal definitions, and problem-solving skills. The MOOC is a research-informed initiative that has been developed to be useful as a whole but also in a modular mode, where videos and activities can be used in the actual classroom to provoke exploratory discussions. Taken together, these characteristics turn the MOOC into a rich educational resource that is capable of supporting teachers in achieving the desired change.*

A particular sample of experienced and technology-enthusiast secondary mathematics teachers engaged in the *Flip Math* project; until now, all teachers have an average of teaching experience of 20 years (five years is the minimum). Furthermore, all of them come with evidence of impressive technological knowledge and skills. The research project turns out to be a project investigating how mathematics teachers with good technological

---

* The interested reader can freely log in the course at www.pok.polimi.it.

knowledge and with long experience in teaching develop new learning formats entailing MOOC videos. Two issues related to the sample emerge at this point: one regards the generalizability of the results of our research, conducted on a purposive sample of experienced and technology-enthusiast teachers, and the other one regards the shift of focus that our research undertook. As regards the former, we maintain that the issues that emerge have a "general" nature, as we will argue in the discussion. As regards the latter, if it is true that instructional videos have learning potentialities (as revealed by promising research findings on teacher education, for example), and if it is true that learning formats such as the flipped classroom are taking the lead in many countries because they are effective, it is also true that it is important to investigate how the teachers introduce and use math videos in classroom. Up to date, and to our knowledge, this phenomenon has received little attention in research. The research focus is not on examining the effects of introducing online learning formats at secondary level on students, but it is on understanding the role of teachers' beliefs, goals, and resources in their decision making during the use of instructional videos in mathematics classrooms. An outcome of the research is the identification of different ways in which experienced and technologically literate math teachers appropriate MOOC videos in their teaching practice.

Technological literacy is not the only factor that influences a teacher's introduction of online resources such as MOOC videos in her classroom, as Anthony (2012) noted. Other individual characteristics such as a teacher's beliefs, or the perceived usefulness of technology, are essential factors to determine both the frequency and the centrality of MOOC video use in mathematics classrooms. We, thus, adopt Schoenfled's (2011) lens and focus on a teacher's orientations, goals, and resources to examine how they influence a teacher's choice of using MOOC videos in her class. It is well-known, in fact, that orientations, goals, and resources play a key role, but how they interact, especially when a conflict between two different goals, or between two contrasting beliefs, takes place, is less researched. The aim of the present chapter is to showcase examples from our extensive data and discuss the role of teachers' beliefs in shaping their choices when they introduce their classrooms to MOOC videos. To this aim, we summarize the conceptual framework that informs our lens of analysis in the next section.

3.2 CONCEPTUAL FRAMEWORK

Schoenfeld (2011) argues that, when people make decisions in well-practiced domains, as is the case for teachers in the domain of teaching, it is possible to model decision making in a quite precise way as a function of knowledge and other intellectual, social, and material resources, goals, and orientations. "Well-practiced" activities are those for which a person has had enough time to develop expertise, knowledge, and routine that shape the majority of her actions. According to Schoenfeld (2011), any person enters into a particular context, such as a classroom, with a specific body of resources, goals, and orientations. For example, a teacher starts her lesson with a specific body of resources, goals, and orientations, but a teacher engages in a new research project with the same specific body, as well. Whatever the situation, a teacher takes in and orients to it. Certain resources become essential and are activated. Goals are either established or recalled (if pre-existing). Consistently with the goals, and within the constraints and the potentialities offered by the resources, decisions are made. An interesting point of Schoenfeld's model is the distinction between familiar and not familiar situations: in the former case, Schoenfeld (2011) notes that actions can be automatic and they can consist mostly of the implementation of scripts, frames, or routines. In the latter case, there is not an established routine; hence decision making is made by a mechanism that can be modeled as a function of the subjective expected values of available options, given the orientations of the teacher. When a teacher joins a research project like *Flip Math*, she finds herself engaged in both familiar (i.e., routine classroom practices) and unfamiliar situations, which can subvert the routine.

The basic assumption of Schoenfeld's (2011) framework is that beliefs and orientations are an essential factor shaping teachers' decision-making, behaviors, and professional development. Schoenfeld proposes to understand beliefs as "perceptions on the part of individuals that shape the ways in which they frame or orient themselves to any particular context, and thus shape they ways they act in that context" (Schoenfeld, 2011, p. 460). This assumption needs to be complemented with additional considerations.

First of all, it is necessary to define what is meant by "beliefs" and "orientations." To this end, Skott (2015) maintains that there are four characteristics of beliefs that seem to be shared by almost all researchers: *conviction*—beliefs can be thought of as knowledge that is true at least in the eyes of the beholder; *commitment*—beliefs are value-laden and relate to motivation; *stability*—beliefs are stable and change only after substantial, new experiences; *impact*—beliefs influence one's individual perception as well as one's practice. In Schoenfeld's (2011) view, orientations include beliefs, but also values, preferences, and tastes. Values are the deep affective qualities which education aims to foster through the school subject of mathematics (Bishop, Seah & Chin, 2003). Values and beliefs are related to each other since there is a presence of underlying values in beliefs: values are conceptualized as beliefs held by individuals to which they attach special priority or worth, and by which they tend to order their lives. As such, values can be thought of as guides to behavior. A category of values is related to an individual's preferences and tastes, i.e., personal values (e.g., when choosing a new car, one may consider the price, or the color, or the design, or functionality). Economic reasons, design, and functional importance are referred to as personal values (see Shimada & Baba, 2015).

The second consideration is that orientations are systemic and contextual. Schoenfeld (2011) posits that in many situations, an activation of certain beliefs triggers the activation of other, related beliefs, and in such a way, a belief system contributes to the choices that teachers make. Skott (2015) suggests seeing the *context* as a constraint on the opportunities for "belief enactment," acknowledging a central role of social interactions that take place in the context.

According to Schoenfeld's framework, teachers' decisions and behaviors also depend on goals and available resources (including knowledge, materials, and personal and interpersonal skills):

> Every sequence of actions can be seen as consistent with a series of goal prioritizations that are grounded in the teacher's beliefs and orientations, and the selection,

once a goal has been given highest priority, of resources intended to help achieve that goal.

(Schoenfeld, 2011, p. 460)

A goal, whether explicit or tacit and unarticulated, is something that a teacher wants to accomplish. Goals can be classified in terms of grain sizes. For instance, a short-term goal can be associated with a single lesson, while a long-term goal can refer to what a teacher wants her students to know as a result of a school experience. In many research projects with teachers, their genuine goals become difficult to observe, since the project's agenda dominates teachers' prioritizations. This one, even if it is the most widespread, should not be the only possibility, as different forms of teacher–researcher interactions may take place (see for example Wagner, 1997). Among the forms of teacher–research interaction identified by Wagner, ours is a case of *co-learning agreements*. In co-learning partnerships, the goals, methods, and principles of inquiry are negotiated openly to maximize the learning and the fruits of learning for both researchers and teachers. Therefore, co-learning agreements essentially reduce asymmetry in the roles of the researchers and teachers. According to Penuel, Fishman, Cheng, and Sabelli (2011), co-learning agreements focus on problems of practice from multiple stakeholders' perspectives, concerns with developing capacity for sustaining change in educational systems, and calls for breaking down barriers that isolate those who design and study educational innovations and those who implement them. Rather than accepting a readymade research agenda, teachers work with researchers on translating both sides' goals into a mutually beneficial agenda. Teachers are invited to work side by side with the researchers in developing, validating, and implementing data-collection tools, and practice collective reflection and data analysis. A particularly important stage in co-learning inquiry is formulating conclusions and implications. Here teachers can bring forth an established sense of what works in education and what makes education work, "a feel for the breadth, depth, and complexity of education as an institution that cannot be picked up by reading about it or observing it" (Labaree, 2003, p. 16). Our claim is that specifically in this form of teacher–researcher interaction it is possible to "see" a teacher's goals.

Finally, Schoenfeld's (2011) notion of resources includes all kinds of "goods" that are available for a teacher, for example, the tools in the classroom, students' knowledge, teachers' knowledge, interpersonal skills, and relations with students. To this respect, focusing on technology integration in general, Anthony (2012) argues that recent research offers insights into why technology integration efforts have not had a greater impact on teaching practice and student achievement: in Anthony's view, a strand of research has emphasized that teachers' individual characteristics, such as technological literacy (a resource in Schoenfeld's view), and constructivist beliefs or the perceived usefulness of technology (orientations with Schoenfeld's words), are essential factors that influence technology use; while another strand of research has focused on institutional conditions, such as school and district settings, as factors supporting or constraining technology use (resources in Schoenfeld's model). Indeed, Anthony (2012) stressed the role of resources (understood in terms of Schoenfeld's views) in shaping technology adoption by a teacher. Moreover, important for our study is Anthony's focus on both technology planning at school- or district-level, and technology integration that teachers enact in their classrooms. Anthony argues that the district-level system of planning and teacher-level system of integration connect in ways that mediate teachers' ability to implement technology with their students. Along this line of reasoning, we pay specific attention to the teacher's ability to utilize the potential of video-integrated lessons for the benefits of students' learning and, adapting Anthony's (2012) dimensions of technology integration, we focus on "centrality" and "frequency" as dimensions that are useful to understand the ways and the extent of MOOC video introduction in mathematics classrooms. Anthony, in fact, understands the centrality of technology use as either seamlessly embedded or peripheral to classroom routines. The frequency of technology use is accounted for as ranging from daily to less than once a month. "Infrequent" technology use occurs when it is used a few times per month, or less. "Frequent" technology use refers to weekly or daily use. Technology use is "peripheral" to a majority of classroom routines when technology is used as an add-on resource for remediation or enrichment. It is "central" when technology use is integral to a majority of classroom routines and is an essential teaching and learning tool. It is so much embedded in the classroom activity that learning would have assumed completely different forms without it.

The focus on frequency and centrality as indicators of "quality" of technology use is widespread in literature, since researchers seem to agree that technology promotes student-centered learning (see Clark-Wilson, Robutti and Sinclair, 2014), and as such it becomes central to know how much technology is employed by teachers, especially in those countries where huge economic investments has been made in order to equip the school with a variety of information communication technology (ICT) tools. A general research finding is that technology integration mainly depends on a teacher's orientations. The general finding on technology use, and in particular on ICT, seems to hold also for MOOC videos. Research findings on teachers' use of technology, and more broadly of innovative pedagogical settings, reveal that there is a considerable gap between the *frequency* of use of these innovative tools and the economic investments that have been made. On one hand, researchers report an increase in the students' academic performance in technologically rich environments (Cuban, 2001; Hannafin & Foshay, 2008; Wenglinsky, 2005); on the other hand, despite access to funding and equipment, a recent study found that less than 35% of teachers in U.S. districts that received specific funding have integrated online learning in their instruction on at least a weekly basis (U.S. Department of Education, 2010). Outside U.S., international surveys indicate low ICT integration in mathematics classrooms, though such indicators mask considerable variation between and within countries and provide little detail into how and why teachers use ICT. For example, in a survey of 42 countries within the Organisation for Economic Cooperation and Development (OECD), on average 32% of students reported that they, or their classmates, performed at least one of a range of seven mathematical tasks on a computer in the last month. A further 14% reported that only teachers demonstrated the use of computers—consistent with a finding of infrequent computer use in mathematics instruction (OECD, 2015). The 2011 Trends in International Mathematics and Science Study (Mullis et al., 2012) reported that only a quarter of students on average used computers at least monthly during mathematics lessons. Like technology in general, also online learning formats in particular are little exploited by teachers and their use is infrequent. As it emerges from this literature review, and as it is cogently argued in Clark-Wilson, Robutti, and Sinclair (2014), there exists a gap between the corpus of research findings that show how technology of different sorts promotes self-directed learning, and the actual practices that are observed in math classrooms. Videos can contribute to promote student-directed instruction, since by their nature they encapsulate both a rather transmissive modality, which might attune to the modalities of use that are more widespread among teachers, and they may require the students to activate resources in order to understand them and to give them meaning. However, to achieve this purpose a priority is to understand the role that teachers' beliefs, goals, and resources play when videos are used in math classrooms. The metric of success for the use of math videos is not the centrality or the frequency *per se*, but the promotion of self-directed instruction in the different scenarios that emerge from our observations.

3.3 METHODOLOGY

In what ways is enhancing self-directed learning a consequence of video use in terms of centrality and frequency? In order to answer to this question, through the lens of Schoenfeld's (2011) model, we explore the roles that orientations and goals can play in the lessons teachers plan and deliver, and in particular whether MOOC videos are central to classroom practices, how frequently are they used, and to what extent non-instructional engagement with the mathematical content takes place. So, in this chapter we aim at answering the sub-question: how are different, maybe contrasting, beliefs and goals related when teachers use videos in their classes? In this chapter, we analyze two lessons, and in the next chapter we analyze other two lessons, which represent a variety of possible scenarios.

3.3.1 The Context of the Research

Flip Math is an exploratory, longitudinal research project where teachers and researchers collaborate closely towards developing pedagogies that are powerful in hybrid learning environments. Teachers join the project on a voluntary basis; hence the sample for data analysis is purposive. Being based upon co-learning agreements, at the heart of the project are three activities: (i) intensive meetings where teachers develop innovative instructional methods, (ii) experimental lessons where the methods are implemented, and (iii) the theory-driven fine-grained analysis which is carried out and discussed with the teachers. Specifically, we use Schoenfeld's (2011) model of resources, goals, and orientations for making sense of teachers' lesson designs and in-the-moment decision-making. Overall, in the last six years we have collected data from a variety of classrooms with students across skill levels. In particular, three teachers, pseudonymously named as Valeria, Nicoletta, and Lorenza, joined the Flip Math project with their classes. In this chapter we consider the case of Valeria, while in the next chapter we consider also the cases of Nicoletta and Lorenza. We present the methodological details concerning the three teachers in this chapter, anticipating some information about the teachers whose data will be analyzed subsequently, because we would like to provide the reader with the full picture regarding our sample.

3.3.2 Participants: The Teachers and Their Students

Valeria, Nicoletta, and Lorenza were three highly experienced secondary mathematics teachers: Valeria has 30 years of teaching experience, she attends professional development courses for in-service teachers regularly, and she is a tutor for beginner teachers; Nicoletta has six years of teaching experience, she collaborates in research projects at the Politecnico di Milano, and she has an outstanding mathematical knowledge; Lorenza has five years of teaching experience, she has a PhD in Mathematics Education, and she regularly participates in international research conferences for researchers in Mathematics Education. Nicoletta and Lorenza are also lecturers at university and are used to online teaching formats like online lecturing, forum interactions, and blended learning. The purposive selection of a group of teachers who have strong technological knowledge, positive attitudes towards online learning, and good teaching expertise allows us to ground our findings on the basis of a rich teaching context, rather than focusing on a teacher's weaknesses.

During the school year 2014/15, when our research was carried out, Valeria was teaching two grade-11 classes (the second-to-last year in high school in Italy), which we refer to as classes A and B. During the school year 2016/17 Nicoletta was teaching a grade-12 class (first-to-last year in Italy), and Lorenza was teaching a grade-10 class. We refer to them as classes C and D, respectively: they will be analyzed and discussed in the next chapter. Valeria's class B, and Nicoletta and Lorenza's classes C and D are described by the teachers as "difficult" to some extent. For Valeria, her class B students are not open to innovation and complain every time she proposes something that is not routine. She, thus, wants to include them in the project to see if some change can happen. Nicoletta talks about class C as a class where the students are not motivated to do mathematics; they rather hate the subject and tend to do the minimum requested by the teacher in order to get good marks. Nicoletta joined the project because her class was already using math videos and she wanted to coordinate with other colleagues. Lorenza says that her students are very nice, hard-working, and collaborative, but have serious difficulties with mathematics. She aimed at conducting research on her students' difficulties, and she wanted to be part of a team of both researchers and teachers. To sum up, we can say that classes B, C, and D are described as "difficult" by the teachers in three different ways:

- Class B does not accept anything new and is not used to video-integrated lessons.

- Class C is not motivated in doing mathematics and is used to math videos.

- Class D has difficulties with mathematics and is not used to videos.

Interestingly, three out of four classes were chosen by the teachers not because the students were "the best possible," but for rather the opposite reasons, namely because the teachers wanted to improve the learning environment for these students, and we believe this is a feature of co-learning agreements that has positive outcomes for both the teachers involved and the researchers, who have the possibility to meet true, actual students.

Valeria, Nicoletta, and Lorenza were invited to use videos from the pre-calculus MOOC course developed at the Politecnico di Milano for students enrolling in the first-year university courses. The mathematical content reflects (and recaps) the mathematical curriculum of the last years in Italian high school, namely logic and set theory, pre-calculus, algebra, analytical geometry and trigonometry, probability, and statistics. The course is structured in six weeks, each one corresponding to one of the above-listed topics. Each week is divided into modules, and each module opens with a theoretical video, which serves the purpose of recapitulating the main concepts; then a series of practicum videos follow, and they serve the purpose of showing typical exercises; at the end of each module there is a quiz. Each week is made of three or four modules. These MOOC videos represented a novelty for all the teachers (also for Nicoletta, who was used to other math videos available online).

3.3.3 Data Collection

After watching the MOOC videos at home, all the teachers were invited to imagine a scenario for integration of some of the videos in their respective classes. Our data come from the meetings with the teachers after they watched the videos, and from video-recorded lessons where the chosen videos had some role. We met Valeria four times before entering her class, where we observed, took field notes, and videotaped two lessons of two hours each for each class. Valeria shared with us all the material produced online before and after the lessons. The videotaped lessons took place in two subsequent days, and on the same day we were present in class A for the first two hours, and in class B for the second two hours. We met Nicoletta in person two times before the lesson, and we also met her on Skype once. She wrote for herself and shared with us many considerations about her class and her goals. Since the three lessons took place in three subsequent weeks, we had time to meet in between the lessons, and she made comments on what had happened. We met Lorenza five times before her lesson, and we shared with her also the theoretical framework we were going to use in order to analyze the data. We observed and videotaped two lessons, in two subsequent weeks, but had no chance to meet her in between them. We met Lorenza once after the lessons. After their lessons, we presented and discussed with Valeria and Nicoletta our theoretical lens of analysis, and they shared various comments.

Our data analyses was concerned with: (i) the teaching context, which refers to how a teacher describes her class, the school, and the pedagogical setting she is used to implement in the class; (ii) the lesson image, namely the lesson plan paired with the teacher's expectations about the lesson; and (iii) in-the-moment decisions that the teachers made during the lessons.

A *teaching context* emerges from the analysis of teachers' resources, orientations, and long-term goals. The aim of this analysis is to understand different scenarios for the introduction of MOOC videos. A teacher can be concerned with the technological infrastructure of the school, teacher's perceptions of students and their abilities, and ideas regarding how videos can be used in a lesson. At this stage, it can emerge how frequently MOOC videos are (intended to be) used by a teacher, as well as which technology that may support their use is available.

A *lesson image* comes into being from the interactions between researchers and teachers after the teachers have watched the MOOC videos and before enacting the imagined lesson. Following Schoenfeld's view, at this phase decisions should be made regarding the mathematical topics of the lessons that will be carried out, the kind of activity, and the (short-term) goals of these lessons. Accordingly, the discussions between researchers and teachers are focused on designing students' engagement with the videos and on the activities that precede and follow the engagement. For instance, mathematical problems and tasks can be created for reaching the lessons' goals. We analyze how MOOC videos are related with the tasks assigned to the students. At this stage, it can emerge whether MOOC videos are central or peripheral to the mathematics classroom activity, namely whether the use of this resource is unavoidable to reach the learning outcome, or whether it is used as a sort of add-on. It is also at this stage that the extent to which teachers want to promote self-directed learning emerges.

A few days after the meetings during which the teaching contexts and the lesson images emerged, Valeria, Nicoletta, and Lorenza went to teach in their respective classrooms. We videotaped the lessons carried out in each class. The videos provided us with data on teachers' *in-the-moment decision making*. The data contain video-recordings, notes of the teachers, students' worksheets, and field notes taken by the observing researchers. We look at matches/mismatches between the lesson image and the actual lesson. We interpret a match as further evidence for the importance of declared resources, orientations, and goals. If there's a mismatch, we will analyze the gap between *declared orientations and goals* and the actual *beliefs in use* and *goals carried over*. As a side note on the reliability of our results, we specify that all the data were analyzed side-by-side by three of us followed by extensive discussions of our interpretations with the teachers, as co-learning agreements entail.

3.4 THE CASE OF VALERIA

3.4.1 The Teaching Context

Valeria teaches in a technologically highly equipped school and has experience in supervising technology-related projects. Valeria told us that in her teaching she often uses various technological devices, such as laptops and an interactive whiteboard, as well as software, such as GeoGebra. She also adopts educational environments like Wikispaces. When we asked her to explain how she uses them, she specified that she devises a learning trajectory for each one of her classes, from grade 9 (the first year at high school in Italy) to grade 13 (the last year). She does not assume that students are able to work with technology "on their own," and hence, she gradually introduces software in grades 9 and 10. At grade 11 she starts assigning homework that involves software, and designs consequent lessons on the basis of students' submissions. In this way, she believes that the students gradually learn to appreciate both mathematics and technology.

Valeria also shared that technology helps her in representing mathematical ideas, giving her spare time and ease in her communication with the students. When we asked Valeria to explain her interest in integrating MOOC videos in her lessons, her response addressed her personal long-term goals and orientations:

It is innovative in terms of didactic methods, and it is an opportunity to use videos produced by experts to develop the teaching around the students. I am very willing to take part in this because it allows me to sharpen my teaching style, to improve myself, and to be involved in new challenges, which is useful for finding out new ways for reducing the gap between students and math.

Valeria watched a number of MOOC videos from the pre-calculus course. Her reaction indicated some of her orientations:

"It looks like a good way to foster the collaborative learning, which is also a keyword in the last ministerial directive."

She concluded that she would need to design new practices for integrating the videos. Valeria, in fact, is used to working with software like GeoGebra, where the students have to manipulate mathematical representations and solve problems. In the case of MOOC videos, the students are confronted with mathematical content that is somehow ready-made and not necessarily related to problem-solving activities.

Regarding her long-term goals related to students' learning, Valeria explained:

I want my students to approach technologies and multimedia with critical thinking, they should be able to use software and applets properly, they should view this kind of videos, which are very dense, getting the main idea and being able to discern the details.

Paired with her belief that MOOC videos provide new learning opportunities for her students, another belief also emerges from Valeria's words, related to differences in her students: she anticipates:

"Not all my classes fit for MOOC videos."

3.4.2 Lesson Image

To recall, Valeria chose two classes for this project, referred to as class A and B. Valeria decided to develop two lessons (one for each class), with a common (short-term) goal: to recall the properties of monomial functions (e.g., x^n) and their inverses, i.e., root functions. Valeria's students learned these topics in lower grades. She chose to use the same six-minute video clip for both classes A and B. In the clip, a lecturer addressed the graphs and definitions of monomial and root functions for natural, integer, rational, and real exponents. The root functions were presented as inverses of monomial functions, and the notions of oddness, evenness, and symmetry were briefly explained.*

3.4.2.1 Class A

Valeria planned to ask the students of class A to watch the MOOC video at home before the lesson and to answer a list of questions that addressed the definitions, properties,

---

* The video can be watched in Italian at https://youtu.be/15wQiw8fOLs.

graphs, and relations among monomial and root functions. At the lesson, she planned to divide the students in small groups and to engage them in solving a challenging problem. More precisely, her lesson plan was to spend the first five minutes dividing the students in groups of four, give each group a sheet of paper with the problem on it, and then leave the students alone doing problem-solving for about 25 minutes. The subsequent 15 minutes were planned to be dedicated to a classroom discussion of the problem; then she planned to show a second video clip on exponential functions for introducing students to a new topic.* The video lasts four minutes, so another ten minutes are left for Valeria's short frontal lesson.

Valeria considered different challenging problems (some of them have been proposed by us), and eventually chose a "paper-folding" problem (see Box 3.1). The mathematical heart of the problem lies at a geometrical progression that emerges when a piece of paper is consequently folded in halves. The connection between the video and the problem should be found in the function $y = x^n$. Valeria wants her students to recall the properties of this function (and of its inverse, the root function), and then to work on a problem that exploits the notion of power (where the basis is fixed and the exponential varies, to introduce the class to exponential functions), and reaches an unexpected outcome: after relatively few folds, the paper becomes really thick.

Valeria explained that she chose this problem because "*it comes from the real life and is very mathematical. Moreover, it shows a non-trivial connection between functions and geometric sequences.*" To discuss the connection with real life and the mathematical nature of the problem is outside the aim of this chapter, as well as it is not the aim of this work to present and discuss the students' answers to the problem. What is central to our chapter is that, in Valeria's view, the MOOC video is a tool that can help the students to recall some math content, and it is expected to be handled by the students alone, at home, without any guidance from the teacher. Valeria's usage of the video clip is quite similar to her usage of the challenging problem: her beliefs and orientations towards class A shape her expectation that the students will be able to learn from the MOOC video by themselves. In fact, the students are planned to be left alone watching the video (at home) and left alone in solving the problem (in class), before a (short) classroom discussion. We can say that this activity, at least in Valeria's intentions, promotes non-instructional engagement with mathematical content, which is one of the key features of online learning at tertiary level. By non-instructional engagement with a learning activity we mean that students are left free to choose what to do, which resources use, and which sources of information resort to, in order to complete an assigned learning task. Instead of it being the teacher who decides (and has full control) on where the classroom activity should go, the students are put in a situation where they have genuine possibilities to choose. This a case of self-directed learning. Are MOOC videos central to the planned activity in class A? To answer this question, we should consider that the first video serves the purpose of recalling math content already "seen" by the students, but they are able to participate in the in-class activity even without having watched it. The second video, planned to be watched at the

---

* The video can be watched in Italian at https://youtu.be/MT9dSVmq2Ek.

end of the lesson, serves the purpose of recapitulating concepts that would have emerged during small group activity. Hence, group work and problem solving are central to the planned lesson in class A, while MOOC videos are peripheral. Moreover, we are prone to conclude that the frequency of video use is high, both because it is planned for both homework and in-class activity, and because in general this class makes frequent use of online resources.

BOX 3.1   The folding-paper problem designed by Valeria

Consider a thin piece of paper. At the beginning fold it in half, then fold the folded paper in half, then again, and so on.

1. How can we describe this situation mathematically?
2. What will be the thickness of the folded paper after 100 folds, if the thickness of the original piece of paper is 0.1 mm?

3.4.2.2 Class B

The students of class B were planned to watch the first video twice during the lesson. After the first time (six minutes), the students would create a table with concepts from the video that were familiar and unfamiliar for them (ten minutes). Then, Valeria wanted to replay particular parts of the video clip to help the students with identifying definitions of the key concepts (another 20 minutes). Afterwards, the students were to be divided into small groups and discuss the concepts of oddness, evenness, and inverse functions. In group discussions the students would have been requested to sketch examples of such functions with GeoGebra and summarize their group work (the remaining 20 minutes). In the case of class B, the teacher *did not plan to assign any homework*: the video would be watched in classroom, and the teacher would have full control of how many times they watch it, and on what it is relevant to dwell on. In class B the students were not planned to be left free to explore the video.

In class B the students are expected to be divided in small groups, but in Valeria's words there's no indication that she is searching for a challenging problem for her class B students (even if, indeed, the assigned task can be perceived as challenging by the students of class B). We have commented that Valeria's usage of the video in class A is quite similar to her usage of a challenging problem. It seems that the same pairing can be made for class B: teacher-guided watching of the clip and teacher-guided filling of the table. However, the MOOC video is central to this lesson plan, since the in-class activity pivots around its first watching and a subsequent replay of some of its parts. The students of class B are expected to work for the entire lesson on the content of the video. But use of MOOC video is infrequent in this class, since the students are not used to online learning resources.

3.4.3 In-the-Moment Decision Making

Valeria started the lesson in class A by asking if the students watched the video and if they had any questions about the concepts that were discussed there. There were no questions, and the class turned to the group work on the paper-folding problem. All the groups answered question 1 correctly.

In one of the groups, the students were working on question 2 and asked Valeria how she came up with the number of 0.1 mm. Valeria redirected the question to the class and asked them to consider possible ways for determining the thickness of a piece of paper. After the lesson, Valeria told us that this question was unplanned. The groups approached Valeria's question differently. One of the groups tried to measure the thickness with standard rulers. When a student in another group noted these unsuccessful attempts, she recalled that *"there exists an instrument for measuring thin things, but I don't remember how it's called."* Then her group engaged in looking for thickness gauges on the Internet and exploring how they work. A student in another group made a connection between Valeria's question and the given problem, and suggested to fold the paper several times until the thickness becomes measurable with a standard ruler. His group liked the idea and engaged in developing a formula for the thickness of a piece of paper as a function of the measured thickness of the folded paper and the number of folds. Another student noted that this solution is not always practical because the number of times that a piece of paper can be folded is quite limited.*

If we focus on the in-the-moment decisions of Valeria in class A, we can comment that the actual lesson resembles her original plans quite well: the students reported that they watched the video at home, then the students engaged in the problem-solving activity, and the unplanned work on question 2 allowed Valeria to invite the entire class to think about a real-life problem from multiple perspectives. We can see a match between declared beliefs and in-the-moment decisions in Valeria's involvement of the entire class in searching for an answer to the second question: the students are expected to solve the problem by themselves, working in groups.

A difference with respect to the planned lesson is that the students did not watch the second video clip on exponential functions. We can infer that another of Valeria's beliefs emerges, namely her appreciation of students' work more than a teacher's "giving knowledge." Instead of stopping the students' activity in order to watch the second video (transmissive pedagogy) and to stick to the plan, she decided to let the classroom activity go on and focus on the mathematical relations between physical quantities (connective pedagogy). Her valuing problem-solving activity prioritizes certain actions, so the MOOC videos in the implemented lesson become even more peripheral than planned.

In class B, after showing the video clip for the first time, Valeria invited the students to create a table of familiar and new concepts, as it was planned. However, the students asked to watch the video again because they did not pay attention the first time. After the second time, they still did not engage in filling in the table and Valeria decided to change the plan: she created a table for the whole class on the whiteboard and replayed particular segments of the video, around ten seconds each, with the concepts and properties that she considered important (e.g., image and domain, graphs). After each video-segment,

---

* We refer the reader to the work of Gallivan for an exploration of the idea of folding paper n times, and an empirical proof for 12 folds. The reader can also see http://pomonahistorical.org/12times.htm.

the students pointed at the central concepts in the segment (practically, repeated the concepts' names) and Valeria explained them by extending the explanations of the lecturer in the video clip. Such interactions continued until the end of the lesson and it did not leave time for the planned group work. The way the lesson in class B turned out to be managed confirms our inference that Valeria has different goals for her two classes, and in particular that in class B she is rather concerned with students' engagement with videos. Also for class B, thus, unplanned decisions were driven by the beliefs and orientations that emerged in the interviews before the lesson start (i.e., the teaching context and lesson image).

3.5 DISCUSSION

If we compare the lesson images and in-the-moment decisions regarding classes A and B, we notice that there are not relevant mis-matches. Valeria's lesson plan allows us to have a "good enough" idea of what actually happened in the two classes, a few days later. Even her decisions for unplanned facts do not take us by surprise: in-the-moment decisions align with her declared goals and beliefs. This situation is interesting since it is not very common: teachers activate different beliefs and thus act differently in different contexts (as Skott, 2015, observes), and different, sometimes contrasting goals emerge. However, this is not the case for Valeria: it seems that Valeria is dealing with a familiar situation.

Valeria's case becomes more interesting, if we consider the striking differences between the two classes: in class A, a high frequency of MOOC video use, despite being peripheral, is connected to self-directed learning that (at least, apparently) could have been taken place even without the introduction of the MOOC video. In class B, we notice very low frequency of online resources, but the MOOC video becomes central to the lesson, where we see a teacher-guided activity where the students have a few possibilities to explore and manipulate the mathematical content that is offered to them. However, it is exactly this striking difference between A and B that can help us find a reason for such a strong match between the planned and the implemented lesson, the beliefs enacted, and the goals emerged in either context: to our interpretation, class A is "too good" and class B is "too poor" in Valeria's expectations, that (with her long and deep teaching experience) she can plan a lesson that resembles more or less exactly what that is going to take place in the actual class. In other words, self-directed learning takes place in class A even without the introduction of a new learning format, which lies in the background. In class B, the MOOC video is brought at the foreground by the teacher, exactly because the class is not used to self-directed learning formats, but the MOOC video is used in a way that is all but promoting non-instructional engagement.

The striking difference between classes A and B allows us to further notice that a hierarchy emerges in Valeria's belief system, namely, her expectations about her students take the priority when it is time to decide the mathematical activity, while the way the video can be used in order to promote self-directed learning comes second. This tells us that the classroom is not passive in the eyes of the teacher: she knows that the students respond to her prompts, and she knows how they respond. However, her (negative) expectations about class B somehow freeze Valeria from proposing something "really new" to them. In this way, the students' different engagement with the task in classes A and B results in being less influenced by the use of technology and more a reflection of the different mathematical attitudes, or motivation to learn. This was anticipated by Valeria in her lesson image, and things went the way they were expected: as a conclusion, Valeria's beliefs about her classes shaped the way the two lessons resulted, and the students' reaction. In order to understand the role of a teacher's beliefs and resources in shaping the use of videos, more nuances need to be added before we can make some conclusions. The cases of Nicoletta and Lorenza are, thus, presented in the next chapter.

REFERENCES

Anthony, A. B. (2012). Activity theory as a framework for investigating district-classroom system interactions and their influences on technology integration. *Journal of Research on Technology in Education, 44*(4), 335–356.

Berk, R. A. (2009). Multimedia teaching with video clips: TV, movies, YouTube, and mtvU in the college classroom. *International Journal of Technology in Teaching and Learning, 5*(1), 1–21.

Bishop, A., Seah, W. T., & Chin, C.(2003).Values in mathematics Teaching–The Hidden Persuaders? In: A. Bishop, M.A.K. Clements, C. Keitel-Kreidt, J. Kilpatrick, & F.K.-S. Leung (Eds.), *Second International Handbook of Mathematics Education* (pp. 717–765), Dordrecht, NL: Kluwer Academic Publishers. Printed in Great Britain.

Bosch, M., Fonseca, C., & Gascon, J. (2004). Incompletud de las organizaciones matematicas locales en las instituciones escolares. *Recherches en Didactique des Mathématiques, 24*(2–3), 205–250.

Clark, M., & Lovric, M. (2008). Suggestion for a theoretical model for secondary-tertiary transition in mathematics. *Mathematics Education Research Journal, 20*(2), 25–37.

Clark-Wilson, A., Robutti, O., & Sinclair, N. (2014). *The Mathematics Teacher in the Digital Era: An International Perspective on Technology Focused Professional Development.* Dordrecht, NL: Springer.

Cuban, L. (2001). *Oversold and Underused: Computers in the Classroom.* Cambridge, MA: Harvard University Press.

Fredriksen, H., Hadjerrouit, S., Monaghan, J., & Rensaa, R. J. (2017). Exploring tensions in a mathematical course for engineers utilizing a flipped classroom approach. In: *Proceedings of the 10th Conference of the European Society for Research in Mathematics Education.* Dublin, IR: CERME.

Gamer, B. E., & Gamer, L. E. (2001). Retention of concepts and skills in traditional and reformed applied calculus. *Mathematics Education Research Journal, 13*(3), 165–184.

Hannafin, R. D., & Foshay, W. R. (2008). Computer-based instruction's (CBI) rediscovered role in K–12: An evaluation case study of one high school's use of CBI to improve pass rates on high-stakes tests. *Educational Technology Research and Development, 56*(2), 147–160.

Hibert, J., & Lefevre, P. (1986). Conceptual and procedural knowledge in mathematics: An introductory analyis. In: J. Hibert (Ed.), *Conceptual and Procedural Knowledge; The Case of Mathematics* (pp. 1–23), Mahwah, NJ: Lawrence Erlbaum Associates, Inc.

Labaree, D. F. (2003). The peculiar problems of preparing educational researchers. *Educational Researcher, 32*(4), 13–22.

Moore, R. C. (1994). Making the transition to formal proof. *Educational Studies in Mathematics, 27*, 249–266.

Mullis, I. V. S., Martin, M. O., Foy, P., & Arora, A. (2012). TIMSS 2011 international results in mathematics. ERIC Digest. Retrieved from ERIC database (ED544554).

Niegemann, H. M., Domagk, S., Hessel, S., Hein, A., Hupfer, M., & Zobel, A. (2008). *Kompendium Multimediales Lernen.* Springer-Verlag.

OECD. (2013). PISA 2012 assessment and analytical framework: Mathematics, reading, science, problem solving and financial literacy. OECD. Retrieved from http://dx.doi.org/10.1787/9789 264190511-en.

Penuel, W. R., Fishman, B. J., Haugan Cheng, B., & Sabelli, N. (2011). Organizing research and development at the intersection of learning, implementation, and design. *Educational Researcher, 40*(7), 331–337.

Schoenfeld, A. H. (2011). Toward professional development for teachers grounded in a theory of decision making. *ZDM, 43*(4), 457–469.

Sherin, M. G. (2003). New perspectives on the role of video in teacher education. *Advances in Research on Teaching, 10*, 1–27.

Shimada, I., & Baba, T. (2015). Transformation of students' values in the process of solving socially open-ended problems. In: *Proceedings of the 39th Conference of the International Group for the Psychology of Mathematics Education* (Vol. 4, pp. 161–168). Hobarth, AU: PME.

Skott, J. (2015). Towards a participatory approach to 'beliefs' in mathematics education. In: B. Pepin, & B. Roesken-Winter (Eds.), *From Beliefs to Dynamic Affect Systems in Mathematics Education* (pp. 3–23), Heidelberg, DE: Springer.

Tall, D. H. (1991). *Advanced Mathematical Thinking*. Dordrecht: Kluwer.

U.S. Department of Education. (2010). *Transforming American Education: Learning Powered by Technology: National Educational Technology Plan 2010*. Washington, DC: Office of Educational Technology.

van de Sande, C. (2011). A description and characterisation of student activity in an open, online, mathematics help forum. *Educational Studies in Mathematics, 77*, 53–78.

Wagner, J. (1997). The unavoidable intervention of educational research: A framework for reconsidering researcher-practitioner cooperation. *Educational Researcher, 26*(7), 13–22.

Wenglinsky, H. (2005). *Using Technology Wisely: The Keys to Success in Schools*. New York: Teachers College Press.

CHAPTER 4

Designing Mathematics Hybrid Classrooms in High School

The Cases of Nicoletta and Lorenza

Chiara Andrà, Domenico Brunetto, and Igor' Kontorovich

CONTENTS

4.1 INTRODUCTION

Online math videos are spreading, and teaching formats like the flipped classroom resort to these resources, especially at university. Our research project aims at understanding the possible uses of math videos in secondary mathematics classrooms and focuses on the ways teachers can introduce and use the videos. In particular, we resort to the lens of analysis provided by Schoenfeld (2011), who understands a teacher's decision making as a selection of goals consistent with her orientations and resources, and we interpret the degree of centrality and the frequency of use of videos in relation to the promotion of student-directed instruction. In the previous chapter, the case of Valeria was analyzed, and it allowed us to partly respond to our research question, that is: in what ways is enhancing self-directed learning a consequence of video use in terms of centrality and frequency? In one class, videos are peripheral but self-directed learning is enhanced. In the other class, videos are central but teacher-paced instruction takes place. A possible interpretation of this is that, in the case of Valeria, beliefs related to the students are stronger if compared to those related to the positive impact of the use of technology in promoting self-directed learning. Hence, Valeria's case allows us to provide a first answer to the question: how are different, contrasting goals and beliefs related? In order to answer the general research question, which involves the ways of enhancing self-directed learning as a consequence of video use in terms of centrality and frequency, the cases of Nicoletta and Lorenza are added. The methodology described in the previous chapter applies also to these two cases.

4.2 THE CASE OF NICOLETTA

4.2.1 The Teaching Context

Nicoletta teaches in a poorly technologically equipped school, and yet, Nicoletta's students were used to watching mathematical videos at home. Nicoletta said that she wants her students to enroll in the whole pre-calculus MOOC on the platform www.pok.polimi.it, and she explicitly added that she expected that class C students' major difficulties would be with logging in and with understanding the organization of the courseware, since they were used to other websites.

In the interview, Nicoletta told us that her major *goal* is to enhance her students' ability to operate and learn from MOOC videos:

> "*I want my students to do not panic if some steps in a mathematical procedure are not made explicit, if one cannot grasp some mathematical concepts, or terms, at first or if different parameters are used.*"

Nicoletta mentions a potential difficulty that can arise when an MOOC lecturer makes use of terminology and symbols that are different from the ones used in a classroom. Nicoletta is aware that videos are not interactive: "*there exists a chance that the students will not engage with videos at home.*" Namely, she is worried that, in case a student did not grasp a concept or did not see a connection between the content of the video and the in-class activity previously done, she would give up. This is particularly important for a teacher like Nicoletta, who believes that her students tend to do the least possible to succeed in mathematics, and hence tend to not complete the tasks assigned if they have a feeling that effort is required. Thus, two long-term goals emerge, namely that class C students become both fluent with the mathematics discussed in the videos, and resilient.

4.2.2 Lesson Image

Nicoletta planned to assign a five-minute-long video to be watched at home. The video recalls some math concepts that had been already learnt by class C students. In it, the graph and some properties of the exponential function $f(x) = a^x$ are explained. The particular cases of e^x and e^{-x} are discussed, and the logarithmic function is also recalled. (The reader may notice that this video was planned to be watched by Valeria's class A in the classroom at the end of the lesson.) Nicoletta planned to assign some questions to be answered at home. In her words,

> I will assign the theoretical video, which recalls definitions and properties, and other two practicum videos which show the solution of exercises. The questions I will assign to my students will enable them to reflect on how a video can be watched, which questions can one pose to oneself, how exercises can be solved.

We can notice that Nicoletta's concern about her students' becoming able to deal with the (new, different) mathematical language of the video emerges in the choice of questions she wants to assign: no further exercises, but questions about "how" to approach the content. Examples of questions posed by Nicoletta are: did you find some mathematical words that are new to you? If yes, how did you deal with them? In case you did not understand a statement in the video, what did you do? Did you notice some difference between the way an exercise is solved in the video, and the way we solve it in class? If yes, which ones?

Nicoletta continued

> "I want to see my students' answers in advance, hence I will collect their work through emails."

Class C homework was expected to be checked by the teacher, even if for them it was not the first time watching a video. We can say that Nicoletta wants to have control of her class C, but we can also say that this practice would allow Nicoletta to prepare the lesson in advance, in accordance with her students' answers. In her lesson plan, the centrality of the video emerges: Nicoletta declared that in class she would "*recall the parts in the video where the graphs of exp(x) and of exp(–x) are shown simultaneously.*" She, indeed, plans to make explicit reference to the video.

> I will ask my students to draw the graph of f(x) and f(–x) for the following functions: a parabola of the form $ax^2 + bx + c$, sin(x) and cos(x). This will prompt the students to notice symmetries in some cases and I will introduce the definition of an even function focusing on the features on examples drawn by the students. The students will work in groups.

Nicoletta also mentions the good quality of graphs in the MOOC videos, and in fact she wants to exploit one of the graphs in the video to introduce the definition of even functions, instead of drawing her own at the lightboard. The MOOC video is central throughout Nicoletta's lesson image: even a new concept is planned to be introduced using the hints offered by the video.

Nicoletta's lesson plan for class C is as follows: she will divide the class in groups of four students each and she will assign a paper to each student (five minutes). Then, the teacher will recall the graphs of y = exp(x) and y = exp(–x) watched in the video at home by the students (five minutes). She will navigate the class while the students do groupwork activity on the basis of the assigned paper (25 minutes). Once the students, in small groups, have drawn the graphs of the parabola, of the sine, and of the cosine, she will recall the part in the video where f(x) and f(–x) are shown and in a frontal lesson she will introduce the notion of even function (15 minutes). In the remaining ten minutes, she plans to answer the students' questions that will arise.

In the implemented lesson in class C, something unexpected happened that is worthy of attention. We now report and analyze it.

4.2.3 In-the-Moment Decision Making

We met Nicoletta 15 minutes before the start of her lesson. She said that very few students did their homework, and she suspected that the majority of them had not watched the video. Her suspicion was grounded in her belief that the class is not motivated and tend to do the least possible in mathematics. A first decision was requested on Nicoletta's side: to either show the video at the beginning of the lesson, or to go on with the planned lesson and not to show the video. Eventually, she decided to start the lesson with the video, saying: "*it will last just for a few minutes, and to show the video won't compromise the lesson.*" We interpret her use of the verb "to compromise" in terms of time: since the video lasts a few minutes, her concern about the fact that she would not have enough time to do all that was planned is relieved. This unplanned decision further speaks to the centrality that the MOOC video has for Nicoletta's lesson: it is not possible to carry it out, without having watched the video.

To our view, however, to show the video at the beginning of the lesson changes the planned lesson in a substantial, even subtle way. First of all, the students were requested to do the homework, and they did not. The nature of the homework was reflective, and the students lost the opportunity to think about their ways of approaching a mathematical video. Nicoletta's goal to exploit the video in order to introduce a graphical approach to functions and to define the concept of evenness is prioritized with respect to her goal of having class C working (more) at home, individually, on a reflective task. Mathematical knowledge is prioritized, in class, with respect to students' individual reflection. The prioritization of this goal recruits resources throughout the lesson, as we now show, and discards other resources such as individual worksheets.

The students were divided in groups of four, but each individual student received a worksheet with some questions on it (as planned). Then, the class watched the video played by the teacher at the smartboard, and on the given, individual worksheets the students sketched the graphs of exp(x) and exp(–x), of sin(x) and sin(–x), and of cos(x) and cos(–x), while Nicoletta navigated the class and engaged in conversations with the students individually. Whilst being divided into groups, the majority of the class worked individually without collaborating with their group peers. It seems possible that the interaction of the teacher with individual students rather that with their groups promoted this behavior. There were two interesting exceptions: a group of four students (two girls and two boys) was really motivated and willing to do their best, so they interacted all the time proposing ideas and checking them within the group. They seemed seriously engaged with the task.

Another group of four boys interacted a lot, almost having fun: they chatted rather than doing mathematics and pretended to be working while the teacher was passing by. After 15 minutes of this kind of "work" in the class, the teacher decided to stop the groupwork, to play again the video from the point at which the lecturer introduced exp(x) and exp(–x), and to comment on the video. Some students, who had previously worked individually, intervened in the discussion. Nicoletta recapped the main features of the graphs of the two functions and then invited the students to go on with the worksheet. The dynamic of the class remained the same for another ten minutes, with the students working individually and the group of two girls and two boys working together. The teacher stopped the work another time and invited the students to find out the general features of the drawn functions. The students were mostly silent, so she drew the functions at the smartboard and in a frontal lesson she introduced the definition of an even function.

We comment that there was congruity between the lesson image and the implemented lesson, even when unplanned decision making was necessary. For class C, self-directed learning was (expected to be) high during homework, but (in the actual lesson) it was low for the in-class part. We also notice a mismatch between the lesson image and the actual lesson in terms of the extent to which the students had the opportunity to engage with the activity. This is partly due to these particular students' unwillingness to work on the assigned tasks.

4.4 THE CASE OF LORENZA

4.4.1 The Teaching Context

Lorenza teaches in a medium technologically equipped school. For example, each class has a smartboard, but the students do not have tablets or laptops. Lorenza wanted to recap exponentials and logarithms, and she wanted to exploit a feature of video-integrated lesson; in her words

> "The advantages of using MOOC are: saving time, better understanding since the students can stop the videos, and favouring the students' self-confidence with technology."

We recall that class D is not used to watching math videos. She adds

> "I also have non-math goals: to favour autonomy, to stimulate curiosity and to provoke critical thinking towards multimedia resources."

Lorenza's goals can be classified into long-term goals within Schoenfled's view, since she also wants to develop critical thinking. Lorenza also shares with Valeria and Nicoletta the awareness that videos are not interactive:

> Video-lessons are attended to at home, where students are comfortable, but at the same time there's a risk they won't work, compromising the efficacy of this pedagogical choice. A drawback is the impossibility to make synchronous questions and to

receive answers from the teacher in the video. This flaw can be dealt with the day after, at school, with their teacher.

Lorenza adds that the teacher is there, in class, to respond to any questions that arise during homework.

Lorenza's reflections on her students make an impression that she sees her students as collaborative:

"It's a class of only girls and they are really cooperative and collaborative with me. Some of them are good in math, but many of them have troubles with the subject."

4.4.2 Lesson Image

Lorenza chose to assign the same videos assigned by Nicoletta at home, i.e., a recap of exponential functions and two practicum videos, where exponential and logarithmic equations and inequalities are solved. Lorenza planned to show class D students how to access the MOOC, in a previous lesson, and to assign them the exercises in the MOOC, both the ones that have a solution provided in the practicum videos and those which are required to be solved in solitude. In class, she planned to discuss with the students their solutions, which she will collect via email in advance. In Lorenza's lesson image, we notice that she plans to spend a lesson commenting on the videos (*"I will discuss with the students their solutions"*), watched at home.

"In class, we will do more exercises."

In Lorenza's lesson image, the video is peripheral since it is planned to be watched at home, then commented on, but the very focus of the lesson would be on new exercises. The frequency of use of videos in class D is low, since they were introduced for the first time in a previous lesson, but the students are not used to them. The students are offered an opportunity to engage in a non-instructional way with math at home, but the in-class lesson image is rather teacher-guided.

4.4.3 In-the-Moment Decision Making

Having asked class D to send homework via email in advance, Lorenza was able to notice that *some* of her students (not *the huge majority*) did not send her their homework. Once in class, Lorenza firstly asked them why. She also asked how the students coped with the assignments: *"How did you feel with watching all these videos?"* To start with a question is a choice that is different both from Nicoletta and Valeria, but it was not unplanned, if we consider that Lorenza wants to start the lesson with a classroom discussion about students' difficulties. The students replied that the videos were clear, but they experienced difficulties with the assigned exercises and asked for teacher's assistance. Hence, the teacher engages the classroom in a rich discussion about "how to do" each exercise. They re-did the assigned exercises at the blackboard, stopping and commenting several times about general properties that emerged, but also on fine-grain details. We can notice that in class D the content of the videos was recalled and "replayed" by the teacher who wrote the exercises on the blackboard.

Class D students proposed the activity they wanted to do, by asking the teacher to redo the mathematical procedures watched at home (this was unplanned by Lorenza), even if the teacher was the one who responded to the questions. The students actively engaged in the discussion, which aligns with Lorenza's opinion about her students' cooperative mood. We also notice that her way of conducting the lesson stimulates the students' critical thinking, since many times during the lesson they were not satisfied with the procedure recapped by Lorenza and wanted also to recap "why to do so." The students took a lot of notes, and in their notes we see many remarks concerning "why" and "how" to proceed, instead of just copy-pasting the exercise that was written (by the teacher) on the blackboard.

4.5 DISCUSSION

Our research addresses the general imperative to understand how our students are "ready" for new-generation learning formats by focusing on teachers' orientations, goals, and resources that shape different ways of using online material in secondary mathematics classrooms. To recall, Lorenza, Nicoletta, and Valeria teach in four classes, and they decided to use the MOOC videos to introduce the same mathematical topic, namely, exponentials and logarithms. They share the opinion that MOOC videos allow a teacher to optimize time and promote the students' self-confidence with technology. They appreciate a feature of MOOC videos, that is, the possibility for a student to stop and re-watch any segment of the video-lesson, but at the same time they know that videos are not interactive, since no one can answer a student's question that may arise. Asynchronous interactions are the sole possibility in the chosen learning environment. In class, both Nicoletta and Valeria aim at recapping some mathematical concepts and at introducing new ones. Both Nicoletta and Valeria intend to split the class in small groups. Differently from Valeria and Nicoletta, Lorenza plans not to introduce new theory, nor to do problem solving, but to do exercises on the basis of class D homework.

Nicoletta conceives her lesson in a way that we can place somehow in between Valeria's classes A and B. Like class A, indeed, class C students are assigned a video as homework. Differently from class A, where the video was not planned to be recalled for the in-class activity, class C is planned to recall the content of the video several times: at the beginning of the lesson, recalling the graphs of exp(x) and exp(–x) and focusing on the symmetries; during the groupwork, drawing other symmetric/non-symmetric functions; and at the end of the lesson, introducing the notion of evenness. Also in Valeria's class B the video is present throughout the lesson, but the kind of work the students are expected to do in class is similar to the work that class C did at home, namely, to reflect on how to watch a video of this sort. To comment on these differences, Anthony's (2012) notions of frequency and centrality come to be useful to us: we would say that the frequency of video-integration is low in class B, where it is used only in that particular lesson because the students refuse innovative teaching formats, while it is high in classes A and C, where the students are used to working frequently with online material. However, in the specific lessons under study, the video-integration for class A was peripheral (since the classroom activity pivoted around a challenging problem), while it was central for classes B and C. The lesson in class B was designed to analyze the content of the video, and the lesson in class C was designed to start from the video watched at home and to bring the class to the definition of a new property of real functions.

In the implemented lesson of Nicoletta, we can notice that the group work activity turned out to be very different from Valeria's class A and to some extent it was much closer to class B's activity, with the teacher having control over what was going on. We further comment that self-directed learning was high in class A, both at home and in class, and it was low in classes B and C.

Lorenza's choice can be seen as an intermediate position between Valeria's class A and class B, too. Namely, in class A, the students are left alone in watching the video and Valeria does not care about possible difficulties that may arise, as if she sees her students able to deal with them; in class B, she wants to control everything and in the class she plays the video and checks how the students deal with it. Lorenza leaves her class D students alone at home (like class A), but she takes into account the possibility that some intervention would be needed in class (like class B). This scenario is similar, to some extent, to Nicoletta's lesson image, in which she planned to assign some work to be done alone at home, but to be sent to the teacher via email so that she knows the possible difficulties arising from the students.

Interestingly, like Nicoletta's class C, in the case of Lorenza's class D self-directed learning was high at home and low in class. However, the students in the two classes reacted very differently to their respective teachers' proposals: class C discarded homework and in class followed the teacher, while class D engaged (at least partly) in the assigned homework and co-participated in the teacher-directed lesson by proposing new directions to be taken, and the teacher followed them. Nicoletta's decision seemed not to be affected by her students' actions and behavior during the in-class activity, while Lorenza was open to changing her plans upon her students' requests. Another difference emerges from a comparison between classes A and D, which both watched the videos at home: class A had been left alone in dealing with the math content, while class D asked for help from their teacher. Videos are played in classes B (and this was planned) and C (but this was unplanned, since the students had to watch them at home). The (content of the) videos had also been "replayed" by the teacher at the blackboard in class D. The reasons why MOOC videos were "replayed" in classes C and D are, however, different: in the former, it was because the students did not watch the assigned videos at home; in the latter, it was because the students *did* watch the videos at home, but asked for the teacher's assistance.

Valeria's case tells us also that in one of her classes self-directed learning is promoted even with a peripheral use of videos. And in the other class, despite central use of video, the lesson is teacher-guided. MOOC videos are central to both classes B and C. In the former, videos are central because the students have to learn how to "use" them, while in the latter it is the mathematical content that necessarily needs to be introduced through the graphs and the terminology employed by the lecturer in the video. We have already commented that in classes A and D the videos are an add-on and different mathematical activities are central: problem-solving and exercises, respectively. In classes B and C the videos are central, yet unavoidable, in two different, contrasting ways. The four cases and the nuances of centrality they bring to the fore show how students deal with the videos and gradually become proficient users of online resources. Namely, the four classes under examination in this chapter can be further characterized by different stages of technology integration (see also Ruthven, 2002): Class B represents a case in which the teacher is mostly teaching how to use the tool, i.e., the video, hence she is concerned with the way the students learn how to manage it, and she proposes an activity that is (inevitably?) teacher-guided and pivots around the sense-making of the video. For this reason, the video is inevitably central to the math lesson. Class D represents a situation where the video as a tool is still a matter of concern for the teacher, and in fact she enacts a teacher-guided lesson where she is involved in troubleshooting, namely in answering the questions raised by her students following their previous, individual activity with the video(s). The video can become peripheral at this stage of co-evolution of technical and conceptual understanding. Going further along the spectrum of stages, we find class C, where the teacher makes use of the video not for the sake of teaching how to approach it, but for teaching mathematics with it. At this stage, the video becomes again central and deeply related to the evolution of conceptual understanding, since the technical one is rather advanced. Finally, in class A the students are so confident with online teaching formats that they do not need assistance from the teacher, and the mathematical activity can go on with a peripheral role assigned to videos. This interpretation, of the introduction of MOOC videos as proceeding through subsequent and interconnected stages, allows us to give sense to the ups and downs in the centrality of video use, and make sense of its different uses. Moreover, it leads us to conclude that MOOC videos should not necessarily be central to the mathematical activity, nor should they necessarily be frequently used, for self-directed learning formats to take place.

All in all, if we look at the four scenarios, and we try to find similarities and differences among them, we can see that the relationship between video use, teachers' views, teachers' practices, and teachers' being in their specific/individual classes is very complex. It is necessary to take this complexity into consideration to authentically understand the phenomenon and to contribute to a technologically rich and deeper-thinking school in the future. Within such a complex scenario, in our four cases we took into consideration the dimensions of frequency and centrality, and the extent to which self-directed learning is promoted by each teacher in each class. Our conclusions, which follow this section and end the chapter, try to understand which scenarios can be understood as successful with respect to the promotion of self-directed learning, and sketch possible future investigations.

4.6 CONCLUSIONS

Our findings are briefly summarized in Table 4.1, from which it emerges that, for example in Valeria's case, frequent use of technology goes along with students' self-directed learning in class A, while infrequent use of technology is paired with rather traditional, teacher-guided lessons in class B. Classes B and C, as opposites of the spectrum, confirm a general research finding that having equipped the school with technology is not necessarily related to the use of technology in classes: Valeria teaches in a well-equipped school but she does

TABLE 4.1   A Summary of Our Findings with Respect to the Categories Outlined

	Centrality of MOOC Videos	Frequency of Use of MOOC Videos	Self-Directed Learning
Class A	Peripheral (only watched at home)	High (general use of technology)	Promoted both at home and in class
Class B	Central (shape the entire lesson)	Low (very traditional)	Not promoted
Class C	Central (home-watched + replayed in class)	High (used to math videos)	Promoted (expected) at home, not in class (implemented)
Class D	Peripheral (meant as exercises)	Low (general poorly equipped)	Promoted at home, not in class

not use technology in class B, while Nicoletta teaches in a poorly equipped school but her use of technology is central and high. Furthermore, Classes B and D relate to cases of teachers who, even if engaged in a specific research project aimed at introducing the use of MOOC videos in their class on a voluntary basis, turn out to use the videos infrequently.

By looking at Table 4.1, which cases can be deemed as "successful" with respect to enhancing self-directed learning? In class A, it is enhanced both at home and in class and, borrowing a metaphor from clinical research, we can conclude that class A represents a "gold standard." In class D, self-directed learning is encouraged at home, and this can be seen as "successful." Classes B and C represent failures in promoting self-directed learning, but grounding a possible explanation on the fact that both classes are "difficult" does not satisfy us, because class D is also difficult but the teacher reached a (partial) success with respect to the aim. Our interpretation goes beyond a focus on the resources (namely, difficult students) and, given that all teachers in the study have positive orientations towards technology, it exploits the notions of frequency and centrality of video use. In both successful classes A and D, videos are *peripheral*. Being used with either high or low frequency seems not to be very relevant with respect to promoting self-directed learning. In both unsuccessful classes B and C, videos are *central*, and the fact that the students were already used to them (i.e., high frequency), or not (i.e., low frequency) seems not to play a crucial role. To draw the conclusion that central use of a resource does *not* enhance *what it is designed to enhance* seems confusing, but we need to take into account the fact that *these particular students* have difficulties with mathematics, they are poorly motivated in doing mathematics, and/or they refuse any innovation coming from the teacher. For this kind of student, a smooth approach with respect to the introduction of a new, somehow disruptive, online learning format seems to work better. A smooth approach to the introduction of MOOC videos assigns to them a peripheral place. In other words, in the cases of failure, the students may have perceived that MOOC videos had invaded their classes in the first time they met them, and as a consequence they have refused to work with MOOC videos. A general conclusion that we can draw is that it is advisable to introduce online mathematics resources in a peripheral way, especially with "difficult" students.

The picture that emerges from the four cases is a picture of teachers struggling to find a way to integrate MOOC videos in their classes, instead of a systematic and well-ordered picture of cases where different choices perfectly work. We recall that they are all expert teachers, and they are all technological enthusiasts. As a consequence, the picture that emerges does not question a teacher's knowledge, and it does not focus on her lack of either mathematical or technological knowledge. It does so *on purpose*. It is as if what makes sense for these teachers is no longer what is clear and precise, but we can use Nathalie Sinclair's words to investigate if they value an aesthetic of ambivalence, namely an aesthetic of simultaneous conflicting feelings, which brings to the fore confusion and loss in mathematics, while putting aside purity, coherence, and connectedness as the very hallmarks of mathematics (Sinclair, 2018). Such an investigation is worth carrying out as a follow-up to our research, since we agree with Sinclair (2018), who argues that this approach to learning phenomena may broaden our understanding of mathematics, and she warns us about the tremendous consequences of narrowing our view to consider only the beauty, certainty, or the systematic nature of the subject. Along this line of thinking, we aim at going on in analyzing "bad cases" that will allow us to unfold relevant issues concerning the uses of online resources in secondary math classrooms, and the meanings that underpin them.

REFERENCES

Anthony, A.B. (2012). Activity theory as a framework for investigating district-classroom system interactions and their influences on technology integration. *Journal of Research on Technology in Education, 44*(4), 335–356.

Ruthven, K. (2002). Instrumenting mathematical activity: Reflections on key studies of the educational use of computer algebra systems. *International Journal of Computers for Mathematical Learning, 7*(3), 275–291.

Schoenfeld, A. H. (2011). Toward professional development for teachers grounded in a theory of decision making. *ZDM, 43*(4), 457–469.

Sinclair, N. (2018). An aesthetic turn in mathematics education. In: E. Bergqvist, M. Österholm, C. Granberg, & L. Sumpter (Eds.), *Proceedings of the 42nd Conference of the International Group for the Psychology of Mathematics Education* (Vol. 1, pp. 51–66). Umeå, Sweden: PME.

CHAPTER 5

Upper-Level Mathematics and Statistics Courses Shared across Campuses

Stephan Ramon Garcia, Jingchen Hu, and Steven J. Miller*

CONTENTS

* *Date*: December 4, 2019. SRG was supported by a David L. Hirsch III and Susan H. Hirsch Research Initiation Grant and the Institute for Pure and Applied Mathematics (IPAM) Quantitative Linear Algebra program. JH was supported by The Andrew W. Mellon Foundation Anne McNiff Tatlock '61 Endowment for Strategic Faculty Support at Vassar College. SJM was partially supported by NSF Grant DMS-1561945. The funding and project management support of the Upper Level Math/Stats Project was provided by the Liberal Arts Collaborative for Digital Innovation (LACOL).

5.1 INTRODUCTION

For students of advanced mathematics and statistics, the liberal arts model offers a deep level of engagement in learning with faculty and peers. Due to practical limitations, small colleges cannot usually offer the breadth of courses available at large institutions with graduate programs. To explore collaborative models that may help enrich curricular offerings, faculty and technologists from several leading liberal arts colleges are experimenting with a consortial hybrid/online course-sharing model. These institutions are partner schools in the Liberal Arts Collaborative for Digital Innovation (LACOL) (`http://lacol.net/`). The goal of these explorations is to increase the wealth and frequency of the advanced classes our students need, both for future graduate study and to delve deeply in the subject.

In this chapter we report on the challenges and opportunities encountered in teaching three different math/stats courses using the hybrid/online course-sharing model:

- Graduate Real Analysis, Fall 2017 (Stephan Ramon Garcia, Pomona College)

- Putnam Problem Solving, Spring 2017 (Steven J. Miller, Williams College)

- Bayesian Statistics, Fall 2017 (Jingchen Hu, Vassar College)

The courses were not completely designed from scratch but rather built upon previous classes taught by the authors, some standard and some partially online. For these shared courses, each instructor opened their course to students across LACOL, sharing lectures, assignments, and other class activities through both asynchronous (e.g., recorded lectures and screencasts) and synchronous (e.g., real-time video-conference of the lectures, online problem-solving sessions, and office hours) means. We viewed these classes not just as isolated courses, but as opportunities to try different techniques and technologies and to determine how to structure future classes.

In this chapter, we report on what we learned. In Section 5.2, Stephan Ramon Garcia from Pomona College reports on teaching Graduate Real Analysis through the LACOL network. In Section 5.3, Steven J. Miller from Williams College discusses experiences, findings, and recommendations on synchronous versus asynchronous instruction. In Section 5.4, Jingchen Hu from Vassar College presents her experiences, findings, and recommendations on how to create and foster an online learning community. Section 5.5 discusses future work.

5.2 CASE STUDY: GRADUATE REAL ANALYSIS (GARCIA)

During the Fall 2017 semester, I taught a graduate-level analysis course at Pomona College. As part of a LACOL-sponsored pilot program, the course doubled as a small-scale experiment to examine the feasibility of bringing high-level course offerings to other small liberal arts colleges. I was motivated by the fact that pure mathematics students at liberal arts colleges often do not have the opportunity to take the graduate-level courses that would make them competitive candidates for elite graduate programs. In addition to a dozen or so local students from Pomona College and Harvey Mudd College, two students from Swarthmore participated in this project, taking the course remotely and asynchronously.

Perhaps the greatest lessons to be learned were on the technical side. Unlike big-budget, professional endeavors, the video recording, processing, and distribution were handled by the instructor. I could not, for example, hire a dedicated camera operator for several dozen lectures spread throughout the semester. Thus, some of my main goals in setting up the course were the following.

(1) Minimizing the reliance on expensive software.

(2) Avoiding the use of auxiliary personnel (e.g., a camera operator)

(3) Minimizing the change to the in-class experience for local students.

(4) Accomplishing this in a repeatable fashion cost-effectively.

I focus here mostly on the logistical and technological aspects rather than the pedagogical aspects of this experience. Indeed, the technical problems that were encountered and some of the solutions that were found may be of interest to a broader audience than the specific pedagogical challenges involved.

5.2.1 Scheduling

The Claremont Colleges, which include both Pomona and Harvey Mudd Colleges, are located in Los Angeles County. Consequently, classes there are scheduled according to Pacific Standard Time. Having synchronous classes during normal business hours for all participants, while avoiding traditional lunch hours, appeared prohibitive.

Another significant barrier to synchronicity was the incompatibility between the Pomona and Swarthmore academic calendars; see Table 5.1. The schedules contain several points of disagreement that would have required significant accommodation on the part of the Swarthmore students.

In addition, we had no funding for a technical-support person to be constantly on site, so having real-time streaming video to and from the remote students would be difficult. Consequently, I decided upon an asynchronous approach (recorded video lectures) early in the planning process.

TABLE 5.1   Fall 2017 Schedules for Pomona and Swarthmore

	Pomona	Swarthmore
Classes begin	Sept. 4	Aug. 29
Fall break	Oct. 16–17	Oct. 13–23
Classes end	Dec. 6	Dec. 12
Finals	Dec. 11–15	Dec. 15–23

5.2.2 Equipment

I anticipated many technical challenges early on and began researching possible solutions during the Spring 2017 semester. Additional tests and tinkering took place throughout the summer of 2017, especially during the weeks leading up to the Fall 2017 semester. I worked closely with Joseph Brennan, Pomona's Director of Media and Classroom Services, throughout the entire process.

Unlike certain other disciplines, in which the instructor lectures with only occasional boardwork, mathematics often requires technical arguments to be spelled out in detail on the board. Although some professors prefer to use Beamer or PowerPoint slides, I felt that a traditional "chalk-talk" approach was best suited to graduate analysis. Moreover, I did not want to reinvent the wheel because one of my primary motivations was to ensure that the in-class experience for local students (most of whom had already taken at least one course with me) did not vary dramatically from what they were used to. I also did not wish the remote students to simply watch slides accompanied by a disembodied voice. Consequently, it was clear that we needed a way to record high-quality video. The selected classroom seats approximately 35 students and features two sets of raisable boards.

Four boards are typically visible at the same time; see Figure 5.1. Put together, the boards encompass approximately 8′ × 18′ (144 square feet), a relatively large area to maintain sharp focus on. In order to utilize all available board space and to keep previous work on the board for the longest time possible, it was necessary to have the camera focused on all four boards simultaneously. This greatly narrowed down the possible cameras that could be employed.

We brought in several models that the Pomona Information Technology Services (ITS) department had in stock. For the initial tests, these were attached to tripods located at the rear of the classroom. However, all of them lacked sufficient resolution to discern fine details on the board. Although most of the writing could eventually be deciphered by the attentive viewer, I could not ask remote students to watch videos of such low quality.

FIGURE 5.1 The classroom features two sets of raisable boards. This is a typical frame from the M4V video files produced for the Graduate Real Analysis course.

One possible option was to write larger than normal. However, I suspected that consciously writing larger than normal for the entire semester would be difficult and that I would eventually revert to my usual writing style. There was also the risk that focusing on "font size" would be a constant mental distraction, something that I could ill afford in such a high-level course. Consequently, the "font size" approach was quickly panned.

Fortunately, these experiments occurred during the Spring 2017 semester, and there was plenty of time to find solutions. Since the in-stock cameras were insufficient for our purposes, we had to purchase a new camera. Through a combination of LACOL and Pomona College funds, a Sony SRG-300H camera was purchased and installed in the classroom; see Figure 5.2. The camera, which cost about 2,000 dollars, can record 60 frames per second at 1080p. Since the camera would be fixed and the lectures would feature no dramatic movement, we decided to record at 30 frames per second to reduce file sizes.

Audio recording proved easier to manage. ITS provided me with a Samson Stage PXD1 microphone system. The instructor wears a clip-on lapel microphone and a transmitter, about the size of a cell-phone, that can fit in a pocket. A small USB stick attaches to the recording computer and communicates with the transmitter. There was only one minor issue on the audio equipment front: a cellphone could potentially cause interference with the transmitter, resulting in poor-quality audio. I had to get accustomed to turning my phone off entirely before class.

5.2.3 Recording

The camera was controlled by a joystick/keypad device, located in small cabinet in a podium near the front of the classroom; see Figure 5.2. Because the classroom was used by several different classes and also served as a student study-room in the evenings, I had to check the alignment of the camera prior to each lecture. Fortunately, only once did I find that the settings had been modified in my absence.

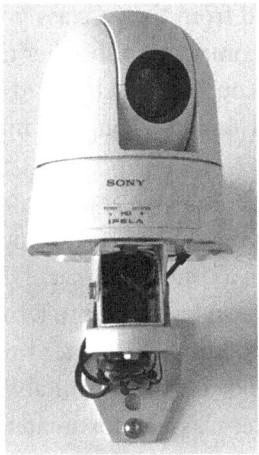

FIGURE 5.2   The camera is attached to the rear wall of the classroom and controlled by a joystick/keypad device located in a cabinet located inside a podium at the front of the classroom.

With the camera and microphone up and running, it was necessary to identify the computer that would be used for the actual recording of the lectures. One obvious solution was to bring my laptop to class and have the camera record directly to it. However, this was undesirable for several reasons. First of all, the power usage necessitated by video recording would require the laptop to be plugged in. This would mean dragging to class an AC adapter, in addition to the microphone, transmitter, chalk, lecture notes, and graded homework assignments, not to mention fiddling with audio and video cables, and adapters every day. Surely at some point during the semester I would forget a key ingredient! Because of the additional hassle and added risk, I decided that using my own laptop was not a viable option.

It was clear that the built-in podium computer (a dual-boot PC/Mac OS machine), which was already connected to the audio-video system, was the best option. However, this introduced an additional wrinkle: how to transfer the large video files from the podium computer while packing up and vacating the class for the next instructor. That important issue is discussed in the following subsection.

Now that the recording computer was chosen, we (Joseph and I) needed to find the appropriate software. Although there were several alternatives, we settled on QuickTime Player. This had the advantage of being simple, easy-to-use, and essentially free, since it comes preinstalled on Mac OS. Moreover, I felt that there was no need to employ fancy video-editing tools or special effects. Thus, QuickTime was perfectly adequate.

The generic podium guest account was used by multiple people, meaning that files could be moved or deleted without my knowledge. There was also the risk of having audio and video settings overridden by other users. Consequently, a special "recording" account was created on the podium computer. This had the added benefit of eliminating the risk of accidentally leaving my personal account logged in during the rush to pack up.

5.2.4 File Transfer

Now that we were able to record lectures at an adequately high resolution, another serious problem emerged. Each lecture video needed to be transferred from the in-class podium computer to either the hosting computer or, at least, my office computer for further tinkering. This process needed to be rapid since a computer science course was scheduled in the same room immediately after my class. After each lecture I had at most a couple of minutes to accomplish this task.

For a typical 1:15-hour lecture, the resulting QuickTime video file was approximately 6 gigabytes (GB). This was far too large to post on the internet directly and large enough to cause difficulty in transferring it from the podium computer in a timely manner.

One of Joe's initial suggestions was to use Box, a file-sharing service similar to Dropbox. Perhaps each lecture video could be posted directly from the podium computer to a Box account? Then the students would be able to view the video files through a suitable link, and the high-quality video could be streamed immediately after class. Unfortunately, this proved untenable for two reasons.

First, the podium computer, despite being hard-wired to the local network via an ethernet cable, was unable to transfer the enormous video files to the Box account quickly enough for me to log out of the podium computer in time for the next class to start.

Second, video files stored on Box and viewed through a web browser are automatically streamed. Consequently, the high-resolution video that we worked so hard to capture was degraded and pixelated when viewed through Box. This would probably be fine for most recreational videos, but for a mathematics lecture in which the fine details on the board needed to be viewed, this proved unacceptable. Consequently, Box had to be abandoned as a viable option.

Joe's second suggestion proved the simplest and most effective. He purchased a SanDisk Extreme Pro USB 3.0 128 GB memory stick. These were significantly more expensive than standard memory sticks and perhaps they still are. If memory serves me, they ran about $75 at the time. The USB 3.0 interface permitted a 6 GB video file to be copied from the podium computer in less than a minute. As part of my "tear-down routine," I would plug the memory stick into the podium computer and copy over the day's lecture video. This permitted me to exit the class room in time for the following computer science class. To ensure that I did not lose or misplace this expensive memory stick, I kept it on a "leash" and in the same bag as the microphone and transmitter; see Figure 5.3.

5.2.5 Video Delivery

Once the raw video files were transferred to the memory stick, they could be copied onto my office machine. This did not resolve the large file-size issue, of course. It merely changed

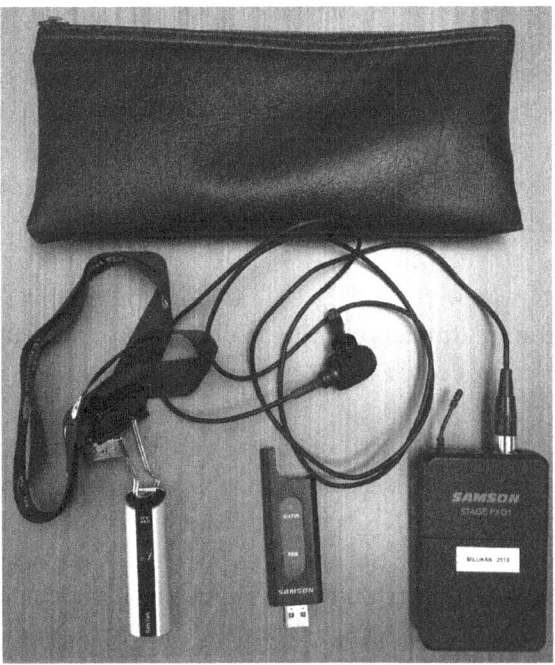

FIGURE 5.3 The only additional objects that I needed to take to class were contained in a small microphone bag: the lapel microphone, transmitter, receiver, and USB 3.0-capable memory stick.

the location of the problem and got things out of the way in time for the next class to use the classroom.

It was clear that the high-resolution video files needed to be compressed without significant loss of fidelity. Moreover, I wanted to accomplish this without purchasing new software and fiddling with advanced tools on a daily basis.

After several experiments, I settled upon VLC. This freely available, multi-format video player is standard fare for Apple users. Although many users might be unaware of this feature, VLC is able to convert between various popular video formats. A little experimentation revealed that the raw QuickTime files, each about 6 GB in size, could be compressed without much loss of clarity to M4V files around 350 MB in size. The reduction in file size was about 94% and rendered the videos small enough to be posted to a course website.

Each 1:15-long lecture took perhaps 45 minutes to compress on my office desktop (an iMac that is several years old). Class ended at 10:50 am, and the compressed M4V version of the lecture would be ready just before lunch. The M4V file, along with the course lecture notes, would be posted on the course website about two hours after the actual lecture. The M4V file could then be downloaded from the course website or watched in the browser; see Figure 5.4.

All of this was done with local computers and with software that was either free (VLC) or included with the Mac operating system (QuickTime). Just as importantly, this required minimal deviation from my usual teaching routine, and it did not significantly affect the in-class experience for the Claremont Colleges students.

5.2.6 Homework

The Claremont Colleges use the Sakai course management system. Similar course management software is available at most American institutions. Each course has a Sakai site dedicated to it, and instructors can post announcements, assignments, and files there. I personally find the system cumbersome and limiting, so I chose to post items on my personal course website. However, I used the Sakai system to keep track of grades and assignments.

The remote students, who were fluent in LaTeX, submitted assignments in PDF form via the Sakai "dropbox" feature (this has nothing to do with the company of the same name). In principle, a remote student could scan a hand-written assignment and then upload it in the same manner.

The course graders (there were two of them, and they alternated assignments) were given teaching assistant access to the course Sakai site. They downloaded the homework assignments and marked up the PDFs using Preview, Acrobat Reader, or other similar software capable of annotating a PDF. The graded assignments were uploaded by the graders to the Sakai site and placed in the remote student's personal directory. In this manner, the remote students could turn in their assignments and receive the graded assignments back entirely online.

The process was smooth, and we encountered no technical problems. In particular, it seemed to be a reasonable way to handle a small number of remote students. Although

Class Schedule			
(Subject to change)			
Date	Reading	Topics	Homework
Tu 8/29	Course Description (PDF) Notes on Cardinality (PDF) Bass, Sections 1.1, 1.2	Foundations Vitali sets, Banach-Tarski Paradox Lecture #1 (PDF) Lecture #1 (m4v)	Homework Template (TeX) Due Tr 8/31: Survey #1 (PDF) Due Tu 9/5: Homework #1 (PDF) Homework #1 (TeX) Solutions #1 (PDF)
Tr 8/31	Logicomix (for those interested)	Foundations Russell's Paradox, Axiom of Choice, partial orders, ZFC, Godel's Second Incompleteness Theorem Lecture #2 (PDF) Lecture #2 (m4v)	
Tu 9/5	2.1, 2.2	Foundations Sigma-algebras, ordinal numbers, transfinite induction Lecture #3 (PDF) Lecture #3 (m4v)	Due Tu 9/12: Homework #2 (PDF) Homework #2 (TeX) Solutions #2 (PDF)
Tr 9/7		Foundations Borel hierarchy, Goodstein's Theorem, Godel's First Incompleteness Theorem Lecture #4 (PDF) Lecture #4 (m4v)	
Tu 9/12	3.1	Measure Theory Measures, null sets, completions Lecture #5 (PDF) Lecture #5 (m4v)	Due Tu 9/19: Homework #3 (PDF) Homework #3 (TeX) Solutions #3 (PDF)
Tr 9/14	4.1	Measure Theory Outer measures, constructing measures, the Cantor set Lecture #6 (PDF) Lecture #6 (m4v)	
Tu 9/29		Measure Theory Cantor set trivia, Borel sets vs. Lebesgue sets Lecture #7 (PDF) Lecture #7 (m4v)	Due Tu 9/26: Homework #4 (PDF) Homework #4 (TeX) Solutions #4 (PDF)
Tr 9/21	4.2, 4.3	Measure Theory Monotone functions, Lebesgue-Stieltjes measures Lecture #8 (PDF) Lecture #8 (m4v)	
Tu 9/26	4.2, 4.3	Measure Theory Lebesgue-Stieltjes measures, regularity properties Lecture #9 (PDF) Lecture #9 (m4v)	Due Tu 10/3: Homework #5 (PDF) Homework #5 (TeX) Solutions #5 (PDF)

FIGURE 5.4 The course website contained videos of the lectures in M4V format.

there do not appear to be any significant technical issues that prevent scaling things up to a dozen or more remote students, there are the matters of money (who pays?) and personnel (who grades?) that would have to be resolved.

5.2.7 Examinations

The local students took examinations in class. Remote students took identical exams that were proctored by a Swarthmore faculty member. This arrangement needed to be set up before the course began, since it required a low-intensity commitment from a remote professor. The completed exams were scanned by the proctor and sent to me via e-mail. This procedure, however, seems less amenable to scaling since it requires each participating institution to have an on-campus proctor. The possibility that an exam might be scheduled on an inconvenient day for one campus or proctor could be uncomfortably high if the number of participating institutions were to increase. Perhaps for advanced classes, "take-home" examinations might be more suitable?

5.2.8 Summary

Our small experiment shows that advanced mathematics courses, even those at the graduate level, can be taught asynchronously online. Perhaps the biggest lesson to be drawn from our experience is: plan early! There are many technical challenges that may arise, and it is hard to anticipate all of them. Do not assume that your current in-class technology will be sufficient. Pre-existing cameras in classrooms might not be able to capture video at the resolution required for remote students. Rectifying this may take time and money, although it is a one-time expense.

5.3 SYNCHRONOUS VERSUS ASYNCHRONOUS INSTRUCTION (MILLER)

An important issue in the creation and execution of shared classes is how the material is delivered: synchronous or not. There are advantages to each. If the lectures are streamed live, the off-campus participants are able to fully participate in real-time. They are able to comment in class, ask questions which guide the lecture, and interact with their fellow classmates. Unfortunately, doing so requires more on the technology side, and also prevents students from taking a class that meets at that block. As many colleges are on different calendars and time structures for the day, these are non-trivial scheduling issues. If the class is entirely asynchronous, many of these issues vanish, though unfortunately now the students are not able to participate in classroom discussions. In this section we report on some of the successes and challenges of these two approaches.

5.3.1 Introduction

For about five years now I have recorded my classes for multiple reasons. The initial motivation was a desire to flip the classroom, either fully or partly, specifically, my Multivariable Calculus class at Williams (where most of the students are freshman, making the transition from high school to college). In college such classes typically meet three times a week for 50 minutes for somewhere between 12 and 15 weeks (though some schools do block off a fourth meeting); in high school classes meet daily for more weeks. Thus there are significantly more contact hours in high school, leading to significantly more time to go through the same material. This gives these instructors a lot of freedom not found in college, such as going at a slower pace and spending more time on worked-out examples. I wanted to be able to assign students videos to watch in future years to move some of the more standard material to home viewing, and free up more class time for value-added lecturing.

Over time, however, my use of recorded lectures changed. Many students found the recorded lectures useful, as they would occasionally have to miss class (either due to illness, family situations, or travel for athletic competitions or job interviews). Further, if there was a topic that was confusing they could watch a subset of the lecture again. They worry less about taking notes during class. Also, this allowed me to assemble a large list of introductory lectures on a variety of topics which could be used to supplement other courses. For example, if a student in complex analysis has not seen Green's Theorem, I can refer them to my lecture where we cover it.

Additionally, Williams is a small liberal arts college; we often only have one section of upper-level classes, and many of these are only offered every other year (if even that frequently). I would often have students wish to take both my course and a class in another department meeting at the same time; by recording the lectures the student could do that. They could still come to office hours and TA sessions; they could still work with their fellow students. The only difference is that they could not participate in class, which means that I do not have the benefit of their perspective in discussion, and they cannot immediately ask questions on material that is confusing or that they wish to see in greater detail.

Over the past few years I have been talking about the rewards and challenges of teaching at a small liberal arts college with many of my colleagues, who have had similar experience, which has led to new thoughts on how to use lecture capturing. One of the greatest problems we all face is class enrollment. Frequently we do not have the manpower to teach all the electives we want, or if we do there is often not sufficient interest on our campus to justify the class.* Many of us have explored solving these issues by sharing classes across campuses. For example, I might teach Complex Analysis at Williams while one colleague teaches Algebraic Topology at their campus and a third teaches Functional Analysis at their institution. We open our classes up to all students from our institutions.

There are numerous problems with making such class sharing work well. Some are administrative issues. What credit do professors get for teaching students at another institution? How much extra work is it to have remote students, and how should the professor be compensated for the additional time demands? What credit do students get for taking a class at another institution? How do the classes show up on their transcript? While these and related problems are important and often challenging to solve (frequently one needs a faculty vote to allow credit for a class at another institution, but a work-around often exists by having a local faculty member at the student's college enroll them in a guided independent study), that is not the point of this contribution. My goal is to talk about two main formats for classes with remote students: asynchronous and synchronous. I'll briefly touch upon the technology and administrative issues, but concentrate on some class structures I've tried, highlighting what works and what the challenges are.

5.3.2 Asynchronous Content

As a professor, an asynchronous class is enormously easier to run than a synchronous one. Many of the administrative issues and challenges disappear or are easily resolved. In particular, if the lectures are recorded and posted online, students can watch at their own pace and at whatever schedule works for them. As colleges rarely have aligned academic calendars (and further are often in different time zones, or some have classes starting on the hour and others on the half-hour), this is tremendously valuable.

---

* This is for the more advanced courses; for the introductory classes the problem is the opposite, where courses such as Introduction to Statistics often turn away as many people as they accept.

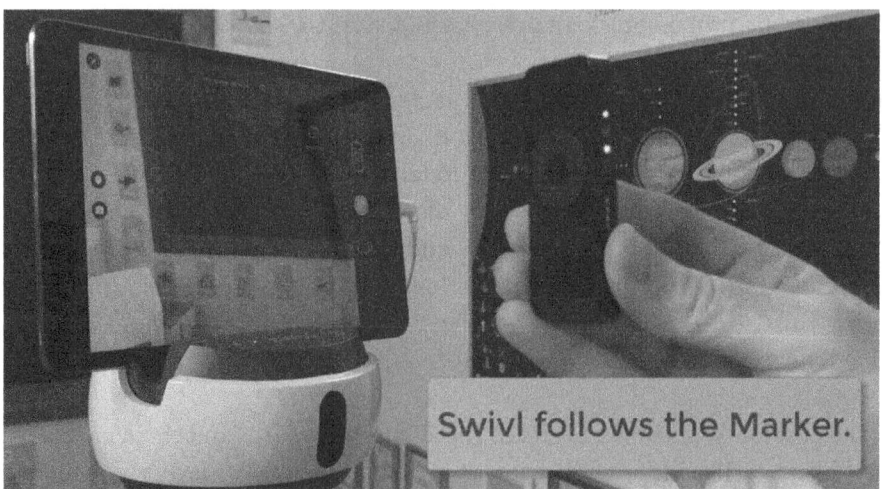

FIGURE 5.5 Swivl system for tracking lectures, image from their homepage (www.swivl.com/how-to-use/). Note the instructor can see what is being recorded in the iPad screen, and the system follows the tracker in real-time.

There are some slight annoyances, as a school that starts later than the host institution could have students submitting say the first homework assignment while the local students are on the second or third, but these are minor issues for grading.* Slightly more severe is that it is harder for students to collaborate across institutions if they are at different places in the course, but this can be mitigated by having some schools do more work in certain weeks to force alignment.

This last point, on collaboration across schools, is worth dwelling on as it goes to the heart of the goal of such course sharing. Is the primary objective to provide content that would not be otherwise available, or to build bonds between students and faculty across multiple institutions? In the asynchronous setting it is easier to work towards the former. While I try to encourage my remote students in such settings to have at least one friend take the course with them, so they have a local study group, this is not always the case. In comments from them over the years the most common remarks are that they are glad to have had an opportunity to see material that otherwise would not be available and overall they are glad they did the course, but that the interactions are not equal to what they have in live classes. The differences range from the challenges of watching lectures on a small screen instead of live in the classroom to feeling more isolated.

On the technical side, it is very easy now to record and post lectures. As Garcia described many of the technologies in depth, I'll just briefly remark here on two options of varying difficulty.

The easiest is to record lectures using an iPad and post them online through YouTube. One can buy a tracking system online for a few hundred dollars, put the iPad on it on a tripod, and just lecture (see Figure 5.5). There is a tracking device one can wear which will

---

* Of course, there is the potential for the other problem: the remote schools have a semester starting first! If the lectures have been recorded from a previous iteration this is not an enormous challenge; if not, one easy solution is to have those students wait till the local school starts.

FIGURE 5.6 Snapshot of a lecture as it would appear on YouTube, recorded using an iPad and the Swivl system. In this model the lecturer wears the tracking device on their shirt, which the base follows. Lecture available online at www.youtube.com/watch?v=NgHIiZUYI6g& feature=youtu.be.

make sure the system follows you, which has the advantage that comments from students won't be detected and recorded. This is important as some students feel uncomfortable asking questions if they know that others will be able to hear them later; for the lecturer, you just have to remember to repeat or rephrase the question. If students are comfortable having their questions recorded and thus being identifiable, you can purchase more advanced systems which have multiple audio input devices, and place those throughout the room.

I prefer to post my lectures online on YouTube rather than through a college course page for many reasons. First, it is simple to go to www.youtube.com/upload and upload a file. You can place tags to make it easily searchable, you can create playlists for classes, but most importantly anyone on the web can see the material. One of the great advantages of this is that you do not need to give remote students accounts at your school. Further, if you want to use material from one class in another you do not have to give current students access to a former class, or move content. The material has a permanent home, and you can use it wherever and whenever you wish. For me, the only real concern is that anything you say is now public domain. In standard and undergraduate elective math courses this is not a big deal,* but in a humanities discussion course where very personal discussions are happening this could be a serious concern.

The biggest drawback to this technology is that you cannot zoom in to the material, and thus the writing can be a little small and not as crisp and clear as one would like (see Figure 5.6); thus it is better for students to watch on a laptop or desktop over a phone or iPad.

The other option I have used is to hire a student to record the lectures (see Figure 5.7). In addition to excellent tracking, more important is the ability of the student to zoom in

---

* That said, there are certain stories that I cannot share with the camera rolling, ranging from certain personal anecdotes to some issues in applying math in practice; these have to be told outside class hours.

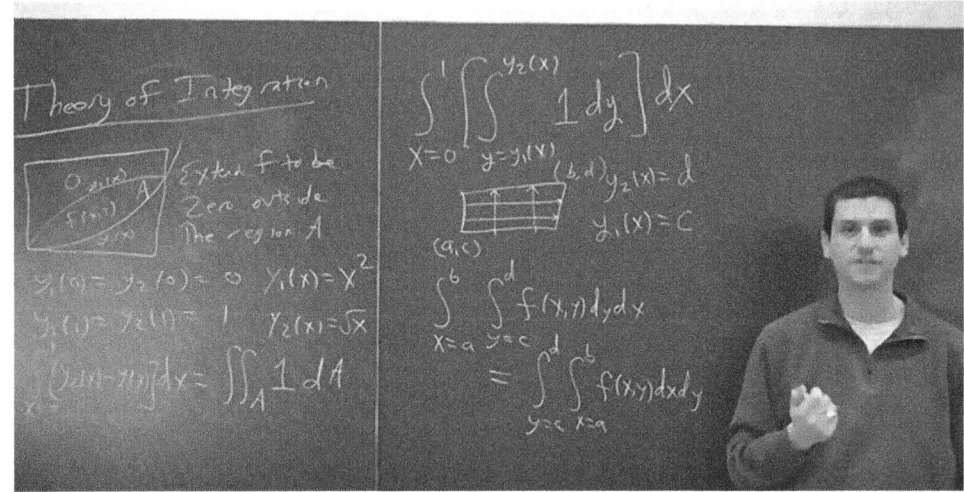

FIGURE 5.7 Snapshot of a lecture as it would appear on YouTube, taken by a student using a high-quality video recorder. Lecture online at `www.youtube.com/watch?v=G9d9lcYevnM&feature=youtu.be`.

on the mathematics being discussed. This makes the lectures easier for students to follow, but requires significantly more local resources. Using an iPad and a Swivl is relatively easy and can be done without too much work for the professor; hiring a student requires funds and often processing time for the video. That said, if funds are available it is worth having the better recording, as it makes the video significantly more useful for remote students and future classes.

In summary, the above are just two of the many options available for running asynchronous classes. The technology is now cheap enough and reliable enough to make this a real option. The minor administrative issues can be overcome without too much trouble; the greatest difficulty is the lack of alignment among academic calendars (though this is also a challenge for synchronous content). Liberal arts colleges pride themselves on close contact between students and professors; this is significantly lessened in this set-up, but with work (such as skyping outside of class, or even better, visiting the remote campuses) personal connections can still be forged.

5.3.3 Synchronous Content

Slightly different technology is required for a synchronous class. To date, I have not done a fully synchronous class. This is due in part to the different academic calendars, but also to students at remote schools (or even the local school!) desiring to take not only my class but another class at the same meeting time. What I have done is taught classes that are mostly asynchronous, but with four synchronous meetings which are scheduled outside the normal time block. Doing so allows real interaction between the remote and local students.

The greatest challenge now is to have the information available to all in real-time. I've used a Microsoft Surface as the input and shared screen technology. At the local

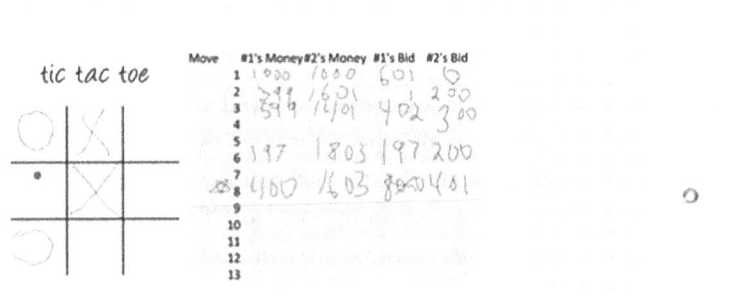

FIGURE 5.8   Snapshot of a shared lecture. Video online at `www.youtube.com/watch?v=DRB` `mLlrztvI&feature=youtu.be`.

school I have my computer connected to a data projector, and thus have the screen viewable by the entire class. For the remote students, there normally are not that many at each institution, and they can just congregate around one connected machine. There are programs now that allow remote users to draw directly on my screen, and thus we can all see the same material at the same time. I've used software available through our class system, Glow, but there are many applications now. The bandwidth requirements are low as most of the time the content has very little change. In such situations I'm not lecturing on the board, but instead I've posed a problem that students are pondering and then presenting.

I've done this in a problem-solving class. Figure 5.8 shows what a shared screen looks like; one of the students is about to move where the red dot lies. The problem-solving class was ideal for synchronous material. I was able to give a short general lecture first, and then the students remotely played a few games* and tried to develop a theory.

These classes were enormously fun, and very different from the standard lecture classes. Stu- dents from each institution chatted with one another before class began. To make the atmosphere more festive and special I arranged for food (typically Dunkin Donuts munch-kins) at each place; this is also somewhat needed as we met at 8 am, well outside normal class hours for today's college students!

Thus, the limiting factor in synchronous courses is not technology (and as the technol-ogy is improving significantly each year, any issues on that end will be short-lived). The greatest challenges are the different schedules, ranging from unaligned academic calen-dars to being in different time zones.

Finally, one concern I am always aware of is making sure that having a class remotely accessible does not negatively impact the experience for the local students. I will lecture slightly differently with screen capture than I would with chalk on a blackboard; however,

---

* The game depicted is bidding tic-tac-toe. Each side starts with $1000, and each turn consists of making a secret bid; whoever bids highest gets the move and gives their bid to the other side; if there is a tie no one moves and you bid again. The game continues until someone wins or all squares are taken.

that is not necessarily worse, just different. Different media have their advantages and disadvantages, and one must adjust one's lecture style to the method used for presentation.

5.3.4 Summary

The technology exists for a variety of shared class experiences, and what we can do will only expand in the coming years. The need is great, especially at small institutions, where we cannot offer all the classes we want with the frequency our students need. Rather than viewing this as a disadvantage, we can see it as an opportunity to forge connections between students and faculty across institutions. In the course of sharing classes, we also assemble a large database of (hopefully!) excellent lectures, which are a valuable resource.

Right now the challenges surrounding asynchronous classes are easier resolved than those for synchronous ones. While both have administrative hurdles, the asynchronous format avoids the headaches arising from different calendars, as well as allowing students to take two classes that meet at the same time. The value of this latter point should not be forgotten; we often have so few classes that are appropriate for upper-level students that we should do all we can to preserve options.

Finally, at liberal arts colleges one of our selling points is the personal relationship between students and faculty. In the synchronous classes I've taught I have made it a point to go to each campus for at least one lecture. There are several advantages to this. First, by visiting for a day or two I get to know the remote students much better than possible through email and skype. Second, and more importantly, it gives the local students a sense of what the remote students are experiencing!

5.4 CREATING AND FOSTERING AN ONLINE LEARNING COMMUNITY (HU)

In this section we report on student experiences in our model of sharing a course across multiple campuses. This approach is unique, and distinct from most of the existing hybrid/online models. Our class meets in a real classroom at one campus, providing synchronous and asynchronous access to remote students from other campuses. To local students, the class is not so different from a traditional face-to-face class, as they can ask the instructor questions during office hours and/or in person, and interact with other local fellow students as they like. However, the experience can be vastly different for remote students, and it is easy for them to feel isolated and left out. We discuss challenges, and experiments and approaches to those challenges. The goal is not only to make remote students feel part of the class and provide them an enjoyable learning environment, but also to create an online learning community involving both remote and local students. The sharing and exchange should not only happen in the instruction and material delivery, but also students' involvement and engagement and contribution to a shared learning community.

5.4.1 Introduction

Vassar College's MATH 347 Bayesian Statistics was offered through the LACOL network in Fall 2017 for the first time. Class met in person on Vassar campus, where local students were present in the classroom during lectures. The lecture delivery was through the conferencing software Zoom.* The course management system was Moodle[†] at Vassar College. Registered Vassar students had a regular Moodle account, whereas remote students had sponsored Moodle accounts created for them by Vassar's Academic Computing Service. The sponsored accounts gave remote students all Moodle functions.

Typically, I as the instructor started the synchronized lecture via Zoom meeting with the classroom desktop computer, and remote students could join class with the Zoom meeting ID. I used an iPad Pro to join the Zoom meeting as well, where lecture slides and other course material were shared and then projected onto the projector in the classroom. The entire lecture (75 min) was recorded, and then posted on a YouTube list of class recordings for anyone to access freely. Therefore, both synchronized and asynchronized access to the lecture was available to registered students.

The course material and interacting tools were all hosted on the Moodle site. For example, lecture slides and R[‡] programming scripts were posted and downloadable from Moodle. Also, the discussion board function of Moodle allows the instructor to create a topic within a module, and everyone has the right to post a thread to the topic.

The course started with four remote students. Among them, two were able to join class in real-time (synchronized access), and the other two were only able to watch lecture recordings due to schedule conflicts (asynchronized access). Eventually, only one remote student stayed in the class, who only utilized the asynchronized access. Nevertheless, the course provided both types of access throughout the semester.

The challenges of keeping every student involved became obvious early in the semester. With a roughly four-to-one ratio of local students versus remote students, there seemed to be no drive for local students to get to know the remote students. To make matter worse, the four remote students came from three different campuses, and their various schedules and availabilities made it difficult to interact even just among themselves. As the number of remote students gradually dropped to one, such communication and engagement challenges only grew greater.

Through conversations with experienced colleagues, readings about online teaching and learning, and a series of trial-and-error experiments, I have identified a few useful tools to create and foster an online learning community. In each of the upcoming subsections, I would like to introduce the tool, illustrate with screenshots when possible, and discuss the advantages, my experiences, and my reflection.

5.4.2 Self-Introduction Posts

The self-introduction posts idea was introduced to me by a blended learning expert in the LACOL network. To me, it was an after-thought, as it occurred to me soon after the course started that I needed to do something to make everyone get to know everyone else, virtually in some cases. The request of a self-introduction post was given to all students in the

---

* For more information about Zoom, visit https://zoom.us/.

† For more information about Moodle, visit https://moodle.com/.

‡ R is a statistical programming language. For more information about R, visit https://cran.r-project.org/.

third week, instead of before the course started or in the first week, which obviously would be more ideal. Still, better late than never.

I started the first post to introduce myself, and then invited every student to make a post on the following aspects:

(1) Your name, year, school.

(2) Your prior statistics exposure (note that we have 22 students from four different schools, so when you talk about your statistics course experience, be more specific than just writing down the course number).

(3) Your prior R exposure (be specific).

(4) Your potential project interest, if you have already started thinking about it. Even if you haven't, anything that would interest you is good to be put down.

After all posts were made, I went in to read each post carefully, and summarized a list of project interest topics and shared on Moodle. This process helped me to get to know the students better.

Overall, I think requesting a self-introduction post is a good practice. It not only allows students to introduce themselves, but also encourages them to reflect on their previous experience, ask themselves what they want to get out from this course, and think ahead about their course project ideas.

There are a few things to keep in mind to maximize the effectiveness of these self-introduction posts. First, do it early, ideally before the semester starts. Second, try to make it more interactive. As you can see in the "Replies" column in Figure 5.9, none of the self-introduction posts were replied to at all. This does not mean that no one read them (well, who knows), but it does mean there was little online interaction activity among the students. If some kind of "reply to one" mechanism can be introduced, that might help improve the actual interaction among students. Third, try to make it more personal. In the current format, the self-introduction posts were made in words. I later received the advice of making a self-introduction video instead of text-based posts, so students can see the face and hear the voice, which sounds very appealing and potentially useful. In fact, I incorporated the video idea into the project introduction posts, which is the second tool that I turn to now.

5.4.3 Project Introduction Video Posts

Like many statistics courses, the Bayesian Statistics course had a project component as a capstone experience. The projects can be done individually, or as a pair of two students. At the end of the semester, a poster session was held for students to present their projects.

The poster session worked in a similar way as any other poster session at academic conferences. All projects were put into two groups. The first half of the poster session was for the first group to present, so the students from the second group were able to walk around, read posters, and interact with the presenters. Then the groups switched for the second

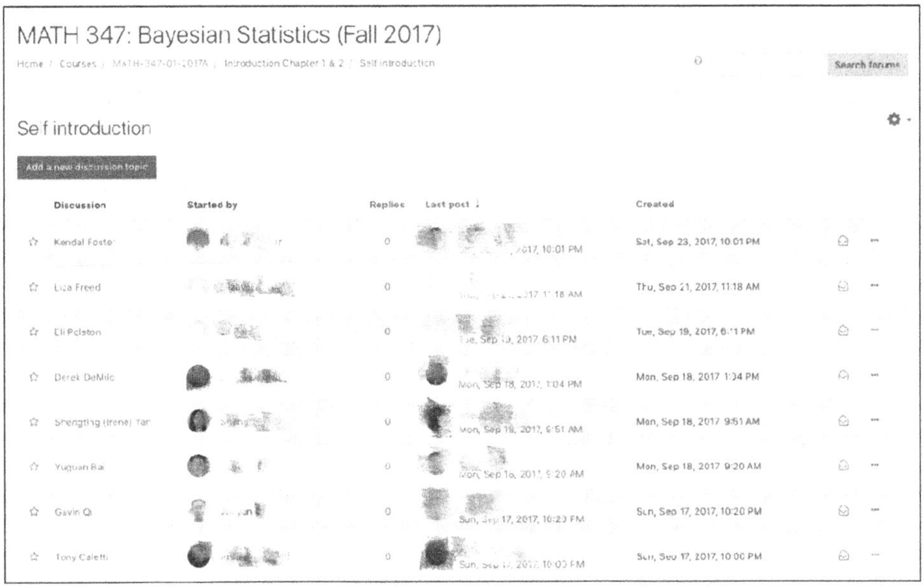

FIGURE 5.9   Students' self-introduction posts (first page).

half. Between the switch and after the second group's presentations, there were allotted time for discussion, summary, and Q&A.

Almost all students were new to the poster session format. To help them prepare their posters, and think about how to present the material from the posters in person (smaller group or even one-on-one, versus a 10- or 15-min presentation at the front of the room to the whole class), I asked each project presenter to prepare a two-min project introduction video and post it on Moodle. All students were required to watch all videos before the poster session took place.

These two-min project introduction video posts served like a pitch talk of students' projects. Many students used their poster slides to go over the project within the time frame. These videos not only helped the presenters to organize their material, thoughts, and ideas, but also gave the other students the chance to get to know the gist of the presenting poster, and help them to navigate the poster session more efficiently and effectively. Figure 5.10 is a screenshot of a sample project introduction video, "Quantifying Prior Opinion."

In particular, this project introduction posts tool was useful to engage the remote student with the local students. The only remote student in the course could not attend the poster session in person, but other students had the chance to learn about the remote student's project, particularly through the posted two-min video. In fact, the remote student created a longer video (around five minutes, acting as if it was the actual poster in person) and posted it on Moodle too. Several local students became very interested in the topic, and they had the chance to learn more about it through these two videos. Overall, the project introduction video posts tool is a very useful and effective practice. It pairs up well with the poster session format in particular. To improve it, in addition to requiring all students watch all videos before the poster session, requirements such as "comment on at least two videos" could increase out-of-classroom interaction to an even higher level.

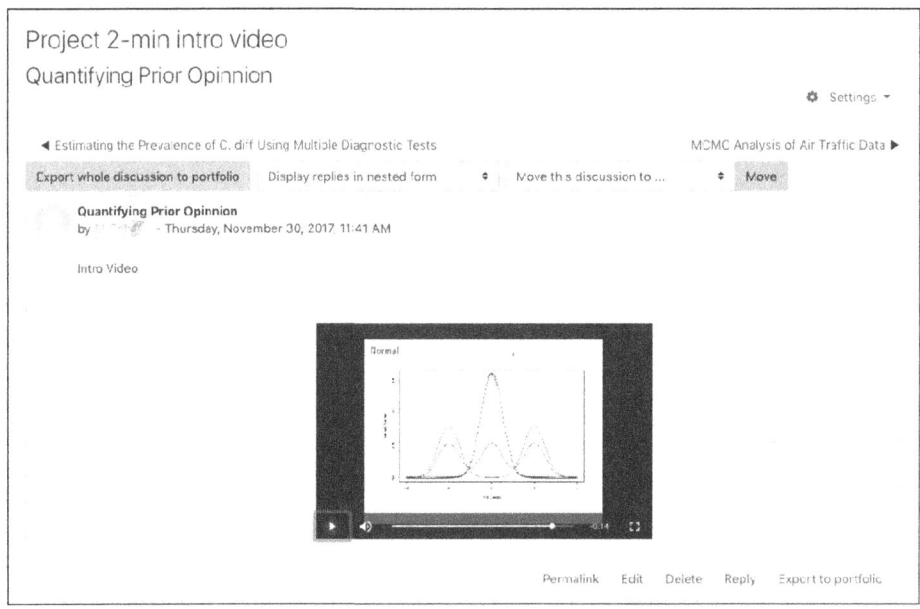

FIGURE 5.10 Sample project two-min intro video on "Quantifying Prior Opinion."

5.4.4 A Learning Environment for Reading a Research Paper

Reading and discussing accessible research papers are important components of an advanced statistics course. When covering the Gibbs Sampler* module in Bayesian Statistics, I chose to have my students read *Explaining the Gibbs Sampler*, a short paper by George Casella and Edward I. George, which appeared in *The American Statistician* in 1992. The paper set up the Gibbs Sampler background nicely, and discussed its features and practices through a series of designed simulations. It is a relatively accessible research paper, even to undergraduate students. Some practices are different from current ones now, which make the paper a good reference for learning and discussing the characteristics of the Gibbs Sampler, and Markov chain Monte Carlo (MCMC)† in general.

In addition to the original paper, I provided a six-question reading guide to the students; see Figure 5.11 for a screenshot of the list of questions. Some questions were related to the verification of presented results, while others were more associated with practices of the Gibbs Sampler, leading to discussion of similarities and differences between practices after the introduction of the Gibbs Sampler (when the paper was published) and practices about 25 years later (what people typically do now).

To engage students in reading and discussion both in class and outside of class, a request of one pre-class post and another post-class post was given to all students. Separate discussion topics for each question on the reading guide were created on Moodle's discussion board (see Figure 5.12). There were two more topics added after the class discussion; one was a supplementary topic raised during class, and the other from colleague's comments. As we can see, the effect of the two posts requirement can be seen in the counts in the

---

* For more information, visit https://en.wikipedia.org/wiki/Gibbs_sampling.
† For more information, visit https://en.wikipedia.org/wiki/Markov_chain_Monte_Carlo.

Fall 2017 MATH 347   Jingchen (Monika) Hu

Reading Guide for Explaining the Gibbs Sampler

Casella and George (1992)

1. [Section 2] How does Gelfand and Smith (1990) suggest to obtain an approximate sample from $f(x)$? How is it different from or similar to the approach we talked about in class? What are the advantages and disadvantages of each approach?

2. [Section 2] The authors claim "Gibbs sampling can be used to estimate the density itself by averaging the final conditional densities from each Gibbs sequence." What is the theory behind this claim? How does Figure 3 support this claim?

3. [Section 3] Write down marginal distribution of y, and verify the conditional probabilities $A_{y|x}$ and $A_{x|y}$. Also, verify $A_{x|x} = A_{y|x}A_{x|y}$ and $f_x A_{x|x} = f_x A_{y|x}A_{x|y} = f_x$.

4. [Section 4] What is a fixed point integral equation in the bivariate case? How does it help illustrate how sampling from conditionals produces a marginal distribution? Hint: check equations (3.5), (4.1), and (4.2).

5. [Section 4] The authors claimed "a defining characteristic of the Gibbs sampler is that it always uses the full set of univariate conditionals to define the iteration." Explain this claim by illustrating how a Gibbs sampler works with k parameters $(\theta_1, \theta_2, \cdots, \theta_k)$.

6. [Section 5] Summarize different approaches to sampling the Gibbs sequence.

FIGURE 5.11   A six-question reading guide for the paper *Explaining the Gibbs Sampler.*

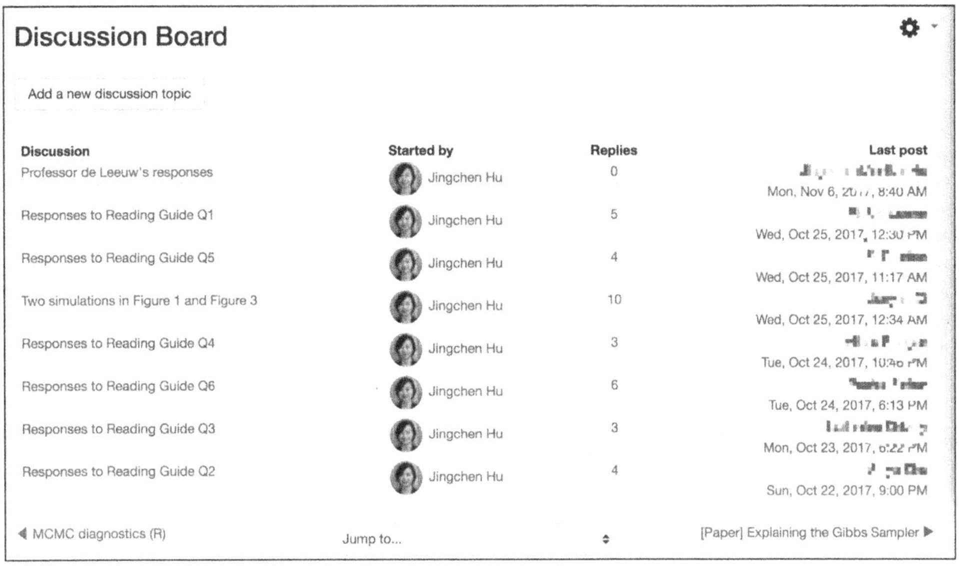

Discussion Board   ⚙ ▾

Add a new discussion topic

Discussion	Started by	Replies	Last post
Professor de Leeuw's responses	Jingchen Hu	0	Mon, Nov 6, 2017, 8:40 AM
Responses to Reading Guide Q1	Jingchen Hu	5	Wed, Oct 25, 2017, 12:30 PM
Responses to Reading Guide Q5	Jingchen Hu	4	Wed, Oct 25, 2017, 11:17 AM
Two simulations in Figure 1 and Figure 3	Jingchen Hu	10	Wed, Oct 25, 2017, 12:34 AM
Responses to Reading Guide Q4	Jingchen Hu	3	Tue, Oct 24, 2017, 10:36 PM
Responses to Reading Guide Q6	Jingchen Hu	6	Tue, Oct 24, 2017, 6:13 PM
Responses to Reading Guide Q3	Jingchen Hu	3	Mon, Oct 23, 2017, 5:22 PM
Responses to Reading Guide Q2	Jingchen Hu	4	Sun, Oct 22, 2017, 9:00 PM

◀ MCMC diagnostics (R)   Jump to...   ⇕   [Paper] Explaining the Gibbs Sampler ▶

FIGURE 5.12   Discussion board for the paper *Explaining the Gibbs Sampler.*

"Replies" column. In those topics with a good number of replies, actual online discussion (e.g., later comments made reference to previous comments) can be identified (see Figure 5.13 as an example).

I went through the pre-class posts before the class discussion to see what students had been thinking and how their understandings were. I was able to refer to some comments during in-class discussion as well, and students' reaction and interaction in class showed their engagement with the pre-class posts, and overall this practice facilitated in-class discussion well.

For the post-class posts, though I was not able to discuss them in detail in later lectures due to time constraints, I was able to respond to their posts directly on Moodle, and also make a summary of the reading guide questions in general.

The online engagement of reading and discussing a research paper complements the in-class discussion, and both practices together greatly enhance students' learning and engagement. I strongly recommend creating a reading guide to help students read and understand the paper. It helps the orientation and flow of discussion too. The pre-class and post-class posts mechanism worked great for the Bayesian Statistics course, and I can imagine variations of this mechanism, depending on the difficulty level and suitability of the paper to the course material, to be effective.

5.4.5 A Learning Environment for Guest Introductory Videos

In traditional face-to-face courses, inviting colleagues or professionals in the community for a guest lecture is a great way to connect course material to real applications. For a

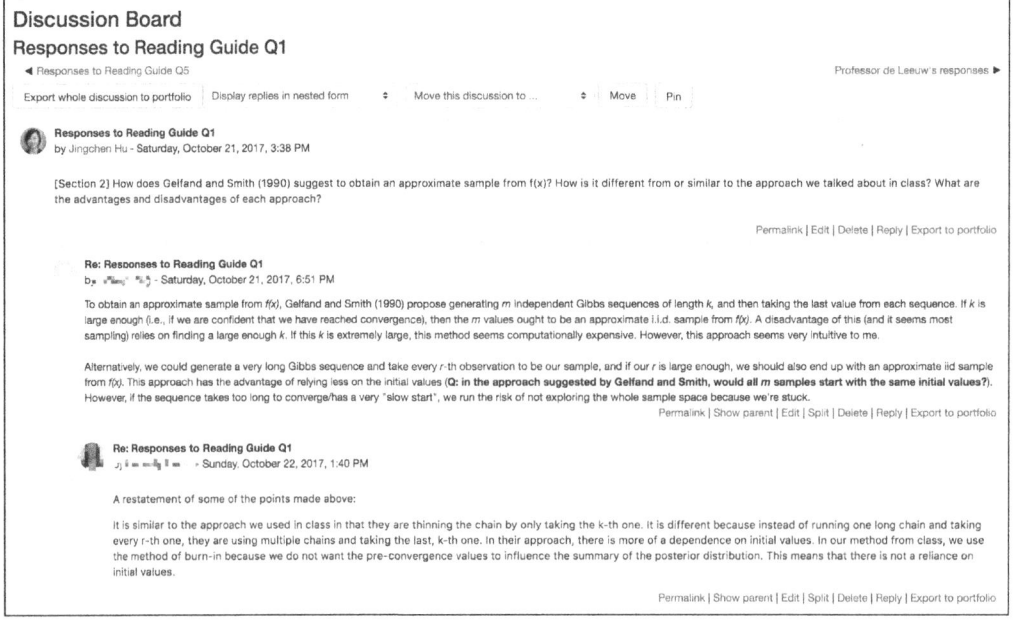

FIGURE 5.13 Discussion board (partial responses to Reading Guide Q1) for the paper *Explaining the Gibbs Sampler*.

hybrid/online course, such a guest lecture can be transformed into a video and moved online. Additionally, a learning environment for the online guest lecture should be created to foster students' learning and engagement.

Bayesian hierarchical modeling* is a Bayesian approach that is the basis of a large amount of applied scientific work, which makes the teaching and learning of this topic potentially interactive and application-oriented. I have been lucky to have a colleague in the Cognitive Science Department at Vassar College who uses Bayesian hierarchical modeling extensively for his research. In particular, he focuses on the development and application of Internet-based tools for the study of human cognition. He had kindly agreed to create a 20-min introductory video to Bayesian hierarchical modeling based on his applied work for my students, so off we started our teaching collaboration.

Creating and tailoring a video for another class proved to be a very challenging task. For a couple of months, my colleague and I met for an hour every other week to discuss the topics, the material, the format of the video, etc. It took me a while to grasp his experimental settings as well as the learning theories in cognitive science. Then it was the period of figuring out a common language in two separate fields (statistics and cognitive science) for presenting the material. There had also been heated discussions about the level of the material and the use of programming languages. Though challenging, the entire process turned out to be a great learning and collaborative experience for me as a statistician. Curious readers could watch the introductory video at `https://www.youtube.com/watch?v=eFmZV67wf4I&list=PL_lWxa4iVNt1TfbsAfv9aW_5KL9rZuAtr&index=30&t=1s`.

To engage students' watching of the video and connect the video's content to our course material, a discussion board on Moodle was created, and a request of one pre-class post and another post-class post was given to all students. Going through the pre-class posts before class helped me greatly to set up the background in class, go about the lecture material, and make reference to the video content when possible.

In addition to posting on the introductory video, there were other topics created for the Bayesian hierarchical modeling module (see Figure 5.14). For example, my colleague had kindly shared his JAGS† script, and it was posted on the discussion board with his permission. Our course just started covering JAGS to facilitate model estimation (as opposed to regular R script that people need to write by themselves), so topics such as sharing JAGS script to replicate results obtained using regular R code from the textbook were also posted. Sometimes, there were inconsistent or non-replicable results that I produced and I could not figure out why. I then went ahead and posted those questions on the discussion board and asked for help, and amazingly students came to the rescue. For that, I appreciated the creation of the online learning environment that makes me comfortable enough to share my own obstacles and concerns. I also believe that compared to traditional face-to-face in-class discussion, the online learning environment made students more comfortable to

---

* For a quick overview, visit https://en.wikipedia.org/wiki/Bayesian_hierarchical_ modeling.
† Short for Just Another Gibbs Sampler. For more information, visit http://mcmc-jags.sourceforge.net/.

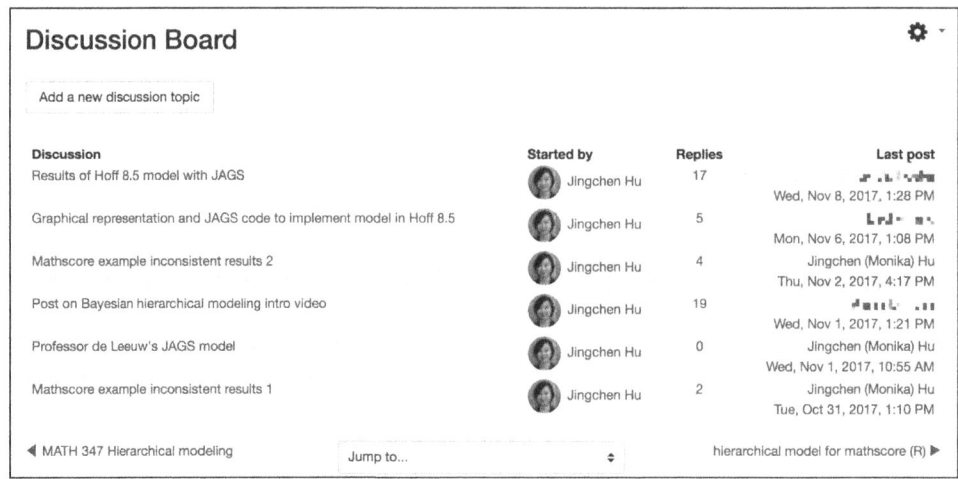

FIGURE 5.14 Discussion board for a guest introductory video on the Bayesian hierarchical modeling topic.

challenge each other or me in an online environment and gave us all more time and space to ponder and discuss.

5.4.6 Summary

In a hybrid/online course, creating and fostering an online learning community is crucial. I have introduced the list of tools that I have found useful: self-introduction posts, project introduction video posts, a learning environment for reading a research paper, and a learning environment for guest introductory videos. I believe many other tools can be developed and proved to be useful.

Based on my experience, the online discussion board and required students' accountability are central to the success. Only if students are actively engaged in the material and the learning community can they connect to each other and to the instructor. I believe the optimal way of enforcing students' accountability highly depends on the instructor's style and the student group. It can be an ongoing, sometimes frustrating, but ultimately rewarding experience for the instructor to create the right dynamic in the online learning community.

5.5 FUTURE WORK

We have run several hybrid/online courses and now have a good sense of what technology is available and how we can use it to engage students from across the country. The next step is to increase the diversity of course offerings. There are several issues that must be resolved. Chief among these are ensuring that faculty and students receive credit for their work and coordinating students at diverse campuses simultaneously enrolling in the same class. Both of these problems, while surmountable, take work.

At many institutions one needs the faculty to vote on course credit for students. The departments or the administration also need to support these efforts. The latter is often challenging, as many schools have strained resources and are concerned that it is too "expensive" to have faculty teach such classes. Fears are often strongest when the course is highly specialized, and involves only a few students from the local campus. There may even be fears that hybrid/online courses will "steal" students away from local courses.

Our hope is that by showing that shared courses can be taught, without an undue amount of work, in a manner compatible with the liberal arts mission that our colleagues will be convinced to add their courses to the mix. Doing so spreads the work across many institutions, and would increase the diversity of course offerings enormously. In such a situation the greater opportunities for students would be clear and help make the case that it is worth the additional faculty time required.

In addition to the more standard advanced classes found at our institution, we hope that a collaborative environment would exist to facilitate creating and offering new courses. One example we have discussed is a class on Financial Mathematics. From numerous conversations with students it is clear that there is an enormous demand for such a course, but most of our institutions do not offer it. One reason for this is that frequently no school has someone who does this type of mathematics/statistics, and creating such a new course is an enormous amount of work. However, if this class were shared across schools, the workload for each involved faculty member would be very manageable. We are exploring a model where three or four faculty members agree to teach a few weeks, thus making the commitment on course development modest and allowing us to build on what we have learned to address a need on our campuses.

CHAPTER 6

Online Statistics Teaching and Learning

Jim Albert, Mine Çetinkaya-Rundel, and Jingchen Hu*

CONTENTS

* Jingchen Hu was supported by The Andrew W. Mellon Foundation Anne McNiff Tatlock '61 Endowment for Strategic Faculty Support at Vassar College and the Liberal Arts Collaborative for Digital Innovation (LACOL).

6.1 INTRODUCTION

Statistics is probably one of the most active fields to embrace and engage online teaching and learning. Numerous massive open online courses (MOOC) have been developed and have attracted a great number of online learners. At the time of writing, there are 346 courses and specializations with the keyword "statistics" on Coursera,* one of the most popular online learning platforms. DataCamp,† a growing online learning platform specializing in data science education, has 159 courses, 22 tracks, and 122 instructors. A broad survey of online statistics education is described in Mills and Raju [1], and a discussion of building an online statistics curriculum is contained in Young et al. [2].

For statistics courses at all levels, teaching and learning online pose challenges in different aspects. Particular online challenges include how to effectively and interactively conduct exploratory data analyses, how to incorporate statistical programming, how to include individual or team projects, and how to present mathematical derivations efficiently and effectively.

This chapter draws from the authors' experience with seven different online statistics courses to address some of the aforementioned challenges. Section 2.1 is an online exploratory data analysis course taught at Bowling Green State University. Section 2.2 is an upper-level Bayesian statistics course taught at Vassar College and shared among ten liberal arts colleges through a hybrid model. Section 2.3 describes a five-course MOOC specialization on Coursera, offered by Duke University.

All of these courses are designed for undergraduate or graduate students with calculus backgrounds. The general aim in this chapter is to provide overviews of these online courses, discuss challenges and approaches, and provide general guidelines for statistics educators interested in online teaching and learning of statistics.

Yang [3] provides an overview of the components of an online statistics course and uses student feedback to gain insight into the particular components that appear effective for learning the statistics material. Everson and Garfield [4] discuss the use of student discussions in an online statistics course and focus on the types of discussion that appear to facilitate understanding of the statistics concepts. Dunwill [5] provides general comments about the challenges of teaching in an online format, and Everson [6] discusses her experiences in teaching online after teaching face-to-face courses in statistics.

---

* For more information, visit www.coursera.org/.
† For more information, visit datacamp.com.

Section 6.2 gives an overview of the seven online statistics courses focusing on the intended audience and the course content. Sections 6.3 through 6.6 focus on particular components of the online course, and discuss how each course addresses the particular component. Section 6.3 discusses the problem of the online course design. How is the instructional content presented and organized, keeping in mind the learning objectives of the course? Section 6.4 focuses on the use of technology in each course. A statistics course will typically include the use of a software package. What types of software are used in each class, and how is the software integrated with the learning of the conceptual material? Section 6.5 describes the different forms of assessment for each course. What types of assessments such as homework or projects are used, and do students have the opportunity to work together on assignments? Section 6.6 describes how the students interact with the instructor in the course and how students interact with other students. How does a student get help in the course? Not all of the presented online courses run smoothly, and Section 6.7 describes the challenges that the instructors faced when implementing each class. Section 6.8 summarizes the general features and challenges of the courses, directed towards the instructor who is developing their first online statistics class.

6.2 DESCRIPTION OF THE ONLINE COURSES

6.2.1 An Online Course in Exploratory Data Analysis

This online course, Exploratory Data Analysis at Bowling Green State University, focuses on the principles of exploring data following ideas from John Tukey's EDA book [7]. The audience consists of graduate and undergraduate students majoring in statistics, and there is a probability prerequisite.

Generally, the main intent of the course is to describe an exploratory philosophy in the analysis of data. One does not wish to impose any assumptions such as normal sampling distributions or equality of variances between groups. Instead, one wishes to explore the data, looking for patterns in distributions and relationships. There are four "R"s in EDA that summarize the general philosophy in data exploration. "Revelation" means that EDA often uses graphical displays in data discovery. "Resistance" means that it is desirable to use statistical methods that are resistant or non-sensitive to outlying values. "Reexpression" means that it is sometimes useful to reexpress variables by a nonlinear transformation such as a log or square root. Last, "residual" means that one usually wishes to look at the deviations from a statistical fit.

Table 6.1 shows the main units for the EDA course. The course begins with a discussion of graphical displays and resistant summaries for a single batch of measurement data. The next general topic is the comparison of batches of measurement data and the use of reexpressions to equalize spreads across batches. Properties of the Box-Cox power family of transformations [8] are explored in Unit 4, and this family is used to perform an appropriate reexpression to make a data distribution symmetric. Scatterplots of two measurement variables are introduced in Unit 5, and Tukey's resistant line is applied as a general method of fitting a straight line to data. In cases where the scatterplot pattern is nonlinear, a running-median smoother is described as a simple way of smoothing a scatterplot to assess the general pattern. Unit 6 explores a two-way table where one summarizes a measurement variable over two categorical variables. Median

TABLE 6.1   Unit and Lectures for the Online EDA Course

Unit	Lecture
1. Introduction	1.1 Introduction to EDA I
	1.2 Introduction to EDA II
2. Single Batch	2.1 Displays
	2.2 Summaries
3. Comparing Batches	3.1 Boxplots
	3.2 Spread Level Plots
	3.3 Comparing Batches III
4. Transformations	4.1 Transformations
	4.2 Reexpressing for Symmetry
	4.3 Reexpressing for Symmetry II
	4.4 Transformations Summary
5. Plotting	5.1 Introduction to Plotting
	5.2 Resistant Line
	5.3 Plotting II
	5.4 Straightening
	5.5 Smoothing
6. Two-Way Analyses	6.1 Median Polish
	6.2 Plotting Additive Fit
	6.3 Multiplicative Fit
	6.4 Extended Fit
7. Counts and Fractions	7.1 Binning Data
	7.2 Binning Data II
	7.3 Fraction Data
EDA Project	

polish is a resistant method of applying an additive fit by the use of a logarithmic transformation; this method can also be used to apply a multiplicative fit to these two-way data structures. The course concludes in Unit 7 by describing methods for binning measurement data, assessing if the histogram has a Gaussian shape, and exploring batches of fraction data.

6.2.2 A Bayesian Statistics Course for Cross-Campus Share

The past decades have seen great methodological, computational, and inferential advancement of Bayesian statistics. While Bayesian statistics continues to gain attention and becomes ever more popular among data analysts and researchers, the topic itself is rarely available to students, especially at the undergraduate level. In most liberal arts colleges, with the staffing constraints, offering a topic course on Bayesian statistics can at most be an occasional luxury. More commonly, such a course is not offered at all.

Vassar College had the chance to offer an undergraduate-level Bayesian statistics course in Fall 2016. The extremely positive experience with a small group of motivated students has encouraged the instructor to think beyond the boundary of a physical college location. Vassar is a member of the Liberal Arts Collaborative for Digital Innovation (LACOL).* Under

---

* LACOL is a partnership of ten liberal arts colleges in the United States, founded in 2014. By leveraging the power of consortial relationships, LACOL focuses on utilizing and adapting emerging technologies to promote excellent and innovative teaching, learning, and research in the liberal arts. For more information about LACOL, visit http://lacol.net/about-the-consortium/.

TABLE 6.2   Section and Main Topics for the Cross-Campus Shared Bayesian Statistics Course

Section	Topics
Inference	Bayes theorem, conjugate prior, posterior distribution, HPD interval, predictive distribution
Computation	Monte Carlo approximation, Markov chain Monte Carlo (MCMC), Gibbs sampler, Metropolis-Hastings alogrithm, MCMC diagnostics, JAGS
Applications	Bayesian hierarchical modeling, Bayesian linear regression, latent class modeling, Bayesian cognitive modeling

the Upper Level Math & Stats Project,* starting from Fall 2017, the Bayesian Statistics course at Vassar College is taught locally at Vassar while shared among the LACOL colleges through a hybrid model.

The Upper Level Math & Stats Project focuses on sharing upper-level mathematics and statistics courses among participating campuses, to supplement existing and probably limited offerings while maintaining the liberal arts flavor. Vassar's Bayesian Statistics course in Fall 2017 is one of the three courses in the pilot study (the other two are upper-level mathematics courses offered by two other member colleges).

The student audience consists of junior and senior students. The prerequisite includes multivariate calculus, linear algebra, and probability. The textbook is *A First Course in Bayesian Statistics Methods* by Peter D. Hoff [9], a book mainly used at the graduate level. The instructor intentionally borrows more applied material from *Bayesian Cognitive Modeling: A Practical Course*, written by Michael D. Lee and Eric-Jan Wagenmakers [10], making the course more accessible to undergraduate students. Occasionally, advanced material from *Bayesian Data Analysis* by Andrew Gelman and others [11] is used to supplement.

There are three general sections of the course: inference, computation, and applications, with main topics in each section listed in Table 6.2. Through the introduction of one-parameter models such as beta-binomial and normal-normal, the inference section covers the inferential basics. Moving to multi-parameter models such as normal with two unknown parameters, computation techniques are covered. Students are equipped with the skills of writing up the Markov chain Monte Carlo (MCMC) sampler when possible, as well as the use of Just Another Gibbs Sampler (JAGS) through the `rjags` R package. Ultimately, through various applications, students are exposed to more advanced models. They are motivated to understand and construct Bayesian models in each application, perform simulation by MCMC with appropriate computation techniques, and answer inferential questions in context.

Participating in LACOL's Upper Level Math & Stats Project is the first time that a Bayesian statistics course is ever shared among colleges from different geographic locations. In addition to motivating and cultivating students' learning of such advanced statistics topics, the instructor needs to redesign an existing face-to-face course to adapt to a hybrid model. Challenges include using software to provide both synchronous and asynchronous access to the lectures, identifying what material is suitable for being moved online, holding online office hours for remote students, coordinating with local faculty liaisons from each remote

---

* For more information, visit http://lacol.net/category/collaborations/projects/upper-level-math/.

campus, and creating a learning community involving all students, among other things. This course uses R extensively for simulations and data analysis, and how to effectively incorporate R programming through a hybrid instruction model is also challenging.

6.2.3 A Five-Course MOOC Specialization: Statistics with R

Statistics with R is a specialization offered on Coursera (`www.coursera.org/specializations/statistics`) comprised of five massive open online courses (MOOCs) designed and sequenced to help learners master foundations of data analysis and statistical inference and modeling. The specialization also has a significant hands-on computing component. The target audience is learners with no background in statistics or computing.

The first four courses in the specialization are Introduction to Probability and Data, Inferential Statistics, Linear Regression and Modeling, and Bayesian Statistics. These courses cover exploratory data analysis, study design, light probability, frequentist and Bayesian statistical inference, and modeling. A major focus of all of these courses is hands-on data analysis in R; each course features computing labs in R where learners create reproducible data analysis reports as well as fully reproducible data analysis projects demonstrating mastery of the learning goals of each of the courses. The fifth course is a capstone, where learners complete a data analysis project that answers a specific scientific/business question using a large and complex dataset. This course is an opportunity for learners to practice what they learned in the first four courses in the specialization.

Table 6.3 shows the modules and associated topics for each of the first four courses in this specialization. Each subsequent course assumes learners have either completed the previous course(s) or have background knowledge equivalent to what is covered in them. Each module is designed to be completed in one week, though learners have the flexibility to extend this if they need to.

Course 1, Introduction to Data, introduces sampling and exploring data, as well as basic probability theory and Bayes' rule. In this course learners examine various sampling methods, and discuss how such methods can impact the scope of inference. In addition, a variety of exploratory data analysis techniques are covered, including using data visualization and summary statistics to explore relationships between two or more variables. Another key learning goal for this course is the use of statistical computing, with R, for hands-on data analysis. The concepts and techniques introduced in this course serve as building blocks for the inference and modeling courses in the specialization.

Course 2, Inferential Statistics, introduces commonly used statistical inference methods for numerical and categorical data. Learners learn how to set up and perform hypothesis tests and construct confidence intervals, interpret p-values and confidence bounds, and communicate these results correctly, effectively, and in context without relying on statistical jargon. Building on computing skills they acquired in the previous course, learners conduct these analyses in R.

In Course 3, Linear Regression and Modeling, introduces simple and multiple linear regression. Learners learn the fundamental theory behind linear regression and, through data examples, learn to fit, examine, and utilize regression models to examine relationships

TABLE 6.3   Modules and Topics for the First Four Courses in the Statistics with R Specialization

Course 1: Introduction to Probability and Data

1.1 *Introduction to data*	Data basics, observational studies and experiments, sampling and sources of bias, experimental design
1.2 *Exploratory data analysis and introduction to inference*	Visualizing data, measures of center and spread, robust statistics, transformations, exploring bi-/multivariate relationships, introduction to inference via simulation
1.3 *Introduction to probability*	Independent and disjoint events, conditional probability, Bayes' rule, introduction to Bayesian inference
1.4 *Probability distributions*	Normal and binomial distributions, assessing normality
1.5 *Data analysis project*	Exploratory data analysis of data from the Behavioral Risk Factor Surveillance System

Course 2: Inferential Statistics

2.1 *Confidence intervals*	Sampling variability and Central Limit Theorem, confidence intervals for a mean, accuracy vs. precision
2.2 *Inference and significance*	Hypothesis testing for a mean, decision errors, statistical vs. practical significance
2.3 *Inference for means*	t-distribution, inference for a mean and for comparing two or more means, multiple comparisons, bootstrapping
2.4 *Inference for proportions*	Sampling variability and CLT for proportions, confidence intervals and hypothesis tests for two or more proportions, randomization tests for small samples
2.5 *Data analysis project*	Inference on data from the Behavioral Risk Factor Surveillance System

Course 3: Linear Regression and Modeling

3.1 *Linear regression*	Correlation, residuals, least squares line, prediction and extrapolation
3.2 *More on linear regression*	Outliers. Inference for regression, variability partitioning
3.3 *Multiple linear regression*	Multiple predictors, adjusted R^2, collinearity and parsimony, inference for MLR, model selection and diagnostics
3.4 *Data analysis project*	EDA and single and multiple regression for movies data

Course 4: Bayesian Statistics

4.1 *Basics of Bayesian statistics*	Conditional probabilities and Bayes' rule, diagnostic testing, Bayes updating, Bayesian vs. frequentist definitions and inference, effect size and significance
4.2 *Bayesian inference*	From discrete to continuous, elicitation, conjugacy, Gamma-Poisson and normal-normal conjugate families, non-conjugate priors, credible intervals, predictive inference
4.3 *Decision making*	Loss functions, minimizing expected loss, Monte-Carlo sampling, prior choice and reference priors, MCMC
4.4 *Bayesian regression*	Bayesian simple and multiple regression, model uncertainty and averaging, decisions under model uncertainty
4.5 *Perspectives*	Interviews with statisticians on how they use Bayesian statistics in their work
4.6 *Data analysis project*	Bayesian inference and regression for movies data

Course 5: Statistics with R Capstone

5.1 *Exploratory data analysis*
5.2 *Basic model selection*
5.3 *Model Selection and Diagnostics*
5.4 *Out of Sample Prediction*

between multiple variables. Model fitting and assessment is done in R, and a substantial amount of examples are focused on interpretation and diagnostics for model checking.

These first three courses were originally offered as a single, much longer, MOOC, for two years, before being split into shorter courses to be bundled up in a specialization. Course 4, Bayesian Statistics, was added to the sequence at this point, in order to make this introductory specialization more complete by adding a different point of view for approaching statistical analysis.

Course 4, Bayesian Statistics, introduces learners to the underlying theory and perspective of the Bayesian paradigm and shows end-to-end Bayesian analyses that move from framing the question to building models to eliciting prior probabilities to implementing in R. The course also introduces credible regions, Bayesian comparisons of means and proportions, Bayesian regression and inference using multiple models, and discussion of Bayesian prediction.

The last course in the specialization is a capstone course. The materials provided for this course are designed to serve as a reminder of learning goals of earlier courses or expand on them ever so slightly. A large and complex dataset is provided to the learners, and the analysis requires the application of a variety of methods and techniques introduced in the previous courses, including exploratory data analysis through data visualization and numerical summaries, statistical inference, and modeling as well as interpretations of these results in the context of the data and the research question. Learners are encouraged to implement both frequentist and Bayesian techniques and discuss in the context of the data how these two approaches are similar and different, and what these differences mean for conclusions that can be drawn from the data.

6.3 ONLINE COURSE DESIGN

6.3.1 Presentation of Content

The EDA course was originally taught face-to-face in the classroom where the instructor would introduce and demonstrate the EDA methods in class and the students would work on weekly data analysis assignments. The online course was designed to follow the same format as the face-to-face version.

1. The lecture material for the class is posted online as PDF documents. Students have had difficulties understanding the material by reading directly from [7], and so the lecture material seems to be a reasonable substitute for the book material. The core EDA material is contained in a series of 23 PDF documents. A particular "lecture" motivates and describes the particular EDA method with an illustration using the R programming language.

2. There are weekly data analysis assignments and no written exams in the course. It is well-known that exams can be challenging to administer in an online format.

3. The work on the assignments is a blend of statistical work such as tables and graphs and interpretation of the results. These assignments are turned in online by the use of R Markdown documents saved in html format.

4. All of the R code in the lecture notes is made available to the students by a collection of R scripts.

In the Bayesian statistics course, regular face-to-face lectures are delivered in the classroom with students present at Vassar College. Every lecture is broadcast and recorded by

Zoom, a video conferencing software.* Remote students can join the lecture in real time with a Zoom meeting ID. Otherwise, they can watch the recorded videos after the videos are posted on the same day of the lecture.

6.3.2 Course Material in Video Form

This hybrid model in the Bayesian statistics course made lectures available in video form. With the flexibility of making videos and the added familiarity of learning through videos on the students' end, other course material has also been turned into video form. This material has conventionally been available as a Word- or PDF document.

For example, when an example is not fully developed during the lecture due to time constraints, a short video on this example is created and made available to students to review if necessary. As another example, when many students are having problems with the same homework question (based on observation from office hour visits), a short video providing hint on this homework question is created and made available.

R programming demonstrations are very suited to video form. By watching a video with a step-by-step demonstration of programming, students are able to pause when needed, see things in action, and practice along the way. Several R programming videos are created for students in this course.

In the Five-Course MOOC, each module includes seven to ten videos roughly four to seven minutes in length. Most of these videos introduce new concepts and the remaining provide additional examples and worked-out problems. The slides that serve as the background in the videos are created in Keynote (Apple's presentation software application) and feature a substantial amount of animations such that text, visualizations, and calculations showed on the slides follow the pace of speech in the videos. Many learners have expressed in their course feedback that these features make the videos more engaging and easier to follow compared to videos in many other MOOCs. Sample videos from the Inferential Statistics course are hosted on YouTube (`bit.ly/2LrO6KZ`).

6.3.3 Learning Objectives

Each module of the Five-Course MOOC featured a set of learning objectives. A sampling of learning objectives from the Inferential Statistics course is shown below:

- Explain how the hypothesis testing framework resembles a court trial.

- Recognize that in hypothesis testing we evaluate two competing claims: the null hypothesis, which represents a skeptical perspective or the status quo, and the alternative hypothesis, which represents an alternative under consideration and is often represented by a range of possible parameter values.

- Define a p-value as the conditional probability of obtaining a sample statistic at least as extreme as the one observed given that the null hypothesis is true: p-value = P(observed or more extreme sample statistic | H_0 true).

---

* For more information about Zoom, visit https://zoom.us/

These learning objectives are constructed using verbs from the revised Bloom's Taxonomy [12] and aim to keep learners organized and focused while watching the videos. The learners are recommended to have the learning objectives handy while watching the videos and revisit sections of the videos and/or suggested readings for any learning objectives that they feel like they have not mastered at the end of the module.

The learning objectives are provided as separate stand-alone documents, and after each batch of related learning objectives are a few simple conceptual questions for learners to check their understanding before moving on.

6.3.4 Suggested Readings and Practice

Suggested readings for the first three courses of the MOOC course came from *OpenIntro Statistics* [13]. This book is free and open-source, meaning that learners enrolled in the MOOC do not need to additionally purchase a textbook. The readings are optional as the videos explicitly introduce and cover all required topics for the course; however many learners have reported in their feedback that they really like having a reference book that closely follows the course material. Practice problems are also suggested from the end of chapter exercises in this book.

For the fourth course on Bayesian statistics, readings are suggested from *An Introduction to Bayesian Thinking* [14]. This textbook has been written by the Bayesian Statistics course development team (faculty and PhD students) specifically as a companion to this course and is also freely available on the web.

6.4 USE OF TECHNOLOGY

6.4.1 R Package

A special R package `LearnEDAfunctions` was written for the EDA class. This package contains all of the datasets used in the lecture notes and the assignments. In addition, the package contains special functions for implementing the EDA computations. For example, the function `rline` computes Tukey's resistant line and the function `fit.gaussian` fits a Gaussian comparison curve to histogram data and outputs the rootogram residuals from the Gaussian fit.

6.4.2 EDA Blog, YouTube Videos

Different methods in the EDA course were used to provide weekly communications with the students. For several iterations of the course, the instructor posted weekly articles on the blog "Exploratory Data Analysis" (`https://exploredata.wordpress.com/`).

In a typical post, the instructor would give an example of the weekly EDA concept and give advice on common problems in applying the interpretation of the EDA method. Blog postings from previous years are made available for the student who wishes to see additional illustrations of the statistical methods. Since some students expressed preference for learning by watching videos instead of reading notes, the instructor added videos at the YouTube channel `www.youtube.com/user/bayesball2/videos`. A particular video would show the implementation of a particular EDA method using R and the `LearnEDAfunctions` package.

This Bayesian course extensively used videos to deliver content. One of the most creative uses of making videos for this hybrid course is to create guest lectures. Conventionally, guest lectures are delivered in the physical classroom. Now, to include remote students, guest lectures can be created in video form and made available to students online. This practice also saves class meeting time when possible. For example, a guest lecture video by a cognitive science professor at Vassar College is created and used as an introduction to Bayesian hierarchical modeling. Students watch the guest lecture before the class meeting, become familiar with the topic by themselves outside of the class, then when meeting in class, the lecture and discussion can follow from the common ground of the material from the guest lecture directly. This practice also exposes students to applications of Bayesian statistics at various stages of their learning.

6.4.3 Shiny Activities

In addition to the weekly data analysis assignments, the EDA class also contains several activities where the student uses sliders and other interactive tools in exploring data. For example, in choosing the "correct" power of a reexpression, the student can choose a value of the power on a slider and see the immediate impact of that particular reexpression in a graph of the reexpressed data.

6.4.4 Use of Technology in Lectures

Instead of writing on the chalkboard or presenting lecture slides on the projector in front of the classroom using a computer, the instructor in the Bayesian course uses an iPad as a participant of the Zoom meeting, brings up lecture slides inside the Zoom software, shares the screen to present the slides to the class on the projector, and uses an Apple pencil to write on the slide or a whiteboard. When doing an R programming demonstration, the instructor joins the Zoom meeting with a laptop, then shares the screen of R/RStudio on the laptop to the projector. Zoom records the video from the projector, and all audio from the lecture. Sometimes, the instructor could use a directional mic on the iPad to improve sound capturing.

6.4.5 Computing Lab

Each module in the MOOC featured a computing lab in R. The objective of the labs is to give learners hands-on experience with data analysis using modern statistical software, R, as well as providing them with tools that they will need to complete the data analysis projects successfully.

The statistical content of the labs matches the learning objectives of the respective modules they appear in, and the application examples (i.e. datasets and research questions) are primarily from social and life sciences. The labs also make heavy use of an R package, `statsr`, which was designed specifically as a companion for the specialization [15]. Two other important aspects of the labs are that (1) they use the `tidyverse` syntax and (2) they are completed as reproducible R Markdown reports. The tidyverse is an opinionated collection of R packages designed for data science, meaning that the grammar used in the packages is optimized for working with data specifically for wrangling, cleaning, visualizing, and modeling data [16]. The choice of the tidyverse syntax for beginners is rooted in wanting to get learners exploring real and interesting data and building informative and appealing visualizations and drawing useful conclusions as much as possible [17].

R Markdown provides an easy-to-use authoring framework for combining statistical computing and written analysis in one document [18]. It builds on the idea of *literate programming*, which emphasizes the use of detailed comments embedded in code to explain exactly what the code is doing [19]. The primary benefit of R Markdown is that it restores the logical connection between statistical computing and statistical analysis by synchronizing these two parts in a single reproducible report. From an instructional perspective this approach has many advantages: reports produced using R Markdown present the code and the output in one place, making it easier for learners to learn R and locate the cause of an error and learners keep their code organized and workspace clean, which is difficult for new learners to achieve if primarily using their R console to run code [20]. Each lab is provided to the learners in an R Markdown template that they can use as a starting point for their lab report. Earlier labs in the specialization include lots of scaffolding, and almost have a fill-in-the-blanks feel to them. As the course progresses the scaffolding in the templates is removed, and by the end of the first course learners are able to produce a fully reproducible data analysis project that is much more extensive than any of their labs. All labs in the specialization are hosted in a publicly available GitHub repository at `https://github.com/StatsWithR/labs`.

6.5 ASSESSMENTS AND ENGAGEMENT

6.5.1 Weekly Data Analysis Assignments

In a typical data analysis assignment of the EDA course, the student works on a particular EDA method using a specific dataset or another suitable dataset chosen by the student. One challenge for the student is to find a suitable data structure to implement the EDA method. For example, if the EDA task is to symmetrize a dataset by the use of a power expression, the student needs to find a strongly skewed dataset that could benefit with a reexpression.

6.5.2 Final Project

In a final capstone project of the EDA course, the students select their own dataset, state questions of interest, and explore the dataset using several of the EDA methods discussed in class. The focus of this project is not on the implementation of the methods but rather on the interpretation of the results in light of the questions that were originally posed.

The Bayesian course has a final project component, and students can choose from one of the following.

- A Bayesian data analysis on a topic of your choosing

- A new Bayesian methodology or theoretical finding

- A Bayesian research paper or a book chapter (choose from a provided list)

Students submit a project proposal after the first midterm exam. They are encouraged to meet with the instructor to discuss their project ideas and their progress along the way.

The final project presentation has two parts: a two-min video on Moodle, and a poster at the poster session. The choice of a poster session is to accommodate a relatively large class (16 local students). However, it mimics real research settings, as it has become common for academic conferences to have a poster session for graduate students and junior researchers. Students in the class overall enjoy being able to talk to audience as a small group. The amount of interaction between presenters and the audience is much more than a regular presentation.

Each course in the MOOC ends with a data analysis project, the focus of which is summarized in Table 6.3, and the specialization wraps up with an extended capstone project. Each student who turns in a project evaluates three other students' work using a peer evaluation rubric. Learners are also strongly encouraged to seek informal feedback on their projects in the course discussion forums. All data analysis projects appearing in the courses in this specialization are hosted in a publicly available GitHub repository at `https://github.com/StatsWithR/projects`.

6.5.3 Homework

Homework for the Bayesian course is on a biweekly basis. The assignment usually consists of a set of derivation exercises to enhance the understanding of Bayesian methodology, and a set of application-based exercises, which require the use of R programming. There are two midterm exams and no final exam.

The teaching assistant for this course holds regular office hours at Vassar. While these office hours are held online too, remote students rarely make use of them. Instead, the instructor meets with the remote students together during a separately scheduled online office hour, also through Zoom.

Homework submission from the remote students is done through scans and emails. Exams for remote students are proctored by the local faculty liaison. Exam papers are sent to the instructor by scans and emails too. All grading is either done by the instructor or the teaching assistant. Graded homework and exams are returned to the remote students by scans and emails.

6.5.4 Case Studies

For the Bayesian course, towards the latter part of the semester when students have been exposed to and gained experience with Bayesian inference, students are grouped to do case studies with real data applications. These case studies are all open-ended. Students are given the chance to freely explore the datasets and come up with their methods and realize their inferences through MCMC computation techniques.

Prior to the case study class meeting, groups need to post their analyses onto the Moodle discussion forum to receive credit. Students in the same group take turns to be the leading writer of the analyses. Such a practice ensures that everyone is prepared to discuss the approaches and findings from the group, and the class meetings usually turn out to be great discussion sessions and ideas bounce back and forth.

6.5.5 Quizzes

Each module of the MOOC course features two sets of multiple-choice quizzes, one formative and one summative. Each question is encoded with feedback that points learners back to relevant learning objectives. Learners can attempt the summative quizzes multiple times with slightly modified versions of the questions.

6.6 INTERACTION

Currently, there is limited interaction in the EDA course. Students communicate with the instructor by means of personal meetings or email or messages sent through the learning management system. There is no regularly planned interaction between students such as an outline chat session, but students are asked to read and review the project presentations of two other students in the class.

In the Bayesian course, to create and foster an online learning community, there are extensive uses of the online discussion forum on Moodle.*

At the beginning of the semester, students make self-introduction posts about their basic information (name, year, and school), prior statistics exposure, prior R exposure, and potential final project interests.

During the semester, online discussion forums are created whenever sharing of information and making comments are needed, and they are for credit sometimes. For example, when covering the Gibbs sampler, the class reads a research paper "Explaining the Gibbs Sampler" by George Casella and Edward I. George [21]. A reading guide for this paper with six questions is provided to the students. Prior to the class meeting, students need to respond to any one question on the online discussion forum to receive participating credit. After the class discussion, students need to make another response to receive credit. Such a requirement not only helps students in reading a statistics research paper outside of class, but also helps facilitate in-class discussion both by more engagement prior to and post class meeting.

There is a final project for students in the course. In addition to presenting their projects in a poster session at the end of the semester, students need to make a two-min video post about their projects on the online discussion forum. Watching a two-min pitch talk prior to the poster session helps other students to arrange their poster visits. These videos also help students to succinctly present their projects in a manner as appealing as possible. While the use of the online discussion forum on Moodle is motivated first mostly to make remote students feel included, it ultimately engages local students much better as well.

In-person student interaction, or the lack thereof, is often a major challenge in online courses. However, in MOOCs the discussion forums are often a major strength of the course. Given that at any point thousands of students are enrolled in the course, even if a small percentage of these students choose to browse the discussion forums, and an even smaller percentage of them interact with other learners on the course discussion forums, this still results in a large number of learners interacting with each other.

---

* Moodle is the course management system used at Vassar College.

Additionally, over the years of the MOOC being offered, a growing number of very knowledgeable and helpful course mentors emerged from the discussion forums. These are learners who took the courses at an earlier time and now volunteer their time to answer student questions and provide direction for new learners. Their contribution is essential to the success of the course discussion forums. In addition to answering student questions directly, they also escalate any questions that would benefit from additional input to the course instructor. This workflow makes it feasible for the instructor to keep tabs on the discussion forums when most needed, but not get overwhelmed by the number of questions posted each day by the learners.

6.7 CHALLENGES

There are several current challenges in the current version of the EDA online course.

- Interaction with the student

  It is beneficial if the student can interact with the instructor and fellow students in an online course. Unfortunately, attempts for interaction such as online chat sessions or online message boards have not been effective in this particular class. Therefore most of the communication is done through email and personal meetings. There is an effort to try out new methods of interaction when they become available.

- Technology issues

  Students can get frustrated with technology issues such as installing software or getting their R markdown files to "knit" properly. It is best to address these issues early in the course so the course is more about the EDA concepts and less about the associated technology.

- Balance of computation and interpretation in assignment work

  In a typical assignment, the student will turn in a Markdown file that blends output from the R system and written text that interprets the R output in the context of the particular applied problem. Since the course is really focusing on the interpretation rather than the implementation of the EDA methods, the assignment should emphasize the interpretation component. Depending on the background, the student may emphasize instead the computation component, but hopefully the students will learn what is expected in future assignments.

Although the instructor of the Bayesian course has been faced with various challenges, advice, suggestions, and sharing from multiple parties have been tremendously helpful. To improve the teaching and learning mode of the hybrid model, lecture videos can be edited and shortened when resources permit. If done properly, the lecture videos can well be a set of learning material for anyone (not necessarily from the LACOL member colleges) who is interested in an undergraduate-level introduction to Bayesian statistics. More thought and consideration on turning course material into video form can further enhance the teaching and learning. While students' interaction can be maintained using an online discussion forum on Moodle, other forms of interaction can be explored and developed to further enhance the overall interaction in the course.

For the five-course MOOC, there are three main challenges with offering this content on an online platform; two of these are associated with the labs and the other is associated with the data analysis projects.

- Autograding: Given that this is a course with thousands of learners enrolled at any given point, human-grading is simply not feasible. The lab assessments are set up as multiple-choice questions. Learners complete the lab exercises by generating R Markdown reports in which they analyze a dataset. Then, they answer a series of multiple-choice questions about the data analysis results. The challenge is that the multiple-choice questions do not assess the full spectrum of the skills we want learners to acquire via these labs—they assess whether they can obtain the correct results using R, however they do not assess mastery of R syntax, reproducibility of their analysis, etc.

- Computing infrastructure: Our preferred method for getting students with no computing background started with R is cloud-based access to RStudio in order to avoid challenges around local installation and to provide a uniform computing environment for all learners. However, it is not feasible to offer a centralized cloud-based solution to all learners enrolled in an MOOC, and hence students have to locally install R and RStudio and the correct versions of all packages they use in the labs.

- As a partial solution for this challenge, we offer students the option to complete the labs in the first course of the specialization on DataCamp (www.datacamp.com), an online learning platform that provides in-browser access to RStudio. This helps students struggling with software installation issues early on in the course to get started with data analysis and go back to tackling software challenges once they feel a little bit more confident with R.

- Peer evaluation: Autograding is not feasible for open-ended data analysis projects, and hence peer evaluation is the only solution for grading of these projects. Even with a very detailed rubric, consistency in grading is difficult to attain, and it is challenging for learners who are just learning the material themselves to evaluate others' work. Additionally, variability in the quality and depth of feedback provided can leave learners frustrated. The option to share their projects on the discussion forums and get feedback can be helpful for some learners, but others are not so keen on publicly sharing their projects.

6.8 CONCLUDING COMMENTS

Although this is current interest in teaching online introductory statistics courses, the online statistics courses described here are directed towards specific groups of students and all of the comments may not be directly applicable to the introductory class. For example, the students in the EDA online course are primarily masters-level or advanced undergraduate students who are comfortable in working independently on assignments and projects, and this particular course design may not be suitable for an introductory statistics class with a minimal mathematics prerequisite. Similarly, the Five-Course MOOC has been designed for a global audience, assuming no common background and varying levels of access to computing resources. The delivery of the course can be simplified for a university audience where assumptions about common access and background might be more appropriate. Nonetheless, there are general common elements to these courses which would be helpful for the instructor who is designing the first online statistics course.

6.8.1 Presentation of Content

Although the EDA course features written instructional content, much of the content delivery in all of these online courses happens in videos. The use of videos makes it possible for several instructors to get involved in the presentation of content. In addition, it is possible for the student to learn from the video at their own pace, replaying parts of the video to help them understand the material.

6.8.2 Interaction

It is important to develop a cooperative learning environment among students in the online course. These courses suggest useful methods for facilitating this type of environment. Discussion forums, as described by the shared Bayesian course and the Five-Course MOOC, are one good way to foster communication between students. Another good opportunity for collaboration is through statistics projects where groups of students explore data on case studies.

6.8.3 Assessment

It should be noted that traditional forms of assessment such as multiple-choice exams play a limited role in assessment for these online statistics courses. Instead, these courses feature data analysis projects, interactive computer lab assignments, and other projects where the student carries out an exploration into a new method or finding that is not covered in the curriculum. However, the necessity for peer evaluation of the projects in the Five-Course MOOC presents its own challenges.

6.8.4 Using Software

All of these online courses use technology or software that may not be familiar to the student. Specifically, since these are statistics courses, one would typically use the R language together with specialized R packages. Introducing this technology creates challenges since the students can vary greatly with their experience with programming in general as well as with the R ecosystem. These courses have presented various ways to mitigate the technology challenges by creating special packages (e.g. the `LearnEDA` package in the EDA course, the `rjags` package in the shared Bayesian course, and the `statsr` package in the Five-Course MOOC) that include all of the datasets and special functions needed for the course. The Five-Course MOOC provides some suggestions to help some of these technology issues such as providing cloud-based access to R via a third-party company, DataCamp. The instructor teaching an online statistics course should think carefully about the use of software, especially from the viewpoint of the student who is inexperienced with technology.

REFERENCES

1. Mills, J. D. and D. Raju, "Teaching statistics online: A decade's review of the literature about what works," *Journal of Statistics Education*, vol. 19, no. 2, 2011.
2. Young, D. S., G. F. Johnson, M. Chow and J. L. Rosenberger, "The challenges in developing an online applied statistics program: Lessons learned at penn state university," *The American Statistician*, vol. 69, no. 3, pp. 213–220, 2015.
3. Yang, D. "Instructional strategies and course design for teaching statistics on- line: Perspectives from online students," *International Journal of STEM Edu-cation*, vol. 4, no. 34, pp. 1–15, 2017.
4. Everson, M. and J. Garfield, "An innovative approach to teaching online statistics courses," *Technology Innovationss in Statistics Education*, vol. 2, no. 1, 2008.
5. Dunwill, E., "Teaching principles transferred to online courses: Strategies to use," *eLearning Best Practices*, 2016.
6. Everson, M, "10 things i learned about teaching online," *eLearn Magazine*, 2009.
7. Tukey, J. W., *Exploratory Data Analysis*. Pearson, 1977.
8. Sakia, R. M., "The box-cox transformation technique: A review," *The Statistician*, vol 41, no. 2 pp. 169–178, 1992.
9. Hoff, P. D., *A First Course in Bayesian Statistical Methods*. Springer Texts in Statistics, New York: Springer-Verlag, 2009.
10. Lee, M. D. and E. Wagenmakers, *Bayesian Cognitive Modeling: A Practical Course*. Cambridge University Press, 2014.
11. Gelman, A., J. B. Carlin, H. S. Stern, D. B. Dunson, A. Ve-htari and D. B. Rubin, *Bayesian Data Analysis*. Chapman & Hall/CRC Texts in Statistical Science, 3rd ed., 2013.
12. Anderson, L. W., D. R. Krathwohl, P. W. Airasian, K. A. Cruikshank, R. E. Mayer, P. R. Pintrich, J. Raths and M. C. Wittrock, *A Taxonomy for Learning, Teaching, and Assessing: A Revision of Blooms Taxonomy of Educational Objec-tives, Abridged Edition*. White Plains, NY: Longman, 2001.
13. Diez, D. M., C. D. Barr and M. Çetinkaya Rundel, *OpenIntro Statistics*. Cre- ateSpace, 3rd ed., 2014. https://www.openintro.org/.
14. Clyde, M., M. Çetinkaya Rundel, C. Rundel, D. Banks, C. Chai and L. Huang, *An Introduction to Bayesian Thinking*. GitHub, 1st ed., 2018. https:// statswithr.github.io/book/.
15. Rundel, C., M. Çetinkaya-Rundel, M. Clyde and D. Banks, Statsr: *Companion Package for Statistics with R*, 2018. R package version 0.1-0.
16. Wickham, V, *Tidyverse: Easily Install and Load the 'Tidyverse'*, 2017. R package version 1.2.1.
17. Robinson, D., "Teach the tidyverse to beginners," 2017. http://varianceexplained. org/r/ teach-tidyverse/.
18. Xie, Y., J. Allaire and G. Grolemund, *R Markdown: The Definitive Guide*. Boca Raton, FL: CRC Press, 2018.
19. Knuth, D. E., "Literate programming," *The Computer Journal*, vol. 27, no. 2, pp. 97–111, 1984.
20. Etinkaya-Rundel, M. C. and C. Rundel, "Infrastructure and tools for teaching computing throughout the statistical curriculum," *The American Statistician*, vol. 72, no. 1, pp. 58–65, 2018.
21. Casella, G. and E. I. George, "Explaining the gibbs sampler," *The American Statistician*, vol. 46, pp. 167–174, 1992.

CHAPTER 7

Statistics for Engineers

Charles E. Smith, Kimberly S. Weems, and Reneé H. Moore

CONTENTS

7.1 INTRODUCTION

Upon the request of engineering faculty, the department created a one-semester course called Probability and Statistics for Engineers. This course consists of about two-thirds statistics and one-third probability and has the largest enrollment among calculus-based undergraduate service courses in statistics. Approximately 100 students enroll in each of the five to six sections taught during both the fall and spring semesters. One distance education (DE) section is available. Statcrunch is the primary software used in the face-to-face sections. The DE section offers the following options regarding software: R, MATLAB, and JMP. The course has evolved somewhat over the last 15 years but an essential component has been the course project. Our goal in this chapter is to show it is feasible to have a designed experiment with two or more factors and its statistical analysis in a *one*-semester course. Some modifications to the traditional way of presenting a one-semester course seem to be required.

In traditional engineering statistics classes, probability distributions and random variables as well as estimation and hypothesis testing are usually presented before regression, ANOVA, and factorial experiments are introduced; for instance, see Navidi (2019), Walpole et al. (2016), and Devore (2015). This course, however, follows a non-traditional sequence of topics: it begins with experimental design, factorial data analysis, ANOVA, and multiple regression; the middle sets of topics include probability, random variables, random samples, and central limit theorem; and it ends with hypothesis testing and confidence intervals. This sequence allows for the students to complete a project, which may be done individually or in groups of two to three students. The objectives of the project are to choose, design, perform, analyze, and describe a completely randomized two-factor experiment using ANOVA or multiple regression and descriptive statistics. Students are encouraged to select a topic that interests them; it does not necessarily need to be related to their field of study. Students submit approximately a 10- to 20-page report including data files. The project is worth 20% of their final grade.

Although this course may be the first college statistics course completed by an undergraduate student, it incorporates material that is usually taught in second statistics courses through its emphasis on experimental design, multi-way ANOVA, and multiple regression.

Before giving the details of the course project components, a brief review of some previous work on course projects and other active-learning methods for engineers is presented. A more general literature review, not focused on course projects, but on best practices for teaching statistics online can be found in Mills and Raju (2011).

Researchers, for instance Brown and Kass (2009) and Garfield et al. (2002), have called for the inclusion of more real data into statistics education. The Guidelines for Assessment and Instruction in Statistics Education (GAISE) College Report (2016) supports the use of real data in teaching statistics and emphasizes the importance of active learning. Particularly, this report suggests the following six recommendations:

1. Teach statistical thinking.

2. Focus on conceptual understanding.

3. Integrate real data with a context and purpose.

4. Foster active learning.

5. Use technology to explore concepts and analyze data.

6. Use assessments to improve and evaluate student learning.

Specific to engineering, Musharavati and Hamouda (2010), Kvam (2000), and Romero and Land (2004) have documented the positive effects of active learning and projects on engineering students' education.

Like Hunter (1977), Khan et al. (2018) emphasize the importance of having engineering students experience all steps involved in an experimental investigation. Also, they underscore the importance of using problem-based learning in statistics courses for engineers where students are self-directed and work in groups to develop strong problem-solving skills. These group projects are supplemented with lectures and tutorials. Roughly three-quarters of their students complete their engineering statistics class online.

Bergquist and Albing (2006) surveyed their engineering alumni to examine the usefulness of statistics in the workplace. Their results suggest that courses should be practical and teach students how to perform experiments with a focus on the statistics involved, what can go wrong in an experiment, and how to handle non-textbook problems. The courses should include a broad discussion about the importance of randomization and encourage students to work on projects that they can relate to, such as "testing optimal settings of a video game."

Bradstreet (1996) argues for teaching statistical reasoning first, with less computation, and then statistical methods, that is, using exploratory data analysis before teaching classical methods. He supports workshop (problem-based and active learning) classes so that students can experience statistics. This order is what is used for the NCSU course considered in this paper.

Stephenson (2001) describes two-semester sequence distance-education applied statistics courses for industry. Emphasis is placed on students' appreciation for statistical thinking with a focus on designed experiments. Their class is taught face-to-face as well as online, with asynchronous delivery of material to DE students. Their class included a data analysis project component which was deemed "the most effect means of actively engaging distance education students in the practice of statistics." Their students worked individually or in groups on a variety of topics that may or may not be work-related. Most of their DE students chose work-related topics

Ojeda et al. (2012) discuss problem-based approaches to teaching statistical thinking in statistics courses designed for several fields, including engineering. The authors stress the need to relate statistics to real-world problems as instructors should avoid "teaching without context." The authors strongly advocated for the use of projects—students learn how to apply appropriate stat techniques, use software, strengthen/develop communication skills, and make conclusions from statistical analyses.

Butt, Fredericks, and White (2004) discuss integrating active learning into statistics for engineers. Weekly two-hour lectures are paired with three hours of lab exercises (in a computer or engineering lab).

Kvam (2000) examined the effects of active learning in calculus-based engineering statistics classes. Primarily, he studied the differences in retention of statistics material between students in a traditional class and a problem-based class. The problem-based classes included four group projects on topics such as DOE, ANOVA, and regression. The challenges of the problem-based course include students being able to think of original topics, stronger students feeling as if they had to carry less prepared students, and the workload for the instructor. The author presented the types of projects used, feedback from students, how students were selected to participate in the study, data comparison, and analysis. The study did not detect any statistical differences in the retention of students from the two teaching methods; however, the data suggested that average students benefit from the active learning.

Grima, Rodero, and Tort-Martorell (2016) discuss explaining variability in an introductory statistics course for undergraduate engineering majors. The course uses a simple circuit and the Wheatstone Bridge to introduce the concept of variability and illustrate the effect of variability. There are short exercises related to circuits in which random variables, probability distributions, functions of random variables, and sampling distributions are covered.

Romero et al. (1995) describe changes in their university system that reflect efforts to design active-learning courses that focus on what students can learn and retain.

Bisgaard (1991) argues that engineering practice and problems should inform how one teaches statistics. He discusses a one-semester engineering problem-solving course that begins with a helicopter experiment. This experiment leads to discussions about blocking, paired comparisons, randomization, and fractional factorial design. Later, data resulting from the instructors' consulting projects are used to create homework assignments. There is a final group project in which students select their own topic and design, conduct, and analyze data from an experiment.

The asynchronous NCSU distance course brings together several of the above-mentioned components. The project is an active-learning experience dealing with real data that the students have chosen and have some interest in, c.f. Ojeda et al., 2012. The project components are taught by a problem-based approach (Kvam, 2000; Ojeda et al., 2012), and all steps of the experimental investigation are performed (Khan et al., 2018) The distance sections of the course tend to have more projects done alone than the face-to-face sections. For example, last semester in the sections taught by Smith the distance section had 21 of 94 students working alone on the project, while the two face-to-face sections had 6 of 97 and 10 of 74 students working alone. The distance section students' projects tend to be more work-related than the face-to-face sections. Later we present some of the students' feedback about their view of the project. The distance section relies primarily on a message board for finding project partners and the instructor trying to pair students submitting similar proposal topics and also emailing students about other students that are geographically close to their location. The distance section has also benefited from pointers by the faculty in the College of Education who teach Massively Open Online Courses (MOOCs) on using innovative approaches and appropriate technology tools (Lee and Stangl, 2017), such as assigning a large number (26) of smaller homework sets.

7.2 PROJECT COMPONENTS

For the face-to-face sections, students submit the parts of their project in stages which allows for feedback throughout the entire experience. These components include the project proposal, executive summary, initial report, and final report, which are described below.

The distance section is handled a little differently, with a detailed project proposal (illustrated below) and a final project following an annotated template. The proposal must be approved by the instructor and the average number of iterations for the proposal is 2.3 submissions. Final projects that are submitted more than one day early receive detailed feedback and the opportunity to correct any errors noted by an initial assessment.

Many students have trouble choosing a topic on their own and struggle to truly understand the difference between a genuine experiment and an observational study. The instructors provide a list of topics and a photo gallery on the course webpage. There are also about 100 sample projects that may be accessed through the libraries' electronic reserves. Students may use ideas from previous projects, but they are required to make some modifications. Several homework questions present previous project proposals and ask them to ascertain whether they are an observational or experimental study and to explain their reasoning. For the distance section, at the project proposal stage, the topic is checked against an archive of previous projects for the course and several websites that contain engineering/science projects in an effort to prevent plagiarism.

In the project proposal, students present the title and a brief description of their experiment. Some recent projects are given in Box 7.1. While a small number of students have access to testing machines in the labs of their major or employer, such as steel or concrete strength, most students choose simpler projects that involve more accessible and cost-effective materials. Details such as variables, treatments and experimental units, analysis methods, etc., are listed in Box 7.2. They estimate how much time will be required for data collection, discuss important sources of experimental error, and describe how randomization and replication will be used. This process allows students to reflect on important topics covered during lectures or videos of them for the distance section, such as the need for a numerical approximately continuous response variable. Students explain whether they will use ANOVA or regression methods and discuss their expected outcomes. Two full models are required so that the students can assess which model is the better fit. The second model is often a log or square root transformation of the response. At least one factor must have three or more levels. The significance of the interaction term between the two factors is also emphasized and provides a clearer understanding for the students of the need for log or square root transformations.

BOX 7.1: SAMPLE PROJECTS

- Distance traveled by three brands of toy cars from ramp at three angles
- Fishing line breakage strength by brand and heat stress testing
- Comparison of the porosity of coffee filters, paper towels, and cotton fabric and their response times with three liquids
- Effect of liquid and environment on egg shell corrosion
- Optimal kicking method and ball type for achieving maximum flight distance
- Effect of packet size and distance on file transfer speeds
- Heart rate differences with respect to swim stroke type and swimming experience
- Tennis ball or superball bounce height or time between first and second impact with varied ball brand and surface type
- The effect of mass and swing length on the period of a pendulum's oscillation
- The effect of number of hidden layers and number of neurons per layer on image classification
- The effect of ball size and drop height on diameter of impact craters in sand

BOX 7.2: PROJECT PROPOSAL FORM FOR DISTANCE SECTION

ST370 DISTANCE—PROJECT PROPOSAL FORM

Fill out at least items 1–9 below and paste into an e-mail to the instructor.
1. Name(s) _____
2. Title and brief description of experiment:
3. The response variable is _____
   and is measured in
   units of _____.

4. Factors:

> Factor 1 is:
> Its levels are:
> Factor 2 is:
> Its levels are:
> If needed Factor 3 is:
> Its levels are:
> What is your "experimental unit"? How many experimental units will be needed to run your experiment?
>
> Draw the DESIGN MATRIX and indicate the random order that the trails will be run (if three replications then EACH cell has three numbers in it, those numbers are the order that combination of factors will be run).
>
> Fill in the names of your factors and their levels; don't use generic level 1, level 2, etc.

	Factor 2			
	level 1	level 2	level 3	level 4
Level 1	|_____	|_____	|_____	|_____ |
Level 2	|_____	|_____	|_____	|_____ |
Factor 1 Level 3	|_____	|_____	|_____	|_____ |
Level 4	|_____	|_____	|_____	|_____ |

> Use "sample" function in MATLAB or R to determine random ordering and put order of runs in the matrix

5. How will randomization and replication be used?
6. How will you analyze the data? With means table and ANOVA (lm with class variables, i.e. with *qualitative* explanatory variables) or regression (lm with *quantitative* explanatory variables) or both?

> [Note ALL project must have: stats output, side-by-side boxplots, mfit (fitted effects), mplot (interaction plots), and rplots (residual plots). You need to also show the model statement for at least two full (main effects and interaction) models in MATLAB or R; typically one has response and other log(response) or sqrt(response); see the four example proposals if you are unclear how to do this.]

7. Briefly, what do you expect to happen?
8. Are there any safety concerns? Will you wear protective eyewear? Gloves? Mask?
9. Any additional comments:

For the face-to-face sections, after the proposal has been approved, students are asked to submit an executive summary. The one-page summary must give enough detail so that the reader can conduct the experiment exactly as intended. Students are required to specify the random order in which trials will be run by completing a design matrix for their factorial experiment. Completing the executive summary requires the students to decide upon all details of the experiment within the first five weeks of the 16-week semester. The initial project report should be as close to the finished product as possible. Students are expected to tie together all of their graphical, numerical, and inferential methods. They include a statement of the practical implications of their study as well as a discussion of further questions raised by the study. Students include at least one photo taken while the experiment was being conducted.

For the distance section, the proposal is submitted after the fifth or sixth week, depending on the timing of fall or spring break. The proposal is more detailed than the face-to-face section and has two or three iterations of feedback with the instructor. The feedback may include doing a pilot or prototype experimental run for the corners of the design matrix to show the experiment is doable and that the levels are chosen appropriately, e.g. it doesn't take 72 hours to get each data point. Having the student send images or video of their apparatus or setup and of a prototype run are a valuable aid for the instructor to provide detailed feedback. The final project submission is usually four to five weeks later using the template shown below. Distance students are strongly encouraged to submit their project or a detailed draft early to check for analysis errors and to circumvent any miscommunication between the student and instructor about the implementation of the actual experiment. Early submissions receive one point per day up to a maximum of seven points.

BOX 7.3: PROJECT WRITEUP TEMPLATE

0. TITLE, NAME, SECTION NUMBER, and TABLE OF CONTENTS
1. An Executive Summary (goals and major findings).
    This part should be about three-quarters to one page and state:
    - What is the question you are trying to answer?
    - Your response variable.
    - Your factors and levels.
    - How you randomized, how many replications you did, and if you blocked on any factor.
    - It should summarize your statistical results: which terms were and were not statistically significant in both your models including interaction terms.
    - Main conclusions.
2. A description of the reason for your study. Why is this study of interest, what did you expect to learn?
3. A statement of how your a priori expectations for the study turned out, that is, how did you think it would come out before doing the experiment?
4. What you did and how you did it (in enough detail that your instructor could replicate it if he/she wished, without having to ask you for more details of exactly what equipment and materials were involved, how they were used, etc.). You may need to include sketches of the physical apparatus used.
5. A list of the raw data you obtained and circumstances surrounding their collection (order of collection, randomization, etc.) and the design matrix. Raw data table should have one column as order of the runs or trials.
6. Appropriate statistical analyses of the data (use graphics as well as numerical summaries). All projects should include stats (statistics by each factor), mfit, mplot, and side-by-side boxplots by each factor.
    For example, if datastructure is called b from b=readfile
    then stats(b.y,b.x1) and stats(b.y,b.x2)
    mfit(b.y,b.x1,b.x2)

```
mplot(b.y,b.x1,b.x2)
bplot(b.y,b.x1)
bplot(b.y,b.x2)
```

Each project needs the two ANOVA or the two regression models or a combination of ANOVA and regression that were stated in your approved proposal. Additional models can be examined as well.

If ANOVA, remember to use the class statement and to include the interaction term. After running each model do an rplot by X_1 and by X_2 where X_1 and X_2 are factors 1 and 2.

If REGRESSION, do at least main effects and interaction terms, model $Y = X_1 + X_2 + X_1 * X_2$; also do a rplot by X_1 and by X_2 where X_1 and X_2 are factors 1 and 2 to see if other terms need to be added to model or if $\log(Y)$ is appropriate. Syntax is:

rplot by X_1

rplot by X_2

For each model, state which terms are significant. Compare this to the results from mfit which indicated the relative importance of each factor and interaction. Give the coefficient of determination, R^2, and standard error for each model. Compare the two or more models; indicate which one has the higher R^2 and if there are differences in which terms were significant. Did the residual plots indicate that the assumption of equal variances across levels was met?

From the mplot and side-by-side boxplots, is an interaction suggested? (Was the interaction term significant in ANOVA or regression?) Is the mean change across levels linear? Are the data in the boxplots highly skewed?

From the means table: which treatment gave the *highest* mean response? The *smallest* mean response?

Give the prediction for these treatments using the mfit results and also from your best regression model if you did regression instead of ANOVA.

7. A statement of the practical implications of your study. What is the take-home lesson from your study? Give the main conclusions.

8. A discussion of further questions raised by your study (that might be investigated in a subsequent experiment). For example, if you were doing the project again, what would you change? What other factors might be considered? Should levels of factors be changed, and did the experiment suggest some new questions to explore?

9. Copy of your *approved* project proposal.

10. The data file as .txt file (with columns being X_1, X_2, Y, and run-order).

Portions of two student projects will be shown to illustrate the types of analysis performed. For the first example, students examined the effectiveness of insulated cups. They produced an interaction plot (also known as mplot) given in Figure 7.1. The two factors were liquid temperature and cup type. They explained that the parallelism of the lines suggests a lack of interaction between cup type and initial water temperature in the change of water temperature. They also noted that initial water temperature has a greater effect on the response than cup type. Later, they produced ANOVA and fitted effects (also known as mfit) results (see Boxes 7.4 and 7.5 respectively) which supported these conclusions. The students determined that an additive model is appropriate. Their results revealed that the cups were better insulators for cold drinks than for hot drinks.

BOX 7.4: ANOVA TABLE FOR INSULATED CUPS EXPERIMENT

ANOVA TABLE:

Source	DF	SS	MS	F-Stat	P-value
Cup	2	74.4	35.7	17.4	< .0001
Water	1	6360.3	6360.3	3108.3	< .0001
Interaction	2	1.56	0.78	0.38	0.69
Error	18	36.83	2.05		
Total	23	6470.07			

BOX 7.5: MFIT TABLE FOR INSULATED CUPS EXPERIMENT

Mfit output:
Overall Mean   20.3708

Fitted Main Effect of Y variable..y, by X variable, Water

Source	N	Main Effect
cold	12	-16.2792
hot	12	16.2792

Fitted Main Effect of Y variable..y, by X variable, Cup

Source	N	Main Effect
A	8	-1.4083
B	8	2.4292
C	8	-1.0208

Table of 2-way Water by Cup Interaction Effects

		Water	
		cold	hot
Cup	A	-0.23333	0.23333
	B	-0.12083	0.12083
	C	0.35417	-0.35417

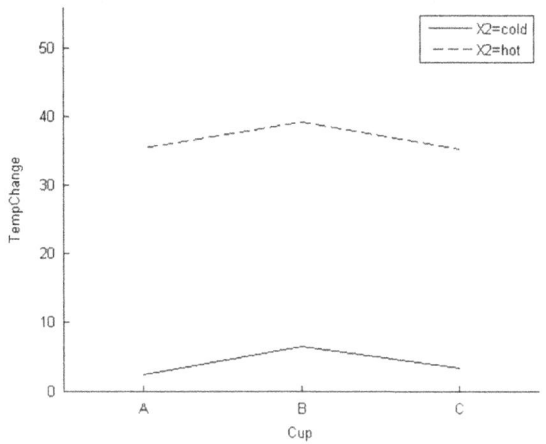

FIGURE 7.1   Interaction plot for effectiveness of insulation project.

Another example project used regression rather than ANOVA. The project had a response variable of time the parachute was in the air. The factors were the height that parachutes were dropped from with levels one, two, and three stories, and the weight attached with levels zero, one, and two binder clips. In Figure 7.2, the experimental units are shown. In Box 7.6 the regression models for responses of time and log(time) are given. The R^2 for the log model was slightly better, and Figure 7.3 indicates that the residual plots for the two models are similar and have roughly equal variances across factor levels. Box 7.7 shows the design matrix for this completely randomized experiment with four replications. The side-by-side boxplots in Figure 7.4 reinforce the conclusions reached from the regression analysis.

BOX 7.6: QUADRATIC REGRESSION MODELS FOR THE PARACHUTE EXPERIMENT

Regression Analysis

time = height+weight+height*weight + height^2 + weight^2
R-square 0.92703
Standard Error 0.69872
Parameter Estimates

Source	Parameter Estimate	Std. Error	t	p-val
Intercept	0.29278	0.96029	0.30488	0.76256000
height	4.02290	1.01350	3.96920	0.00041557
weight	0.51917	0.62169	0.83509	0.41027000
height*weight	-0.54625	0.17468	-3.12720	0.00390420
height^2	-0.28417	0.24703	-1.15030	0.25910000

log(time) = height+weight+height*weight + height^2 + weight^2
R-square 0.94696
Standard Error 0.11008
Parameter Estimates

Source	Parameter Estimate	Std. Error	t	p-val
Intercept	0.5082800	0.151280	3.359700	0.00213810
height	1.0372000	0.159670	6.495600	3.5204e-07
weight	-0.0751230	0.097941	-0.767030	0.44906000
height*weight	0.0023737	0.027519	0.086258	0.93183000
height^2	-0.1498500	0.038918	-3.850400	0.00057498
weight^2	-0.0935350	0.038918	-2.403400	0.02263000

BOX 7.7: DESIGN MATRIX FOR THE PARACHUTE EXPERIMENT

Design Matrix

	No Weight	Weight 1	Weight 2
3rd story	32, 22, 34, 35	6, 3, 16, 11	30, 33, 7, 28
4th story	17, 14, 8, 5	29, 21, 25, 31	27, 26, 19, 15
5th story	1, 36, 23, 2	4, 18, 24, 13	9, 20, 10, 12

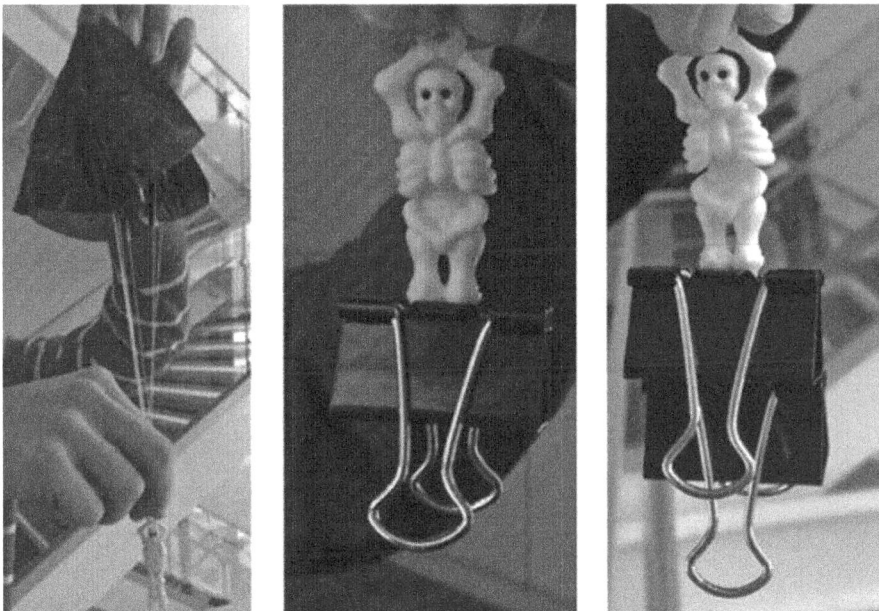

FIGURE 7.2   Experimental units for parachute experiment with zero, one, and two binder clips.

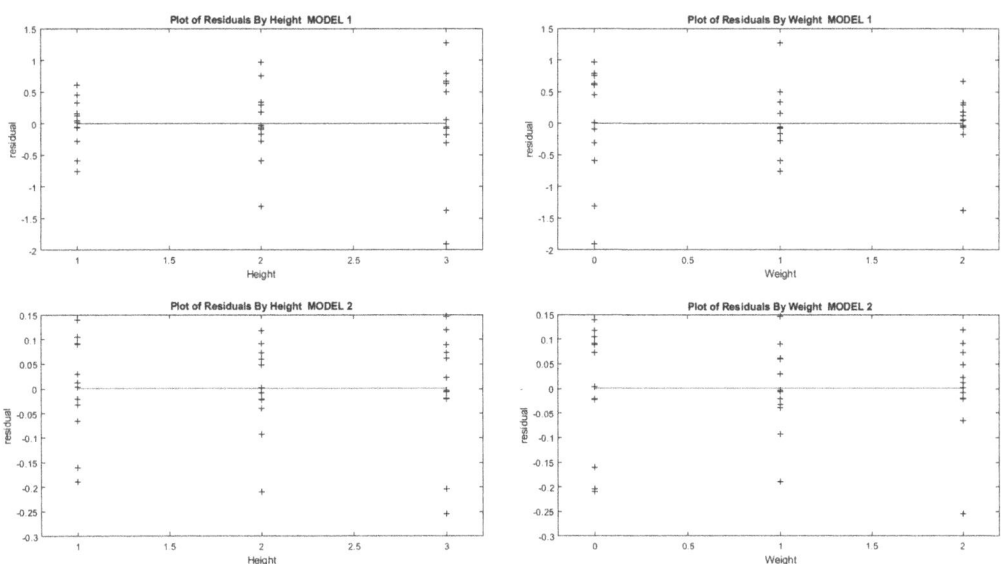

FIGURE 7.3   Residual plots for regression models in parachute experiment.

After submitting the project, students are asked to evaluate the contributions of each person, including themselves, toward the completion of the project. These evaluations are done anonymously and submitted electronically. Lazar et al. (2009) noted that the peer evaluations "give the students a chance to engage in self-criticism, and to give their perspectives on how the team functioned." On a few occasions, students have submitted

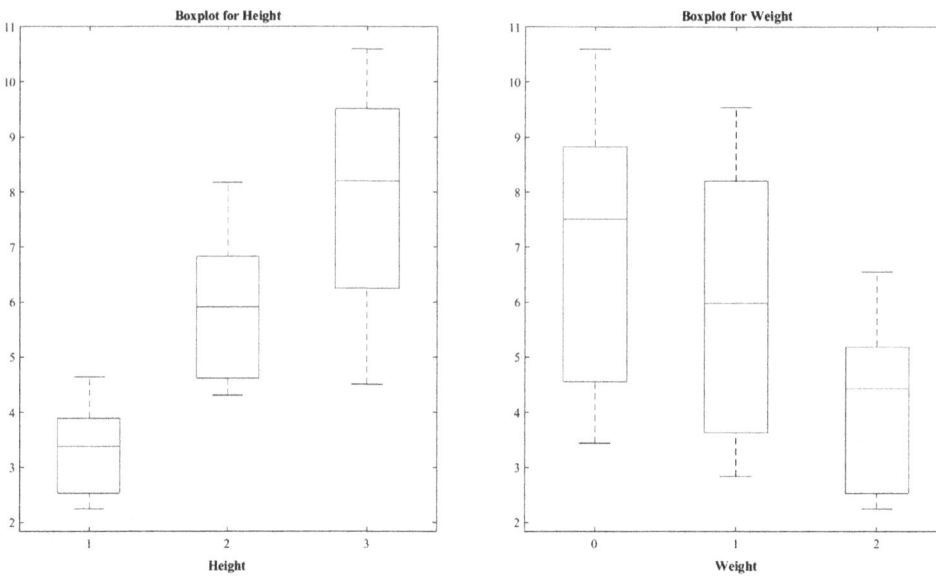

FIGURE 7.4   Side-by-side boxplots for the parachute experiment.

complaints in person, but the instructors have observed that most teams prefer to work out differences among themselves. Students are more likely to give honest, constructive feedback via the peer evaluations completed at the end of the project. The feedback is usually positive.

A sample of the "reasons for the study" comments in the project writeup provides some insight into how the students regard the project requirement for the course.

- "All three members of our team have taken an engineering dynamics class and learned about free falling objects, so we decided to test this concept with varying weights and heights of falling parachute men."

- "We started this experiment inspired by collapsing of the structure especially the collapse of the Tacoma Narrows Bridge on the morning of November 7, 1940."

- "This study was of interest because of the idea on a larger scale. How are wooden structures affected by material altering substances on an everyday basis?"

- "A month ago at a family outing to a local golf course, Asa left his putter at home and had to use his other clubs to make his shots. This did not go over well because all the other clubs have a lot of differences to the putter."

- "This experiment was chosen by our group, because it related to our common interest in astrophotography, and it gave us a chance to explore the factors involved in the formation of the craters we so often see. We varied the size of our projectiles while maintaining a fixed mass, in order to see how crater size related to the buoyancy/density of projectiles."

7.3 COMMON MISTAKES

Many of the common mistakes involve improper implementation of software, misreading of software output, or omission of required elements. Also, graphs and tables are often presented without proper or any discussion.

Besides the somewhat routine errors listed above, the most common mistake is improper use of randomization. Often, there is confusion between randomization and simple random sampling. For instance, related to the insulation example, a student might say, "we randomly selected a cup from the package to use in our experiment" and consider this process as randomization. Some students, particularly computer science majors, argue that randomization is not necessary in their experiment.

Another major stumbling block is the interaction between factors as indicated in the interaction plot, fitted effects table, and ANOVA table. Rather than discussing the importance of main effects and interaction, some students solely focus on factor combinations that support their a priori expected results. A means or interaction plot that is not parallel and an ANOVA table that says that interaction is not significant presents a dilemma in their report for some students until they examine the size of the standard error in their model.

The distance course project webpage includes a section on the 13 most common errors in projects. An email reminding the students to read this section is sent to the students several times in the two weeks before the project is due.

7.4 PERFORMANCE

In this section the performance of distance vs. lecture students is compared for sections taught by the same instructor. The median of the course average for students in the distance section over the last five semesters is: 86.77 (n = 94), 84.40 (n = 99), 82.12 (n = 98), 82.14 (n = 101), 82.14 (n = 98). The corresponding median for the lecture section in four of the last five semesters is: 84.69 (n = 97), 79.41 (n = 89), 81.84 (n = 73), 79.43 (n = 90), 83.84 (n = 92).

The percentage of the class receiving a letter grade of A over the same five semesters for the distance section is: 42.6, 37.4, 36.7, 25.7, and 40.8, and for the lecture section is: 26.9, 29.2, 30.1, 25.6, and 37.0.

Both measures indicate that the distance section, in terms of final grade average and percentage of As, is doing as well as, if not better than, the lecture section.

Several comparisons of distance vs. lecture traditional vs. lecture flipped have been made for business statistics (Dutton and Dutton, 2005) and for a statistical literacy course (Gundlach et al., 2015), but none for calculus-based engineering statistics that we are aware of.

7.5 CONCLUSION

The capstone project is an important component of students' learning experience in the engineering statistics course. Most students produce a very good report, earning an A grade. In fact, a student project from the distance education section won a prize in the

Undergraduate Statistics Project Competition (USPROC). The discussion of the project details with group members clarifies and solidifies the homework and lecture or video concepts. In addition, the project promotes creativity and allows students to relate statistics to their extra-curricular or academic interests.

There are some challenges. The assessment is time-intensive for the instructor. Also, the selection of project partners may be difficult, particularly for the online sections. Using a message board and having evening or virtual brainstorming sessions have helped with partner selection. Also, some distance students are at a common location such as the interning students at Havelock, NC, for the Mechanical Engineering Department. The 14 students there last semester all formed groups rather than working alone.

In summary, it is feasible for students to perform a two-factor ANOVA or regression experimental project in a single semester. Having the first half of the course emphasize statistical reasoning over computation seems to be required to meet the one-semester time constraint (Bradstreet, 1996).

More details can be found at the course's project website: www4.stat.ncsu.edu/~bmasm ith/NewST370WEB/project.html

BIBLIOGRAPHY

Bergquist, B. & Albing, M. (2006). Statistical methods – Does anyone really use them? *Total Quality Management*, 17(8), 961–972.

Bisgaard, S. (1991). Teaching statistics to engineers. *The American Statistician*, 45(4), 274–283.

Bradstreet, T. E. (1996). Teaching introductory statistics courses so that nonstatisticians experience statistical reasoning. *The American Statistician*, 50(1), 69–78.

Brown, E. N. & Kass, R. E. (2009). What is statistics? *The American Statistician*, 63(2), 105–110.

Butt, S., Fredericks, T. & White, B. (2004). Elements of an activity-based statistics course for engineers. Paper presented at the Annual ASEE Conference, Salt Lake City, Utah. https://peer.asee.org/13463.

Devore, J. L. (2015). *Probability and Statistics for Engineering and the Sciences* (9th ed.). Boston, MA: Cengage.

Dutton, J. & Dutton, M. (2005). Characteristics and performance of students in an online section of business statistics. *Journal of Statistics Education* [Online], 13(3).

GAISE College Report ASA Revision Committee. (2016). Guidelines for assessment and instruction in statistics education college report, http://www.amstat.org/education/gaise.

Garfield, J., Hogg, B., Schau, C. & Whittinghill, D. (2002). First courses in statistical science: The status of educational reform efforts. *Journal of Statistics Education*, 10(2).

Grima, P., Rodero, L. & Tort-Martorell, X. (2016). Explaining the importance of variability to engineering students. *The American Statistician*, 70(2), 138–142.

Gundlach, E., Andrew, K., Richards, R., Nelson, D. & Levesque-Bristol, C. (2015). A comparison of student attitudes, statistical reasoning, performance, and perceptions for web-augmented traditional, fully online, and flipped sections of a statistical literacy class. *Journal of Statistics Education*, 23(1).

Hunter, S. (1977). Some ideas about teaching design of experiments, with 25 examples of experiments conducted by students. *The American Statistician*, 31(1), 12–17.

Khan, S., Khadem, M. & Piya, S. (2018). Teaching statistics to engineering students – An Australian experience of using educational technologies. *Sultan Qaboos University Journal for Science*, 22(2), 120–126.

Kvam, P. H. (2000). The effect of active learning methods on student retention in engi-neering statistics. *The American Statistician*, 54(2), 136–140.

Lazar, N. A., Reeves, J. & Franklin, C. (2009). A capstone course for undergraduate statistics majors. *The American Statistician*, 65(3), 183–189.

Lee, H. S. & Stangl, D. (2017). Design and implementation of professional development MOOCs for teachers of statistics. *AMSTAT News* (Special issue on Statistics Education), September, Available online: http://magazine.amstat.org/blog/2017/09/01/pd_teachers/.

Mills, J. D. & Raju, D. (2011). Teaching statistics online: A decade's review of the literatureabout what works. *Journal of Statistics Education* [Online], 19(2).

Musharavati, F. & Hamouda, A. (2010). Project-based instruction: Creating excitement for learning how to design manufacturing systems. In *Second International Conference on Education and New Learning Technologies*, Barcelona, Spain, pp. 2695–2706.

Navidi, W. (2019). *Statistics for Engineers and Scientists* (5th ed.). New York, NY: McGraw-Hill.

Ojeda, M. M., Sahai, H. & Khurshid, A. (2012). Disseminating significant learning in statistics service courses. *Science Journal of Mathematics and Statistics*. doi:10.7237/sjms/103.

Romero, R., Ferrer, A., Capilla, C., Zunica, L., Balasch, S., Serra, V. & Alcover, R. (1995). Teaching statistics to engineers: An innovative pedagogical experience. *Journal of Statistics Education*, 3(1). doi:10.1080/10691898.1995.11910481.

Romero, J. & Land, M. (2004). Group learning, contextual projects, simulation models and student presentations in enticing engineering statistics students. *Proceedings of the Statistics Education Section of the American Statistical Association*, Toronto, Ontario.

Stephenson, W. R. (2001). Statistics at a distance. *Journal of Statistics Education*, 9(3).

Walpole, R. E., Myers, R. H., Myers, S. L. & Ye, K. E. (2016). *Probability and Statistics for Engineers and Scientists* (9th ed.). New York, NY: Pearson.

PART 2

Student Interaction

Encouraging Higher-Order Thinking in Online and Hybrid Mathematics and Statistics Courses

Larry Copes

CONTENTS

8.1 WHAT IS THIS CHAPTER ABOUT?

What sequence of isometries moves the figure on the left to fit exactly on top of the figure on the right (Figure 8.1)?

Student 1 Answer

> To move the figure on the left to fit exactly on top of the figure on the right, you will need to flip the figure at least once. To do so, I flipped the figure up just once. The colors are all on the same place already so I would not worry about it. When you flip the figure, you will have to rotate it clockwise at least 35–40 degrees to match the right figure. When you are done rotating, the figures should match.

> Another way to do it is, flip the figure sideways and keep rotating either counter clockwise or clockwise until it matches the right figure.

> I was also wondering: What if there was a limit of how many moves we can do to figure A to match figure B?

Student 2 Reply to Student 1

> Might you have used mathematical names for the steps? You know, rotate, reflect, and translate.

Student 3 Reply to Students 1 and 2

> Actually, the response does use the term "rotate." But don't you also have to shift (translate) the figure to the right to lie exactly on top of the other one?

Student 4 Reply to Student 1

> I like that you gave two methods and explained them. Your last question might be how to do it with the fewest moves. I think it might be done with one reflection, but I'm having trouble seeing where the reflection line is.

Do you find this snippet of a conversation among students as appealing as I do? Would it surprise you to learn that the conversation took place in an online forum for a hybrid mathematics course?

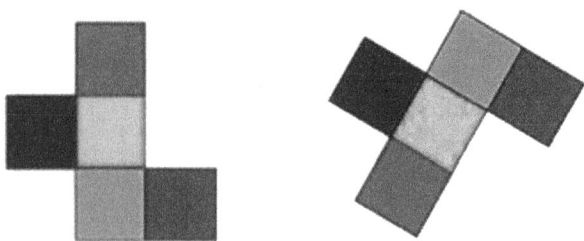

FIGURE 8.1   From Copes, Challenge 39 in Odyssey *Mathematics for Elementary Teachers F14.*

Why do I like it? There are numerous reasons.

- All of these students seemed to understand and be able to apply the concept of an isometry.

- Student 1's response shows a willingness to describe two approaches, even though the question could be interpreted as asking for only one.

- Student 3 realized that Student 1 didn't fully answer the question.

- Student 2 wanted more use of mathematical terminology.

- Student 3 challenged Student 2's claim that mathematical terms weren't used.

- All of the students seemed to be thinking for themselves, without relying on an authority such as a teacher.

- Students 1 and 4 posed an extension question: How to do it in the fewest moves?

- Student 4's willingness to engage with Student 1's musing wins my mathematical heart (and makes me grab a pencil and paper).

This kind of reasoning goes beyond rote memorization and applying algorithms to solve problems with single answers. It is often called *higher-order thinking*, which I abbreviate as HOT.

What questions might you have about HOT? In the rest of this chapter I'll address some of them, in the context of online and hybrid mathematics and statistics courses.

8.2 WHAT IS HOT?

You may recognize some of thinking revealed in the conversation above as representing different categories in Bloom's (1956) *Taxonomy of Educational Objectives*. In fact, that's the reason that I extracted this conversation from the online forum. HOT starts with Bloom's structure but goes beyond it.

Bloom is best-known for his classifications of kinds of thinking (as modified over time; see, for example, Westbrook, 2014):

- Remember (describe, identify)

- Understand (explain, interpret, predict)

- Apply (classify, illustrate, use)

- Analyze (see patterns, compare, contrast, critique)

- Evaluate (assess, rate, conclude, justify)

- Create/synthesize (integrate, plan, imagine, hypothesize, find analogies) (Figure 8.2)

In the opening conversation, Student 3's careful reading might fall into the Analyze and Evaluate categories, and Student 4's questioning could be classified as Create.

The conversation, however, illustrates intellectual mindsets as well as skills. For example, Students 1 and 4 embrace multiple answers and pose a question to extend the inquiry. Student 4, when admitting to have failed in solving a problem, demonstrates self-awareness. I was also struck by the willingness of Students 2, 3, and 4 to take the intellectual risk of critiquing the response of a peer. Moreover, the communication was clear.

These positions go beyond what are often called *higher-order thinking skills* (HOTS) or *critical thinking skills*. I'm using the term HOT to encompass not only the skills but these other intellectual characteristics:

- Comfort with multiplicity and ambiguity
- Independence of authority
- Curiosity and questioning
- Self-awareness
- Ability to note inaccuracies
- Risk-taking
- Communicating effectively

Hence the opening conversation illustrates some HOT. But let's take a step backward and ask, *is it even good to be teaching HOT in mathematics and statistics?*

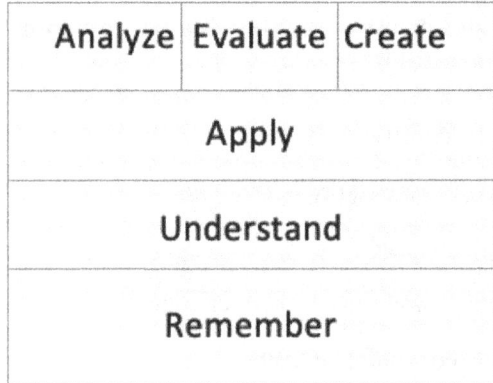

FIGURE 8.2   Bloom's taxonomy.

8.3 WHY TEACH HOT, AND WHY NOT?

Because this chapter is about teaching HOT, you might rightfully conclude that I believe teaching HOT should be included in mathematics and statistics courses. Here are some of my reasons:

- HOT is necessary for solving novel problems.

- HOT is used to determine whether answers by calculation make sense. A common example, from the National Assessment of Educational Progress, cites the many students who calculated that a fractional number of army buses were needed to transport soldiers. (For example, see Schoenfeld, 1988.) The field of statistics is especially notorious for being cited to defend dubious claims that rely only on calculation.

- HOT is important for understanding mathematical and statistical concepts well enough to apply them. How many of us teaching elementary combinatorics have heard the complaint, "I don't know when to add and when to multiply?"

- The knowledge of terminology and theorems and even techniques from probability and statistics doesn't by itself keep us from getting jerked around by politicians and marketers; we must be able to apply that knowledge to note inaccuracies in all we encounter.

- From early school days, students have been taught that *mathematics* means *arithmetic*. In fact, the phrase *do the math* refers to calculations. But mathematics and statistics go well beyond calculations. HOT is what professional mathematicians and statisticians do.

It's not just me. Really! The U.S. National Research Council is one organization that cited cognitive research backing up its claim that HOT is important even in elementary school. (See Resnick, 1987.)

Unfortunately, so often in mathematics and statistics courses we focus only on Bloom's Remember or perhaps Understand. Some courses venture into Apply, but few address Bloom's higher-order skills or HOT as a whole.

We justify our restraint by claiming that one can't analyze without knowing facts. Before doing higher-level thinking, one must build a strong foundation of lower-level thinking. This is *mastery learning*. (See Guskey, 2010.) In fact, Bloom himself was an early adherent of this theory. One has to walk in order to run.

This claim of walking and running reminds me of our daughter as an infant. She was a very fast crawler. Who knows why she even tried to stand? She certainly couldn't walk faster than she could crawl. Anyway, she was no sooner steady on her feet than she was running. Running was why she learned to stand and walk.

I am also reminded of the community chorus in which I sing. The conductor pushes us hard to be excellent. Even when we first read through a new piece of music, he's critiquing our pronunciation, our phrasing, our dynamics, and the emotions we're projecting. Sometimes I want to scream, "Just let me learn the notes first." I mutter to myself, "I have to walk before I can run." But then I realize that, while we go over and over the music to improve our pronunciation of vowels, we're learning the notes also. The goal of running makes us learn to walk.

As Westbrook (2014) says,

> The different skills can and should be used in a more integrated way. For this reason, it can be helpful to consider them as a circle, with no start or finish, and where the skills can be integrated in any order.

To do my own HOT, I'll point out that not everyone agrees on its importance. Although outdated now, an earlier platform of the Republican Party of Texas (2012) cites one argument very clearly:

> We oppose the teaching of Higher Order Thinking Skills (HOTS) … [which have] the purpose of challenging the student's fixed beliefs and undermining parental authority.

Without addressing the question of parental authority, I'll claim that challenging students' beliefs can be a good thing. This argument actually supports my belief in the value of HOT.

Another challenge to teaching HOT does not deny its importance but only its practicality. How do we as instructors assess HOT? How do we find time to teach it? And how would we teach it if we did have time?

I'll address the first and second of these questions in Section 8.8. For now, let's consider the "How" question.

8.4 HOW TO TEACH HOT?

Much has been written about teaching HOT in face-to-face situations. Some of these ideas, such as using appropriate tasks for students, can easily move online. In this section, I'll start with those suggestions and move on to other teaching techniques; in Section 8.5 I'll consider how to transport these techniques into online and hybrid courses.

So, first, here are tasks for students, organized by Bloom's categories.

8.4.1 Analyze

Let's begin with a challenge that might induce thinking in Bloom's Analyze category—looking for assumptions being made—even while practicing understanding a statistical graph:

Challenge 1: In Figure 8.3 there is a graph taken from a web site. Without getting into moral or legal issues about abortion, what is your response?

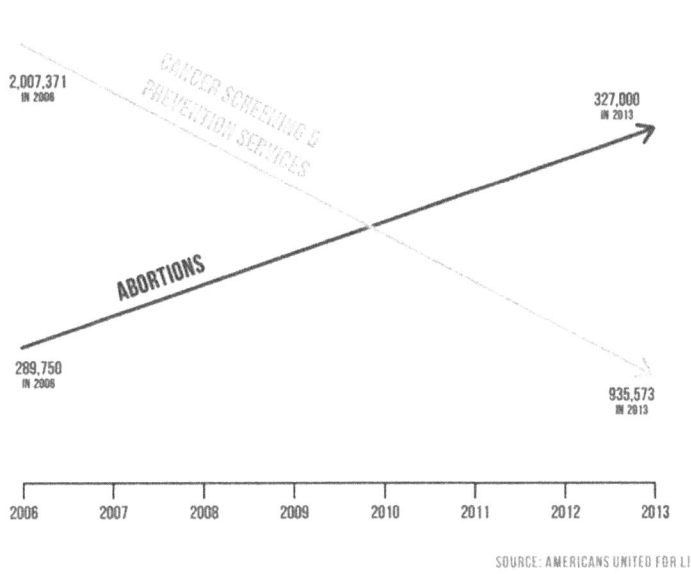

PLANNED PARENTHOOD FEDERATION OF AMERICA:
ABORTIONS UP – LIFE-SAVING PROCEDURES DOWN

SOURCE: AMERICANS UNITED FOR LIFE

FIGURE 8.3   Schield, Challenge 9 in Odyssey *Augsburg 2019A Spring MIS 264.*

A lower-order prompt might have been, "How is this graph misleading?" That phrasing assumes that the graph is misleading rather than inviting students to consider the assumptions being made.

8.4.2 Evaluate

A question requiring only lower-order thinking is:

Challenge 2a: Fill in the blank: Two geometric figures are congruent if _____.

To address this challenge, all that's needed is to Remember (or look up) the definition given by some authority, such as the text book. The kind of thinking required might be categorized as Apply if several figures are given with the task of identifying those that are congruent.

A more HOT-oriented challenge might be:

Challenge 2b: Online or in text books, find at least three different definitions of *congruence* (in a geometric sense) and critique them.

or

Challenge 2c: One common definition of *congruence* is

Geometric figures are congruent if they have the same size and shape.

A student once drew a 6 × 3 rectangle and a 5 × 4 rectangle. Because the rectangles have the same size (perimeter 18) and shape (rectangle), the student claimed they were congruent. Do you agree?

Each of Challenges 2b and 2c asks students to Evaluate. The concept of congruence can be replaced with virtually any mathematical or statistical notion.

8.4.3 Create/Synthesize

Synthesizing often involves comparing and contrasting. Here's a HOT example:

Challenge 3: Please describe differences among:

- The identity equation $x^2 - 4 = (x + 2)(x - 2)$
- The conditional equation $x^2 - 4 = 0$
- The equation $y = x^2 - 4$ to be graphed
- The function $f(x) = x^2 - 4$

Lower-order tasks might ask for calculations or applications involving any one of the four objects.

Beyond Bloom's categories, we can address other aspects of HOT.

8.4.4 Comfort with Multiplicity and Ambiguity

A straightforward task in elementary mathematics might be:

Challenge 4a: Please calculate the perimeter and area of a rectangle with length 6 and width 3. Do the same with a rectangle of length 5 and width 4.

(Yes, the task might be broken into parts and the caution added to show work. In this chapter I'll condense some of the problem statements. Later I'll say more about justifying results.)

This task requires lower-level thinking to Apply known methods for calculating areas and perimeters.

Name	Pets
Deeab	1 dog, 2 cats
Beeab	3 dogs, 1 salamander, 2 snakes
Leeahem	hamster
Sanha	cat
Casajes	None
Losemar	3 cats
Reelar	2 dogs
Moon	1 dog, 10 fish
Rasa	1 rabbit

FIGURE 8.4   Copes, Challenge 2 in Odyssey *Teaching Secondary Mathematics F15.*

Here's a version of the problem that requires more HOT:

Challenge 4b: What are the base and height of two rectangles that have the same perimeter but different areas?

Challenge 4b asks for the same calculations as Challenge 4a—with even more practice—but also encourages HOT with multiple answers. (Moreover, attempts to generate two appropriate rectangles require experimentation that may lead to deeper comprehension of the concepts of area and perimeter.)

Another example requires multiple interpretations:

Challenge 5: Here's a table of completely fictitious data about some people and their pets (Figure 8.4).

What's the most popular pet among these people?

A lower-order question might ask only for Understanding the table: *What kind of pet is owned by the most people?* or *What's the largest number of pets owned by anyone in this group?* The HOT question is intentionally ambiguous to require students to see that the question can be interpreted in either of these ways.

8.4.5 Self-Awareness

Tasks conducive to multiplicity might be extended to help with self-awareness: They might ask for a student's favorite approach (justified) or answer. For example, not just

> **Challenge 2b**: Online or in text books, find at least three different definitions of *congruence* (in a geometric sense) and critique them.

But perhaps

> **Challenge 2c**: Online or in text books, find at least three different definitions of *congruence* (in a geometric sense). Either choose one that is your favorite or create one you prefer. What is it about *you* that makes it your preferred definition?

Thus tasks can be used to encourage HOT. But teaching HOT goes well beyond the tasks. Also important is interaction in face-to-face classes.

My usual approach is to pose a challenge on which the students work in groups while I observe and circulate. I'll ask selected groups to share their ideas. Then I'll summarize, introducing mathematical terms and drawing connections to previous topics. This is standard problem-based learning (PBL) (Schmidt et al., 2011), first introduced in medical schools and adopted in business curricula (Scherpereel and Bowers, 2006), Malaysia (Win et al., n.d.), Singapore (Chun and Wong, n.d.), and elsewhere. It resembles the Socratic method of questioning through scenarios. (See Murray, n.d.)

In PBL, students see each other's ideas. As Schinkten (2017) points out, when students encounter a multiplicity of answers or approaches, and when they interact with their peers, they can begin to develop awareness of their own thinking.

Can this kind of interactive teaching transfer to online courses? That's a question for Section 8.5.

8.4.6 Questioning

The topic of questioning is frequently addressed in mathematics education: What questions should teachers ask? How should they be phrased? In contrast, little attention is paid to teaching students to ask questions that are not only for clarification. (See Brown and Walter, 2005.) As Schinkten (2017) says,

> We need people who can ask the right questions … Start by placing more emphasis on the importance of questions. Foster this with class discussions, online forums and assignments asking students to generate questions.

I add a questioning component to my PBL teaching. In the course of a class discussion, and/or explicitly after the summary, I ask the class to generate extension questions. These are questions that we don't answer, though I might mold some of them into challenges for later class periods. Early in the course, students need a lot of help with this activity, so I introduce some question stems, such as:

> **Why ... ?**
>
> **What other ... ?**
>
> **What if not (WIN)... ? What if ... ?**
>
> **How many ... ?**
>
> **Is there a different way . . . ?**
>
> **Under what conditions ... ?**
>
> **Is it possible to...?**
>
> **What do we mean by (WDWMB)... ?**
>
> **What assumptions...?**

Over time, the prompts fall by the wayside. Even though many of the students' questions show limited understanding, I display them. Eventually many students seem to feel some freedom in asking questions without worrying about answers.

A question we must address in Section 8.5 is *Are there ways to transfer this teaching to online situations?*

8.4.7 Independence

Independence is a controversial goal. Colleagues have objected, "But students shouldn't be thinking independently until they know enough to be experts." Some have even whispered, "But I went into teaching so I could be the center of attention."

To the latter, I would whisper back, "Get over it." In a more benign moment, I might inquire, "Can you boost your self-esteem by feeling rewarded when your students come up with ideas on their own?"

My response to the former objection is to ask, "Does graduating from high school or obtaining a higher degree magically bestow an ability to think independently?" We need to help our students, like our own children, to learn to run on their own.

What kinds of tasks might we give to encourage independent thinking?

Please recall:

Challenge 4b: What are the base and height of two rectangles that have the same perimeter but different areas?

Consider this slight rephrasing:

Challenge 4c: Are there two rectangles that have the same perimeter but different areas?

I asked a tiny focus group of one how Challenges 4b and 4c differ. She thought that the latter was more inviting. It would draw the student into an investigation rather than looking for an answer already known to the questioner. In my words, it would encourage more independent thinking.

For teaching HOT, some experts (e.g., Boostrom, 2005) stress the importance of the student–teacher interaction as I "circulate" among groups in my class.

What I have observed—have you as well?—is that students will wait until I arrive at their group and then turn to me for help. Even if I only ask questions, any response I make increases dependence on my authority instead of emboldening the students to think for themselves.

I have learned that my best reaction is to deflect a student's question to the group, preferably by simply looking at another student or, if necessary, by asking for another student's opinion. Students quickly learn that they must rely on their peers or themselves to answer their questions. Sure, some just sit there and wait for their peers to have ideas. Some conclude that I am incompetent for not knowing the answers. (One very astute student concluded that I was brilliant.)

Is it possible to accomplish this kind of redirection online? That's yet another question for Section 8.5.

8.4.8 Risk-Taking

To develop psychological and emotional comfort, in a face-to-face class I try to reward all student contributions, with at least a "thank you" and often with enthusiasm. I wonder, though: How much of the students' goal is to please the teacher? How independent is it to exhibit questioning, or self-awareness, or even independent thinking, only to impress an authority? Perhaps doing so is all we can ask. But can we do better?

And is it even possible to accomplish online? Yep. Section 8.5.

8.4.9 Communicating

The goal is clear communication, orally and in writing. I'm sometimes criticized for "trying to teach English in a math course," but I focus primarily on clarity. My response of "I don't understand" often leads both the student and me to a comprehension different from what I had initially guessed.

Online teaching must somehow reward and perhaps teach clarity of communication, even in mathematics and statistics. Can it?

It's time for Section 8.5!

8.5 CAN HOT BE TAUGHT ONLINE?

We arrive at the much-heralded Section 8.5, in which you (unless you're using HOT) expect answers to the questions raised earlier.

Many writers (for example, Clark and Mayer, 2012; Oh et al., 2018; McDonald, 2018) have addressed teaching HOT online. All of the articles that I've read claim that it's more difficult to accomplish online than in face-to-face interactions.

When I first started teaching hybrid and asynchronous online courses, I used standard learning management systems—Moodle, WebCT-Blackboard, D2L—provided by my institution. By assigning appropriate tasks, I could begin to address Bloom's Analyze, Create, and Evaluate. Having students read various approaches to a concept or solutions to a problem and asking which they preferred also led to some self-awareness.

Unfortunately, I found no ways to encourage the rest of HOT: Questioning, risk-taking, and independent thinking. Moreover, student participation was tentative and minimal.

McDonald (2018) suggests,

> In addition to elaborating on the discussion question, you can engage students by addressing their responses. In doing so, you can ask various types of questions to get students to address problematic argumentation, elaborate on incomplete ideas, or think through the implications of their assertion(s). In other words, you can help students use their initial response as a springboard into deeper and meaningful thinking.

That's all good stuff, but posing the questions for the students did not encourage them to question on their own or to wean them from my authority.

So I began to dream of an online forum with several characteristics:

- I would pose challenges that encouraged Bloom's higher-order skills and self-awareness.

- To encourage risk-taking, students would post anonymously.

- To encourage self-awareness, students would critique each other's work, rating each post on the basis of explicit criteria that rewarded clear communication, addressing the challenge, and questioning.

I hope you have asked, "Wouldn't such an anonymous forum degenerate into the incivility we often see online?" In answer, my fantasy forum would include

- Part of the rating system could be on civility. Posts would lose status if they were uncivil.

But would this kind of forum be more engaging than the forums I'd already used?

- Perhaps for engagement my dream forum could take the format of a multi-player game, in which students gain "power" for clear communication and posing extension questions.

Moreover, if the teacher's contributions to the forum were also anonymous (and didn't state explicitly, "I am the teacher, so Heed my Truth"), I might be able to ask MacDonald's suggested questions without students depending on me as authority.

The points earned could be part of a student's grade despite the anonymity if I alone could see which student contributed what post. But would that really be necessary? Could

I, like the students, give ratings based on the quality of a post without knowing who its author was? For grading purposes, would it suffice for me to know how many points each student had earned? Or perhaps for helping students who were foundering, I could know how many posts each student had made.

Could there be such an online forum?

8.6 MIGHT THE O2S FORUM ALLOW TEACHING HOT?

As a test, my colleague Benjamin Cooper and I adapted a game-like forum called Odysseys to Sense (O2S) (2019) that we'd originally designed to cut down on trolls in online media.

How does O2S work? I'll provide a few of the many details. You can find information about availability in Section 8.9 of this chapter. If you want to know more specifics about its functioning, please contact me.

8.6.1 Challenges and Responses

Through the course, the instructor puts up *challenges*. For example, below is a challenge similar to that of Challenge 4b above. (All text in this font, complete with errors, is copied directly from Odyssey forums.)

> **Challenge 6: Areas and Perimeters**
>
> What are the base and height of two rectangles that have the same perimeter but different areas?
>
> What are the base and height of two rectangles that have the same area but different perimeters?
>
> *Copes, Challenge 52 in Odyssey Mathematics for Elementary Teachers F14.*

Students then post, anonymously, *responses* to the challenge, ideally before the deadline the instructor has specified. Here are two of the 16 responses to this challenge posted by students in this course:

Response 1

Part A. Rectangle 1: Rectangle has a base is 5ft Height is 5ft. The perimeter is 5ft+5ft+5ft+5ft= 20feet. The area is 5ftX5ft = 25ft squared

Rectangle 2: Base is 7ft height is 3 ft. The perimeter is 7ft+3ft+7ft+3ft=20ft. The area is 7ftX3ft=21ft squared

The perimeters are the same but the areas are different.

Part B. Rectangle 1: Rectangle has a base of 5ft and a height of 4ft. The area is 4ft X 5ft = 20ft squares. The perimeter is 4ft + 5ft +4ft +5ft =18ft.

Rectangle 2: Rectangle has a base of 2ft and a height of 10 ft. The area is 10ft X 2ft =20ft squared. The perimeter is 10ft + 2ft +10ft + 2ft = 24ft.

The areas are the same but the perimeters are different.

Note: Creating the 2 rectangle with the same area was easier for me than creating the 2 rectangles with the same perimeter, possibly, because finding the area involved multiplying 2 numbers and finding the perimeter involved adding 4 numbers.

Ext: What if the rectangle were a box or cube? What if we were using decimals or fractions? What if there were labels with inches and feet instead of just feet?

Response 2

When the perimeter is the same we can find that the area can be different. For example i the picture that I uploaded [Figure 8.5] the perimeter is 12 cm, but the area for 1 rectangle is 8cm^2. Base=4 and height=2 the 2nd rectangle has an area of 6cm^2. Base=6 and height=1.

With these two rectangles, I first draw them to have the same perimeter assuming that my measurement is equal. Having same perimeter it does not mean that we will have the same area, it will depend on how we draw the rectangle.

The recatangles with same area of 18 cm^2 have a different perimeter. the rectangle 1 has a perimeter of18cm.

Base=6 and height=3

the 2nd rectangle has a perimeter of 23 cm.

Base=9 and height=2

So I will say that either the base or the height will determine the area of the rectangles.

Extension: How the perimeter or area will change if there is a different shape? for example a polygon.

You ask, "What are the 'ext' in Response 1 and the 'extension' in Response 2?" These are extensions—either questions or connections that go beyond the challenge. Why would a response include them? The student is trying to satisfy reviewers of the post.

Reviewers? After the student posts a response, but before it actually appears in the forum, its author must *review* a number of software-selected responses that have already

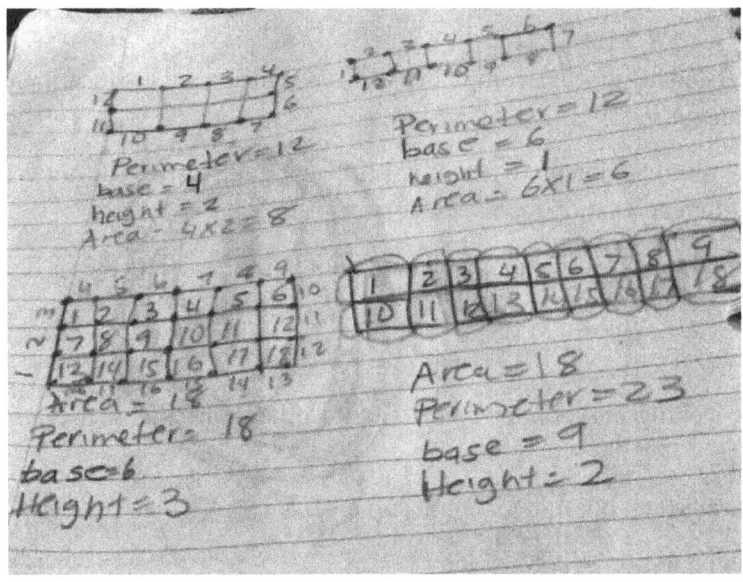

FIGURE 8.5   All student posts are used by permission of the Institute for Studies in Educational Mathematics.

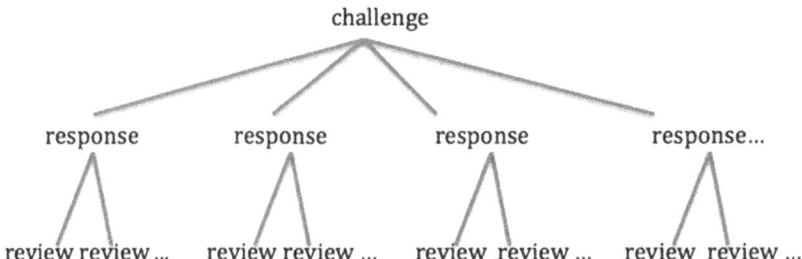

FIGURE 8.6   O2S challenge, responses, and reviews.

been posted (with exceptions for the first few responses). This choreography is intended to diminish group-think. After a response is posted, the student may review as many of the other responses as desired (Figure 8.6).

A review consists of a numerical rating and a comment explaining the rating. The rating is given by checking options on the criteria that appear at the O2S site (used by permission of the Institute for Studies in Educational Mathematics):

Responsiveness

Acceptable, comprehensive, professionally written answers to ALL questions (+1)

Partially acceptable or less than comprehensive or unprofessionally written answers to SOME questions (+1/2)

Minimally acceptable or unacceptable or missing answers to SOME questions (+0)

Explanation

Valid, professionally written, comprehensive, AND insightful justification/explanations of ALL answers (+1)

Inadequate, unprofessionally written, not fully comprehensive, OR not very insightful justification/explanation for SOME answers (+1/2)

Minimal or missing justification/explanation for SOME answers (+0)

Extension

Thought-provoking question, or connection to other concepts, that extends the challenge (+1)

Superficial extension question or connection (+1/2)

No extension question or connection (+0)

(All rating criteria copied by permission of the Institute for Studies in Educational Mathematics.)

These criteria encourage work toward some of our HOT goals:

- **Responsiveness** encourages Bloom's category of Remember.

- **Explanation** encourages Bloom's Understand category.

- **Extension** encourages questioning and, to some extent, Bloom's Create.

- All encourage clear communication.

Coupled with challenges that address self-awareness, multiplicity, and Bloom's Apply, Analyze, and Create, the system seems to show potential for HOT teaching and learning.

As an example, one of the two reviews given to Response 2 above assigned it a rating of 3 with this comment:

Review of Response 2

i must admit i found your explanation just a tad bit confusing and for the 2 rectangle for the problem of same area but different periemeter, you said your perimeter = 23 cm, with a base of =9 and height =2, but 9+9+2+2= 22 not 23. Good try though i kinda see where you were going with it.

I'm sure that you could write a better review!

Each response has a *score*, a weighted average of ratings given in reviews. Responses automatically receive 1 point. The maximum rating is 4 points, so that's the maximum score. This rubric mimics often-stated grading standards:

- D for trying

- C for meeting expectations

- B for exceeding expectations

- A for excellent

The score of the first example response was 3.75. The score of the second response was 3.0.

The game-like aspect of the forum comes from amalgamating a student's response scores and participation into a *power*. The power affects the weight the student's ratings of others have.

But what if others—teacher or students—disagree with the rating or comment posted in a response?

8.6.2 Critiques

Participants can reply to each other's reviews through *critiques* (Figure 8.7).

Students can critique critiques in an unlimited chain. Like reviews, critiques include comments, as well as ratings based on criteria:

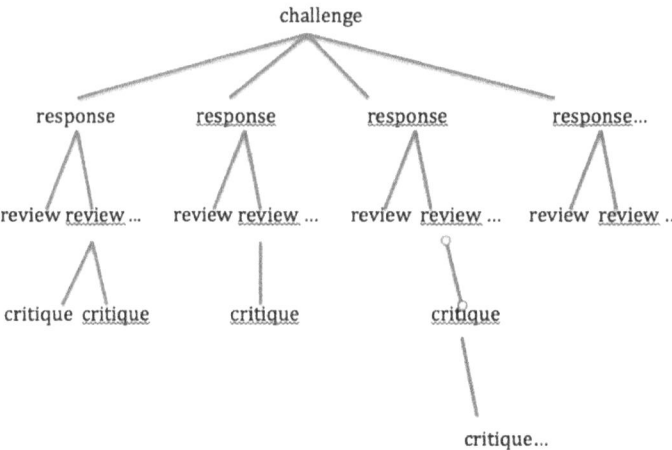

FIGURE 8.7 O2S challenge, responses, reviews, and critiques.

Accuracy

   O Rating about right (+2)

   O Rating off by a point (0)

   O Rating off by more than a point (–2)

Interest

   O Comment very helpful or thought-provoking (+2)

   O Comment somewhat helpful or thought-provoking (0)

   O Comment minimal (–2)

Civility

   O Civil enough (0)

   O Uncivil (–2)

   O Unusually civil (+2)

The **Accuracy** rating encourages work toward Bloom's Evaluate. **Interest** encourages helpfulness and self-awareness, and **Civility** helps tame trolling impulses that might arise from the anonymity conducive to risk-taking.

When a review is critiqued, its score is adjusted according to the rating given by the critique, affecting the score of the original response.

8.6.3 Dynamics

Changes appear immediately on every page opened to the site. Students can view all posted comments, including ratings, but the only scores and power they see are their own. (They also see the median power of the class.)

Although students can participate in the discussion of one of their own responses, the ratings they give do not affect the scores of any posts in that conversation.

What about us instructors, or teams of instructors? Our power is automatically high, so that our ratings carry maximum weight. We have more privileges than the students in that we can display reports on each challenge and on the Odyssey as a whole, seeing each student's scores, powers, and amount of participation. We cannot, however, identify the author of any particular post other than our own.

8.6.4 More Examples

Here are two examples that might clarify the full flow. The first is similar to Challenge 2 above, quoted exactly from an Odyssey shared by two statistical literacy courses taught by the same instructor.

Challenge 1 Revisited: Planned Parenthood Graph

The attached graph [Figure 8.8] was presented by US Representative Jason Chaffetz (Republican) to the US President of Planned Parenthood during her 2013 appearance before the US House Oversight Committee. Planned Parenthood provides health care services and information to women. These services include cancer screening and abortions. Copy at www.StatLit.org/images/2013-Planned-Parenthood-Services.jpg.

Challenge: In what ways is this graph misleading? Be specific. Assume that the numbers shown are correct for each service provided. Do not comment on or argue about the morality or legality of abortion.

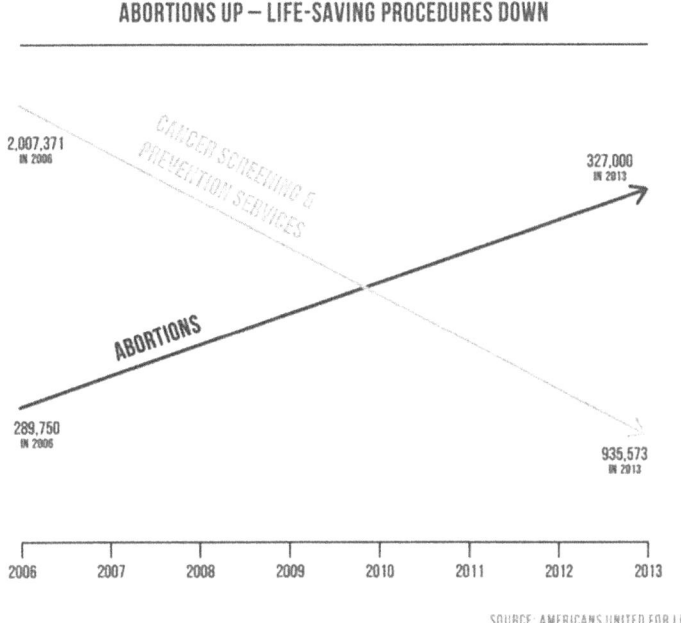

FIGURE 8.8   Schield, Challenge 9 in Odyssey *Augsburg 2019A Spring MIS 264.*

Students posted 46 responses to the challenge. Here's one that was chosen at random:

Response
The graph regarding Planned Parenthood services is misleading in the way it presents its data. The graph shows that the number of abortions increased from 2006 to 2013 while the number of cancer screenings and prevention services decreased. The two lines on the graph imply that the increasing abortion amount and the decreasing cancer screening/prevention services amount are numerically similar, but they are not. The number of abortions increased by 37,250 - an increase of 12.9%. The number of cancer screening/prevention services decreased by 1,071,798 - a decrease of 53.4%. The pink line (cancer screening & prevention services) should be much more steep than the red line (abortions) in order to accurately depict the data.

This response received ten reviews. Here are a few:

Review giving rating 4
I really liked how you included the percentages in your response it was helpful in seeing your point.

Review giving rating 3.75
I agree with this analysis. It is also important to consider why these two variables were compared and who was doing the comparing. The group which created the graph most likely has a certain bias and wanted to portray their desired outcome with a simply glance at the graph, as most people are not expected to look at it closely and see the numbers you outlined in your response.

Review giving rating 4
This graph even lacks labeling the units, its assumed that the figures are suggesting the amount of services rendered but given these congressional graphs I see I wouldn't be surprised if it is funding.

The ten review ratings, as weighted by the powers of the reviewers, gave this response a score of 3.9.

A second example of a challenge comes from an Odyssey for a geometry course for prospective secondary mathematics teachers. Fifteen students were enrolled.

Challenge 7: The Surfer
Here's a problem, adapted from Harold R. Jacobs, *Geometry*, Freeman, 1974:

You are shipwrecked on an island whose shape is an equilateral triangle. You're a surfer, and the waves are great, so you're in no hurry to be found. You want to spend your time on the beaches along the sides of the triangle. And you want to build your hut to have a minimum sum of distances to those beaches. Where should the hut be?

Experiment with The Geometer's Sketchpad or other dynamic geometry software. As you do so, try to record all of your thoughts (relevant to the problem!) and what you tried. Your challenge is to summarize those thoughts in order and reflect (with the advantage of hindsight) on what a better approach might have been.

Copes, Challenge 15 in Odyssey Problems in Secondary Geometry

The challenge elicited 15 responses. Most responses went into detail and gained scores of approximately 3.5 (out of 4). I'm showing you this response because you might find the conversation revealing and even amusing:

Response
I didn't need sketchpad for this.

1- I'd put my hut on the vertex of a triangle so I'd have immediate access to two beaches whenever I woke up. Zero travel time to two beaches is the perfect solution.

2- Surf.

The ensuing dialog resulted in a response score of 2.366:

Review giving rating 3
Your idea of putting the hut on vertex of a triangle and that you only have access to 2 sides of the beach, what is the third side? I think that is your first thought. I encourage that and my first thought is the same as yours. My appraoch is the hut wil be located at the intersection point where all 3 lines meet to create a balance for each side ans each corner of triangle. What do you think?

Review giving rating 3
I definitely agree that this would work, but would it be any different from putting it anywhere else on the island?

Critique
Of course it would be different. I wouldn't be living on two beaches (unless I chose one of the other vertices).

I figured since no matter where you choose the total distance is going to be the same, so I went with maximum number of beaches for zero travel. Roll outta bed and party on!

Review giving rating 1
What about the third beach? Would putting your hut on the vertex of the triangle minimize your distance traveled to the third beach?

Critique
Because it doesn't matter where you put it, it's always the same.

I found being at two beaches always to be preferable to only being at one or zero. That third beach isn't very nice anyway. Too many jellyfish.

Critique

How do you know it's always the same? Justification is always key. Besides, you still have to get to the third beach even it it has too many jellyfish. The waves are killer.

Critique

"How do you know it's always the same?"

Honestly? Common knowledge. I've seen the problem 1,000 times. The problem didn't ask for justification. It asked for my thought process. That was my thought process. I chose immediate access to two beaches.

That was all. Maybe the problem wasn't rich enough. (I'm not sure if I'm too internet savvy because I wan't to put a little winky smiley here. I realize I put out the bare minimum effort, but again, this problem asked me something that was to me was the equivalent of asking me, "What is the capitol of Wisconsin? How did you decide that?"

A challenge for you, dear reader: How would you contribute to this conversation? Can you do reply without revealing that you're the instructor?

8.7 DOES O2S WORK?

Now the big question: How effective is the O2S game-like forum for teaching?

We have some data, because since 2009 nine faculty members in six institutions of higher education have used O2S for over 1,100 students in 68 courses, many of which were online or hybrid. Using the forum besides mathematics, mathematics education, statistics, and statistical literacy courses have been courses in children's literature and an online module in veterinary ethics.

In fact, the only published research on the effectiveness of O2S concerned its use in that ethics module for 200 veterinary students over two years. Comparing the module of case studies presented one year without O2S and the next two years with this forum, Kustritz and Copes (2013) found evidence that the O2S students had significantly better understanding and retention of content than students not using O2S. Apparently HOT improved Bloom's Remember category.

Other research has addressed the question of student satisfaction, based on end-of-term surveys in statistics and mathematics courses. These studies (Schield, 2011a, 2011b, 2014; Schield and Copes, 2011) indicate that students had mixed reactions, but the O2S forum seems to have been especially attractive to students accustomed to defending their opinions in non-quantitative courses. Because adults should be able to support claims with evidence in quantitative fields, these data can be interpreted as an indication of growing pains.

For our question about the effectiveness of the forum for teaching HOT in mathematics and statistics courses, I must turn elsewhere. I looked at a variety of those courses and rated, on HOT criteria, a randomly chosen sample of the anonymous posts. I found that the Odysseys for some courses showed impressive improvements in student posts, but other Odysseys did not.

What factors were different? My investigation led to several observations.

- The most effective challenges for meeting the HOT goals seem to have several characteristics:

  - They encourage multiplicity: Multiples in perspectives (like "The Surfer" example), multiples in answers (as in the "Planned Parenthood Graph" example), or multiplies in approaches (as in the challenge about isometries in the opening conversation).

  - They encourage the use of external resources. (For example, from a statistics course: "Please read about the friendship paradox in sociology. How might the friendship paradox relate to statistical sampling?")

  - Challenges do not say "Defend your answer" or "Show your work" or the equivalent. Justifying opinions is built into the forum's culture via the rating criteria, toward the goal of instilling the habit of doing so.

- Odysseys with several challenges per week are more likely to produce HOT than those with only one challenge per week.

- HOT goals seem to be met only in Odysseys in which we instructors are active in the forum, at least at the beginning of the course. We mold the culture of the forum as we anonymously post responses of both high and low quality, review them, and give low ratings to posts that don't meet the criteria. If we are not active in an Odyssey, student posts do not seem to demonstrate much HOT; for example, the class as a whole apparently will ignore the **Extension** criterion, not having any examples of how it might be met. (You may have noticed that this was the case with the Planned Parenthood Graph challenge.)

8.8 HOW CAN O2S BE USED EFFECTIVELY?

First, I'll consider the questions raised about time and effort needed to teach HOT and about assessment. Based on my experiences and those reported by instructors through the years, I have learned:

- Designing challenges that require HOT can be time-consuming. But if we already have a collection of them, then entering them into an Odyssey requires only copying and pasting. Challenges may also be imported from our own earlier Odysseys.

- The time required to set the standards and mold the environment by entering our own responses and reviews can be extensive at the beginning of a course, though not necessarily more so than actually grading papers.

- In addition, our interaction can diminish as students gradually practice HOT and take over the role of maintaining the desired class culture. If student posts are saying what we would say, there may be no need for us to add anything.

- With appropriate challenges, a student's power is a fairly accurate assessment of HOT.

I can also add these observations:

- If we, working anonymously, can place ourselves in the mindset of a student with HOT who is first encountering the course ideas, then our comments can be understood. Consequently, students seem to believe that it's possible for them to think the same way rather than be awed, or at least silenced, by our authority.

- Playing the role of intelligent student makes it more fun for some of us than responding as authorities. Others of us, however, may be reluctant to hide our hard-won expertise and to give up the limelight.

- On the reverse side, if we accidentally post something less than brilliant, it will be anonymous. If students don't point out our error, we have yet another opportunity to model civil, constructive criticism.

I'm sure that you, in practicing your own critical thinking skills, have other questions. Here are my current answers to a few:

- Not being able to associate individual posts with their authors can at first be disconcerting to us instructors. We find ourselves having to base our opinion of each post on its content, unaffected by any knowledge of the writer.

- Many students are uncomfortable with having to deal with multiplicity or being "graded" by their peers. The forum does not replace the encouragement we must provide for learners who are encountering not only new concepts but also more mature ways of experiencing knowledge, thinking, and learning. (As needed, I point out to students that I'm still responsible for assigning actual grades.)

- O2S allows us to address misconceptions more immediately than through the process of collecting and grading and returning individual submissions. If responses and reviews show widespread misconceptions, we can address them in a general comment to students (or, for hybrid courses, as early as the next class session).

- Similarly, O2S can be integrated into a course by our written or oral comments giving advice about how to practice HOT when posting. To make our points, we can cite individual (and anonymous) posts.

- Contrary to expectations, only rarely could students identify authors of the posts. An extreme case was an Odyssey in which a student and his mother were both enrolled. The mother told the instructor at the beginning, "Of course I'll be able to identify what he writes. I've seen his writing all his life." But she couldn't, even though there were only five participants.

I must add a comment about the possibility of trolling. Incivility in the forum is extremely rare; over the years only one review in any of my courses ever qualified as remotely uncivil.

There were more problems with students' "gaming" the forum. Some time ago, we noted the appearance of many identical and vapid reviews, such as "Good job." A small amount of power is given for participation, and a few students apparently were copying and pasting many times to increase their power. The software was modified to reject, with feedback, reviews that were too short or too similar to others by the same author, and the practice seemed to dissipate. In the spring of 2018 it re-emerged in one course, however. So from then on credit for reviews in each challenge decreased to 0 after a few posts beyond the required minimum.

I'll conclude this section with an excerpt from a critique in a course for prospective secondary mathematics teachers. After reviewing others' responses to a challenge, a student demonstrated notable synthesis, self-awareness, and risk-taking:

The sound you hear is my palm slapping my forehead.

How many hints do I need to make this connection? I feel a bit stupid.

I'm going to stop feeling stupid and just marvel at the connections among these three apparently-different things

8.9 IS O2S AVAILABLE?

You can visit the forum (Odysseys2Sense, 2019) at no cost. You can see discussions of sample topics and can access Odysseys that the instructors have made public over the years. You can view a video of an actor thinking aloud while making posts.

Anyone can set up an Odyssey. In order to maintain the website, there's a small cost. The most popular option is to charge each student $5 US for five months of posts. (Students may make ten posts—responses and reviews—for free to get started.) Alternatively, the instructor can choose to pay the $5 per student up front. (Other options for payment, more relevant to the forum's use by online news media, are available as well.)

If you wish, you can contact me for sample instructions to students for enrolling in an Odyssey, as well as other advice and answers to questions.

8.10 WHAT CAN WE CONCLUDE?

To summarize, Odysseys2Sense (O2S) is an online forum adapted for teaching higher-order thinking (HOT) in online and hybrid academic courses. Students anonymously respond to challenges posted by the instructor and discuss the responses through posts. To each post, students give a numerical rating based on set criteria that reward careful higher-order thinking, clear communication, and civility. These ratings affect other students' power and influence, adding a game-like feel to the forum.

Apparently O2S can be a helpful tool in teaching HOT if the instructor:

- Posts challenges that require HOT.

- Posts several challenges a week.

- Integrates the forum with the rest of the course.

- Is engaged anonymously in the forum, molding its culture to adhere to the rating criteria and to exclude superficiality.

To date, no published research has determined whether or not the use of O2S actually correlates with improvement in HOT. I am comfortable, however, in claiming that this forum can be used as a tool consistent with HOT goals. Furthermore, I can testify from experience that participating in it can be more fun for instructors than traditional grading.

REFERENCES

Bloom, B. S. et al. 1956. *Taxonomy of Educational Objectives: The Classification of Educational Goals. Handbook I: Cognitive Domain.* New York, NY: David McKay Company.

Boostrom, R. 2005. *Thinking: The Foundation of Critical and Creative Learning in the Classroom.* New York, NY: Teachers College Press.

Brown, S. I. and Walter, M. 2005. *The Art of Problem Posing,* 3rd edition. Mahwah, Lawrence Erlbaum Associates.

Chun, H. and Wong, P. nd. *Preservice Teacher Use it to Present Scenarios for Problem-Based Learning.* https://www.tp.edu.sg/staticfiles/TP/files/centres/pbl/ pbl_huchunphilipwong.pdf. [Accessed 25 January 2019].

Clark, R. C. and Mayer, R. E. 2012. "e-Learning to Build Thinking Skills." In *e-Learning and the Science of Instruction,* 3rd edition (eds. R. C. Clark and R. E. Mayer). New York, NY: Pfeiffer (John Wiley and Sons).

Guskey, T. R. 2010. "Lessons of Mastery Learning." *Educational Leadership* 68(2), 52–57.

Johnson, L. and Lamb, A. 2011. *Critical and Creative Thinking - Bloom's Taxonomy.* https://eduscapes.com/tap/topic69.htm. [Accessed 25 January 2019].

Kustritz, M. V. R. and Copes, L. 2013. "Use of a Civil Discourse Web Site for Ethics Training." *Journal of Veterinary Medical Education* 40(1), 45–51.

McDonald, D. 2018. *Critical Thinking in the Online Classroom.* https://ctl.learninghouse.com/critical-thinking-online-classroom/ [Accessed 25 January 2019].

Murray, J. n.d. *Teaching Strategies About the Socratic Method.* https://www.teachhub.com/teaching-strategies-about-socratic-method [Accessed 8 February 2019].

Odysseys2Sense. 2019. *Odysseys2Sense™: A Game-Like Online Forum.* North Haven, CT: Institute for Studies in Educational Mathematics. http://www. odysseys2sense.com. [Accessed 25 January 2019].

Oh, E. G., Huang, W.-H. D., Mehdiabadi, H. A. and Ju, B. 2018. "Facilitating Critical Thinking in Asynchronous Online Discussion: Comparison Between Peer- and Instructor-Redirection." *Journal of Computing In Higher Education* 30(3), 489+.

Republican Party of Texas. 2012. *2012 State Republican Party Platform.* https://www.empowertexans.com/wp-content/uploads/2012/07/2012-GOP-Platform-Final.pdf [Accessed 25 January 2019].

Resnick, L. B. 1987. *Education and learning to think.* Washington, DC: National Academy Press.

Scherpereel, C. M. and Bowers, M. Y. 2006. "Using Critical Problem Based Learning Factors in an Integrated Undergraduate Business Curriculum: A Business Course Success." *Developments in Business Simulation and Experiential Learning* 33, 13–21.

Schield, M. 2011a. *Teaching Statistical Literacy Using Odysseys2Sense™: A Unique Web Discussion Forum*. Paper Given to the Joint Conference of the Mathematical Association of America. http://www.statlit.org/pdf/2011SchieldMAA.pdf [Accessed 2018 June 26].

Schield, M. 2011b. *Teaching Teachers Statistical Literacy Online*. NNN. http://www.statlit.org/pdf/2011SchieldNNN.pdf [Accessed 2018 June 26].

Schield, M. 2014. *Odyssey: A Journey to Lifelong Statistical Literacy*. Invited Paper for 9th International Conference on Teaching Statistics (ICOTS). http://www.statlit.org/pdf/2014-Schield-ICOTS.pdf [Accessed 2018 June 26].

Schield, M. and Copes, L. 2011. *Odysseys: Teach Critical Thinking*. Poster Session at Research Conference, Minneapolis, MN: Augsburg College. http://www.statlit.org/pdf/2011SchieldCopesPosterOdysseys.pdf [Accessed 2018 June 26].

Schinkten, O. 2017. *How to Get Students to Engage in Higher-Order-Thinking*. https://learning.linkedin.com/blog/education/how-to-get-students-to-engage-in-higher-order-thinking [Accessed 25 January 2019].

Schmidt, H. G., Rotgans, J. I and Yew, E. H. J. 2011. "The Process of Problem-Based Learning: What Works and Why." *Medical Education* 45(8), 792–806. Abstract available from https://www.ncbi.nlm.nih.gov/pubmed/21752076. [Accessed 25 January 2019].

Schoenfeld, A. H. 1988. "When Good Teaching Leads to Bad Results: The Disasters of 'Well-Taught' Mathematics Courses." *Educational Psychologist* 23(2), 145–66.

Westbrook, C. 2014. *Teaching Critical Thinking Using Bloom's Taxonomy*. http://www.cambridge.org/elt/blog/2014/04/18/teaching-critical-thinking-using-blooms-taxonomy/ [Accessed 25 January 2019].

Win, N. N., Nadarajah, V. D. V and Win, D. K. 2015. "The Implementation of Problem-Based Learning in Collaborative Groups in a Chiropractic Program in Malaysia." *Journal of Education Evaluation for Health Professionals* 12, 17. https://www.jeehp.org/DOIx.php?number=143 [Accessed 25 January 2019].

CHAPTER 9

Tools for Communication and Interaction in Online Mathematics Teaching and Learning

Shay Kidd

CONTENTS

ONLINE LEARNING HAS BECOME a regular part of educational opportunities at many post-secondary and K-12 schools. The rapid growth of these online environments has left a gap in the "how to" aspects of teaching in the online environment specific to mathematics (Engelbrecht & Harding, 2005). While there has been some gap filling in general, this chapter provides several examples specific to the gap of how to teach mathematics online and how to help students learn mathematics online. Personally, having eight years of experience in teaching mathematics online to K-12 students, three years of teaching post-secondary mathematics, having been an online student for my master's in mathematics education, and attended numerous meetings or conferences that had a focus on online teaching, have helped me fill some of this gap. This chapter is written with the purpose to help others fill the gaps that they find.

This chapter is a case study that offers many insights into the teaching and learning of mathematics from largely a K-12 perspective. Sourcing to research for any of the larger ideas is provided, but where there is not yet sufficient research, the content can be considered as specific to the author's experience. This case study is one of reporting the phenomenal experiences of myself and other online instructors to then bring out the important aspects of teaching mathematics within the online format. These experiences have been gathered from my own experience or the experience of colleagues over the years that has been shared with me.

In this sharing of experience, I want to first take a few lines and help house this as an example of the pedagogical content knowledge (Shulman, 1986). The content of mathematics is clear from the title of the book. "Designing online courses is substantially different from designing courses for traditional courses" (Akdemir, 2010, p. 50). The pedagogical aspect will have a larger focus on online teaching. Some examples use the broader brush of online teaching and then focus in on the mathematical aspects. With this in mind, the first major section is about the types of data that can be found in online courses as compared to the face-to-face classroom. These data are helpful to develop a relationship in the online environment that can be the foundation for teaching mathematics (Eichhorn, DiMauro, Lacson, & Dennie, 2019). Additionally, tools and supports that can be used are provided, starting with the locations and people, followed by a sample of online tools that have been found to be helpful. Knowing each of these sources, as well as how to use that source, will help instructors gather data they are more comfortable with in the teaching of mathematics.

9.1 TRACKING DATA ON STUDENTS FOUND IN THE LEARNING MANAGEMENT SYSTEM (LMS)

A student on my online summer course was emailing back and forth with me in a rapid manner with a series of quick questions. The student stopped emailing, though there were still questions unanswered, and I went back to work in my office. A few minutes later there was a knock on my door and the student walked in. Though the student was taking the course online, they were on campus some of the time during the semester. The face-to-face interaction was their preferred method and they were only taking the course online because they were not able to always attend face-to-face. In our short conversation, I learned so much about the student who had only been an online profile previous to our conversation. Getting to know students requires more work and knowing where to find the information in the online environment.

One of the most important things to know about students is the amount of time that they are spending in a course (Akdemir, 2010). The learning of mathematics takes time. A significant feature of many LMSs is the ability to track the time of students within the course and content. This ability to track when and for how long a student is accessing a course and what they are accessing helps the instructor in knowing what priority the course and content are. Students who mistake an online course to be easier may neglect the available resources that the course provides and make attempts to complete all graded assignments and not take the time and energy to learn from the resources provided. When this happens students often find out how much they do not know. Some students respond to their lack of knowledge as the instructor's fault (Bambara, Harbour, Davies, & Athey, 2009). That fault may be viewed as making the course too difficult or that the instructor did not communicate the specific resources required to pass the graded assignments.

Using the online tools of requiring students to view or complete some component of the course is a good way to ensure that the student's lack of knowledge is not due to a lack of opportunity to learn. The old adage of "you can lead a horse to water, but you can't make it drink" describes well the issue of students not learning, even with many great learning resources available. Instructors need to ensure that every opportunity for learning has been made available to the students. An LMS can help to track each of these opportunities to learn and how much time students use them. When students return to a page multiple times, it can mean that there is a key piece of information that they use there or that the information confuses them. As a student, I would refer to certain examples in the course every time I had some similar problem. The LMS tracks the number of visits to pages as well as how long they are viewed. Several short views are often more about referencing the material. Longer visits are efforts to understand.

Multiple attempts on assignments can also be a good source of information. Some instructors limit the number of attempts while others allow students to attempt as many times as they will take the time for. Both strategies have advantages and disadvantages. The limited attempts help to have students focus more on the individual attempts and put their best foot forward. Some students do this even with unlimited attempts available. If this approach is used, be sure that students have clearly seen all the problem types assessed. This is their time to shine and show what they know. On the other hand, multiple attempts allow the assignment to be a learning experience for the students. Feedback given in the assignments is used to improve their score on the next attempt. The assignment is still part of the learning process rather than the end result (Akdemir, 2010). How students use their attempts can also be seen in the LMS data. Multiple attempts done back to back show a rushed pace for completion rather than an effort to improve. When instructors can compare the multiple attempts of students, trends in the answers to similar questions can be found. Are the students moving closer to the answer? Are they using the feedback? Are they making all the same mistakes? This helps as the instructor works to make a difference for the students and the communication that is provided. How long students take on assignments and tests can tell an instructor if students are putting in the effort or not. Also, if it is too right and too short, that might be a flag for cheating.

Some of the expectations for a course are built from tools within the course. Students need to use a specific tool to get the correct answers. Example: Loan calculator. Within one of the courses that I have taught several times is the calculating of loans by using a spreadsheet file created and loaded into the course. Due to the technical issue of using different devices or not having the capability to open a spreadsheet, students did not use the tool provided. Since the students used different tools, the instructions for performing the calculations did not match with the tools that the students attempted to use. Much confusion and many questions were had by students as they tried to problem solve their self-created problem. I've had parents take their students to professional loan officers to check their calculations. Some students have even worked to point out the fault of the course and that all the calculations are incorrect in the course. All of these other tools helped to provide answers well within the margin of error for the calculations. The difficulty is that the students and LMS system expected answers to match exactly. Exact calculations require exact use of tools and the tools be the same. The requirement of a single tool felt like a bottlenecking of resources to meet the needs of the course to the students. This is an important balance that is required with an online course as there are multiple tools available to students on the internet. If instructors allow the use of a wide range of tools, then they must accept the wide range of answers that will come from each of the tools. For my course, the choice was made by course developers to use one tool so that the instructor could be an expert on that one tool and method and be most helpful to the students. Instructors need to select their methodology for the course.

Tracking of where a student goes in the course is also nice to see if they have looked at the resources that were required. If they have not looked at the resource, then it becomes clear why they have the question. If they have looked at the resource, then the instructor knows to answer the question differently. Students may be skipping the learning resource of a topic as they feel they already know it. If they are continually getting incorrect answers, it is important to look for error patterns to determine if the students have a misconception (Ashlock, 2006). Being able to view the multiple attempts of students and the multiple types of questions on a specific topic can help provide the necessary data. With all the work of students saved and tracked in the course, the amount of data available is ready for analysis.

9.2 HOW TO ENGAGE STUDENTS IN AN ONLINE ENVIRONMENT

Engagement of students is often a key factor evaluated within a face-to-face classroom. Engagement is readily measured by eye contact, body language, and interaction of students. In the online environment, engagement happens in different ways. The engagement of a student should not be only within the confines of a course and content but should extend to other interactions. In the same way that a teacher builds relationships with students outside of the classroom interactions, an online instructor should engage students outside of their online course. One of the first steps to meaningful interactions is to understand the tools available for communication.

The available methods of communication for students include, but are not limited to, phone calls, emails, texting, announcements, lesson content, video, and face-to-face interactions. The teaching of content in an online environment does not necessarily eliminate the possibility of face-to-face interactions.

The additional methods of communication that are listed previously should all be readily accessible for a good online instructor. The use of a telephone call is often a method of communication to engage parents of K-12 students. Clear specific communication shows a respect for the parents' time but can also come off as gruff for not taking the time for "necessary" small talk and pleasantries. Phone conversations need to have a clear purpose for the receiver. With K-12 students, a call to their parents or home number for each student to get some background information about them and their situation helps to gather information that is readily accessible in face-to-face situations. Start by asking for the parent or guardian by name, as this information is provided in the data management system. Then ask if they have the time to talk about their student. Comparing a phone call with the parents that come by the classroom during a back to school night helps to show the time and place that these conversations would happen, but with the online environment. During a back to school night, parents come and happily share the information about the history of their student in mathematics. This type of information gathering can happen in an online assignment that students fill out within the course as well with post-secondary students.

The three questions that are always the focus of this phone call are to ask about the student's experience in online learning, the student's experience with mathematics content, and why they are taking this course. Within these three questions, there are several details that will help the instructor to better understand the students. The final question, about taking this course, provides a wealth of information about the students. That there is a typical order for taking mathematics courses and there is a need to complete courses in this order are considered in this answer. The need to retake a course or to take the course out of sequence are often part of the answer provided. There is also the consideration of why they are taking this course in the online format. Though this question is not directly asked, it is often answered. If students are catching up or working ahead of their general education tells much about the students. Any difficulties with mathematics content, or school in general, are brought up in this answer. Ability levels and knowledge levels provide a good set of data about the students.

Another important data set is student profiles within a Learning Management System (LMS). These profiles may have some system-created information or may need to be built by the student. An early assignment to build this profile with the necessary information to help make connections can be a nice tool in interaction efforts. Profile pictures that are actual pictures of the student are helpful to visualize and recognize the students and their work. Allowing the picture to either be a picture of the student or a picture that represents them is a nice way to work around students who may not be willing to share their personal appearance in a new environment.

Often within the profile, there are places that students can include information like their major, interests, and other information. Some students have used this as a social networking opportunity and written pages of information. Others fulfill the minimum requirements of the profile and never think about it again. As students are looking for other students to interact with the development of these personal profiles tells much about if the students are willing to interact in the online environment. Of course, an instructor's profile should be more than the minimum, but also not be excessive.

9.3 COMMUNICATING USING ONLINE TOOLS

With a set of background data to build from, instructors can then move into their communication with the students. Students engage in electronic communication so regularly that their expectations for engagement are very different than the expectations of many instructors. For students, the most important information is going to be stated clearly and concisely. With communications being limited by the number of characters, like Twitter, lengthy messages are not even possible. Understanding that longer communications disengage students, the difficulty of getting enough information to students now requires greater frequency in electronic communication.

Many mathematical concepts require a large base of pre-knowledge for students to then learn the next step. The ability to shorten a communication to a comparable tweet of 140 characters was once compared to a single slide of a PowerPoint slide show. The trouble is that this is not the first slide in the presentation. Placement of the information helps students to understand the content of the slide. Additionally, it is helpful to provide the links to the previous "slides" of information so that the information is available but not all included and overwhelming to the students.

9.4 SET TIMES FOR COMMUNICATION

Communication can be done in multiple ways with students in the online environment. Students often expect communication at the moment that they are communicating. Google provides thousands of responses within a second for any search done. This can be one of the biggest difficulties of working in the online environment, that is, most of the work is done out of sync or asynchronously. Unless instructors are going to be available all the time, setting up regular times for synchronous communication is necessary. This can be done in several different ways and using multiple means of communication.

A common method for approaching the need for synchronous communication is the use of office hours. Having online office hours on a regular basis serves much the same purpose as it would with face-to-face courses. It is a time that students know they can find the instructor in a set location and the instructor is ready and willing to help them. This "location," and how students reach it, needs to be clearly communicated to the students. It may be that students can count on an instructor to be at their computer and responding to emails within minutes. Similarly, the instructor is by the phone and will answer the phone calls of students. Finally, it may be that there is a virtual room that students can come into and ask the instructor or other students questions about the course. Understanding the timing of students and when they need feedback is one aspect of knowing students that affects their relationships within the course (Akdemir, 2010).

One common type of feedback that many students require in mathematics is the check-in that their process is being done correctly. These simple check-ins for students as they work a problem in the early stages of their learning help them to see that they are following the methodology that the instructor provides well enough that they will then get the correct answer. The tutoring requests that turn into watching a student work and giving positive reinforcement on their work are an aspect of the face-to-face class that students miss as they move into the online environment. In place of the instructor being the only provider of this feedback, small groups can meet and check in to share the work that they are doing. This can group students who may come to the instructor with a question and the instructor can then answer many students' questions all at the same time.

The interaction within the classroom can include the written messaging option, calling or use of the computer microphone, and video conferencing. The comfort level of the instructor usually dictates what office time includes. An additional benefit of a virtual room for students to come to is that of recordings of these office hours. These recordings can be edited to create a resource in the course. When several students have a question on an assignment or content item it is nice to create the resource that answers that question and have it ready in the course. Then future questions are just a reference to the resource rather than having to repeatedly respond to the same question. These resources are great for helping to meet the needs of students. Then as students are struggling with different assignments an instructor can help to personalize what resources to refer them to. All the resources are available, but as an instructor reaches out to a student with a specific prescription of resources for them to use, they feel a connection and realize that the instructor is a human who is working to help them with their specific needs.

A word of caution about using online rooms when teaching mathematics: Inputting mathematical symbols or equations can be very difficult. Ensure that instructors are familiar with any mathtext tools available so that they can explain to the students how to use them. Some of these tools are based on some other software package that may not be available on all systems. To help work around this, a whiteboard tool can be shared. If the virtual classroom does not have such a whiteboard tool, it can be worked around by sharing screens and opening a drawing program, like paint. Students can do this same process if they are learning what a symbol is.

One particular difficulty I've had was when a student asked what the e symbol was. No more context was given. I assumed the student meant e for the natural log. I jumped into a long explanation of the natural log. After a few minutes, the student responded that this was all information that seemed to be far from the rest of the content of the lesson. They followed up with questions about what the number on top and on the bottom were. This took a minute for me to process as e is typically only involved with exponents, numbers above. I thought about the subscripts with logarithmic functions. As I was typing out another long description for all these parts of logarithmic functions the student shared a picture of what they were trying to describe: . As this picture popped up on my screen, I felt a wave of frustration at all the time and energy I had just wasted explaining the wrong symbol to the student. Since then I have worked to make sure I fully understand and even see what the students are asking as I approach my office sessions. Getting the visual of what is being discussed is essential to make sure that instructors have effective communication with students.

9.5 COMMUNICATING OUT OF SYNC

Apart from the synchronous communication opportunities that the instructor schedules, there is much asynchronous communication that happens in the course. Every part of the course that the students can access within a time frame of the course is asynchronous. It is something that the instructor, or content developers, put into the course and that can be used by the students at a time that works for the student. Most of the resources and experience in fully online courses are asynchronous. That is one of the major reasons that students take courses in the online format (Deal III, 2002; Li & Irby, 2008) They do not have the regular time to connect with a teacher, that fits the schedule of the school, teacher, or 20 other students. These many other "communications" are the bulk of course content and online interactions. They are often just referred to as resources in the course. Many of the different types will be discussed specifically later on in this chapter.

With the sequential nature of mathematics, adaptations to some of these resources help students to see the content as more than just several disconnected resources. When possible, having a specific sequence that can be referred to and help guide the students through creates this cohesiveness. Leading into a new lesson with a reference to previous content and learning is a best practice for helping to prime students' minds in all learning situations (Skemp, 1987) and works well in the online environment. Referencing solving systems of equations by graphing before starting on substitution is a little thing that helps the students to see order in the course, even if it is not in sync.

A consistent difficulty that students have expressed in online courses is a feeling of isolation (Bambara et al., 2009). Even when they have a great instructor who interacts with them regularly, the students long for student-to-student interactions. Like most interactions in the online environment, student-to-student interactions take effort to make happen. The instructor is usually the one who needs to create these opportunities for students.

Planning out each of these details of the course helps students to get to know their instructor. When an instructor develops a course, they have a connection to the content and it is a representation of them personally (Eichhorn et al., 2019). One colleague expressed feeling vulnerable when asked to give access to others to see how well their online course worked. This colleague had put in much personal time and effort for their students to see their teaching style and personality represented. Courses should have a personal aspect to the instructor, even with a standardized curriculum. If a course does not have a personal touch, then why have a person connected to it? After revamping a course several times, I received a new standardized course with many of my personalized updates included. It was a bit disheartening to me to see my personal work applied to the masses, but then I realized that I must be doing something right if my school wanted my work to be used by all students. Being recognized for creating good online resources is important. Be sure to use other great ideas and to credit them. Setting up a course is the first step in engaging students. Additional suggestions are included in the following sections.

A common interaction for online course work is the use of a discussion board. The discussion board has been seen as an essential tool for student engagement (Eichhorn et al., 2019). A discussion board is a tool in the course where the instructor provides a prompt that the students will create replies to, often called threads. Each student should create a new discussion thread so that other students can respond directly to them. In mathematics, discussion boards require open discussion prompts. The discussion cannot be focused on the right answer to a problem. As these prompts are created, the goal of interaction and a discussion needs to be the focus (Engelbrecht & Harding, 2005). Consider the different levels of questions and even the wording (Wiederhold, 1995). A discussion prompt needs to be very open in the responses that students can provide. A correct answer focus ends the discussion once the answer is found and limits the interaction.

What is expected of the student in these discussion boards also needs to be very clear. The minimum expectations for discussion boards that I have used as a student and a teacher are one thread that responds to the instructor's prompt and then at least two replies to other students' threads. The openness of the prompt needs to consider that every thread can then be carried on with a discussion. One good prompt that I've been able to use in my courses has been to provide students with a system of equations. The prompt provides the system of equations in a story problem format so that students may use a variety of approaches to the problem aside from the typical writing of equations. Students have learned the graphing, substitution, and elimination methods for solving systems of equations. The prompt has students choose a method and work out the problem. The replies are to ask a student who selected a different method and ask what made their method easier or more difficult. This wider range of discussion is not easy for many students in mathematics. The early prompts will need to help students through step by step.

Additionally, the timeline of different parts is very important. Students need to have enough discussion threads to choose from. To help with this, I've often scheduled out the time frame for original discussion threads and a different time for replies. Usually the replies time frame can begin right after the original threads. This should create a time frame in which all can put out their original posts. Then when they come back during the reply phase, they have all possible threads available and the students have the widest range of options to pick from. A major issue that I've dealt with in all LMSs that I've worked with is that there is not a separate time frame for each of these parts of the discussion board. It is a single grade and a single due date. Even with an explanation of the need for two separate parts to the discussion board and two separate times to get into the course, students work to do the minimum and will participate on only the last day to attempt to get points. They do not give anyone a chance to respond to their original thread and no one can respond to their replies. It is a large disservice to the interactions of the course. It is a hurdle that needs to be laid out early and a clear expectation repeatedly reinforced.

The expectations of the replies should also be very clear in these discussion boards. As the goal of these is to have students interact, replies need to be more than a comment, but rather add to the discussion. Thinking about a bad social interaction, it may be someone who only answers with single words or vague comments. The discussion takes more work than it is worth. Students need to continue the discussion with additional information or questions. Again, the instructor should be an example of this, but clear expectations and some examples should also be provided.

Another opportunity for student-to-student interaction comes from group work (Engelbrecht & Harding, 2005). Group work has always been a sore spot for me as a student. The group projects or presentations that I have had to do as a student were graded on a group basis. While this does simplify the grading process for the instructor, considerations should be made for when a group does not work well. As part of that, instructors should think about the method for grouping students. If the course is set up with the ability to have students select their own groups, this can be useful to start out. This is a time when the initial interactions would greatly influence how students select other group members. This can provide interesting information to the instructor about connections that have been made or if there are difficulties in the course that they should be aware of between students. These influences would then be considered when the instructor selects the groups for students on later projects. Even just an exchange of papers between students online can be a very helpful thing. I have had great success in having two students, who both understood parts of a mathematics concept but could not put the whole thing together, suddenly understand as they shared between the two of them. As a key reason that students choose to take a course online is the inability to meet the schedule needs of the school, instructor, or other students, instructors should be clear about what scheduling requirements are included in the course for any group projects at the start. I use the peer review setup in which students need to leave a file by a certain time and then have a time frame to peer review and return the file. This allows the flexibility in schedules that students prefer.

The grading of group work needs to be done individually whenever possible. All the information on grades should be filtered through the instructor to ensure that grading is fair and appropriate. If the end product of the group work is poor, but the majority of the group did their part and contributed greatly the instructor needs to provide individual grades that represent this. In like manner, one person may make a great end product while the majority of the group failed in their part of the group work. Careful consideration of what is graded and how it is evaluated should be done at an early stage of planning.

9.6 COMMUNICATION THROUGH THE COURSE CONTENT

Communication in an online course is easier and harder. Communication is a key part of all learning. It is the method by which instructors package their knowledge and deliver it to students. In a face-to-face environment, communication can be done with kinetic, auditory, and visual methods. Having this wide range of communications is not what makes the online learning environment different, rather it is the interactions of these communications. In fact, the methods of communication in the online environment can be better than those on the face-to-face environment in certain aspects. Face-to-face communications are received at the broadest ability of the receiver. This means that everything that the receiver can see, hear, and feel is part of that moment for learning. This allows the interaction of people on the broadest levels. Each person communicates, even unconsciously. These interactions are the basis for all communication at the earliest stages of life. Communications in an online environment are filtered.

The online environment is created through the communications that people are willing to share. This puts major limits on the methods of communication and what information is actually shared. As a result of the limited communication that we receive but are not used to the human mind works to fill in any gaps in the communication. This filling in of gaps happens as we read books and work to provide emotion to the words on the page. In this same manner, actors provide meaning and emotion to their lines in a script. The more information that is provided to them the more they can fill in the gaps to provide the whole communication that was originally intended.

Misinterpretations of the meaning of communication occur often as the context of the communication may be unknown or different from the one that the communication is received in. This gap of environment can be a change of location or a change of time (Aboujaoude, 2011). Some astounding piece of information is common news at a later time. The day that World War II ended was a monumental day and was deeply felt by all who received the news. Now to read the history lesson about the end of World War II has little direct meaning for the world's population. In this same manner, the day the end of World War II was announced was a day of victory or defeat based on the geographical location. Even the mental and emotional states of each person who heard the news affected how it was received.

In the teaching of mathematics, there is often a removal of context to help solve problems. When this is done the parts of the problem have to be boiled down to their most specific and pertinent information to solve the problem. Mathematicians do this type of context removal very smoothly and naturally after some practice. In like manner, the message or the information that the instructor is working to send to students needs to be delivered without any additional information. This is the first step to good communication in the online environment. Sometimes this is achieved as all the information from a classroom environment is repackaged to the online environment, and other times it is the only step done and students feel a lack of relevance and context that takes away the meaning of the knowledge.

This simplified message needs to only be the first step. For some students, it may be enough for them, but many will need more than raw knowledge. They will need more context, application, and meaning to have a lasting memory or use of the knowledge. The minimum knowledge that is published in the online course needs to be required by all students. Additional direct applications should also be included so that students receive the knowledge and can use it. Additionally, the relevance of the knowledge is what helps students understand the meaning. Each of these additional aspects of the communication should be optional for students to use. Instructors need to provide all methods of communication in the online course for students to access, but not require all parts to be accessed. This will overlabor many of the students, and they will shut down or quit the course. This is an important balance that needs to be had in the tools and resources within an online course.

One aspect of communication that is easier is that it can be standardized. Standardizing communication helps the instructor to provide the same level of information and expectations for all students (Akdemir, 2010). Sending out an email or providing an announcement to all participants of a course means that all participants have access to the communication. The priority of these may be different to different students. The multiple methods for communication are necessary to meet the preferences of all students. By being able to provide the same written communication in the multiple methods helps to provide the same expectations. The recording of all these communications in the online environment is also helpful to instructors. I've often had questions from students, parents, or schools about some detail that I can simply refer to a previous communication. I have often even re-sent the original communication. As an instructor, I have kept my email inbox as a to-do list for keeping up with my students' questions. As someone who is nagged by seeing these each time that my email is opened, I am told that I respond to emails much quicker than most other online instructors. Sorting and filing the communications that are received as an instructor is helpful in keeping up with student needs.

Instructions need to be perfectly clear on what to do, but not give away all the parts of how to do it. Providing written instructions can be great and should include as much information about the expectations as possible. Rubrics are instrumental when there is more than a right answer required. In addition to written instructions, video demonstrations are helpful. Each of these different methods has to be built by the instructor or content developer, even if not all of them will be used by the student. The more that can be built ahead of time the fewer issues that will need to be addressed in the moment. A colleague recently expressed their preference for teaching online as it involves much setup work and then simpler maintenance during the course. I would generally agree with this feeling for time allocation for online mathematics courses.

All methods of communication need to say the same thing. Due dates in every part of the course need to be the same or the work of the instructor will have to go to clarifying the communication more than teaching. Consistency across the course communications is an important thing that instructors should pay attention to before students have access. When I have taught a course that has been preloaded with content, checking all the communications is a top priority. Some previous due dates or variation of expectation from my own are quickly found by students. Even my own reuse of course content requires the update of due dates. The reuse of a previous resource may also require the checking of links and surrounding information.

9.7 USING LOCAL SUPPORTS

One aspect of mathematics that math teachers often have engrained in them is to remove the context of a problem. The context is often what helps the answer make sense. In that same light, each student comes from a specific context that helps them, their knowledge, and their abilities make sense. Teaching in a school that has multiple feeder schools is a good way to understand this. Multiple teachers have taught the group that is found in any course. Each teacher made a skill or concept a high priority in their classroom. Students now have different priorities because of the different teachers that they had before.

One difference in priorities is what technology is available to students. One interesting thing that I have had to develop through the years is a checklist for what technology the students have available to complete the course. Ensuring that students have all the right technology for success should be an early and high priority of the instructor. I have certain assignments in my course that require the students to use an online graphing tool, take a picture of their graph, and then upload it to the course. Having a practice assignment to get all of this troubleshooting out of the way early can be a nice approach. Mathematics often requires the use of special symbols or particular ways of writing things. Ensure, before the students are working with content, that these technical aspects are not going to get in their way. The use of Flash is not an option on certain devices. If there is an online tool that can only be used on a certain type of device, be sure that this is clear at the beginning of the course. If printing and scanning are part of the workaround for symbols, be sure that this is upfront as well.

9.7.1 Schools

Online courses mean different things to different schools. Some schools use online math classes as a supplement for students who are above or below where their mainstream group is at. The courses are not supported by time or personnel. The students have to do all the work outside of the school time, and that makes the online instructor the most important resource for the students.

Other schools use online mathematics courses as part of their curriculum and supplement the courses with an in-house assistant. "Assistant" can mean many different things. This assistant may have no content knowledge and be there to make sure that students are working. The assistant can be a teacher who allows students to work in their room during a prep hour and is a content expert but does not have the course time to teach. The assistant may be a full teacher who has not received their license, and so the students are getting the course from a certified teacher and being helped with daily lessons from another teacher.

Knowing what supports from the school are available to each student helps to determine some of what can be expected from each student. It also helps the instructor know who they will be hearing from often and who they should talk to as local support rather than the student directly.

9.7.2 Parents

Parental expectations for an online mathematics course vary as much as for students. With that in mind, the support that parents provide will vary too. Some parents will work every problem right along with their child. Others will never even know the student is in an online mathematics course. In the K-12 system, it is important to keep parents in the loop of expectations and coordinate with the school supports too. Alignment between school supports and parent supports can be difficult. I have had parents who knew that the students had time at school to do the online coursework and then never expected that there would be any time required out of school for the students. I often have to follow up with the question of if their student has homework from their face-to-face classes. That same expectation should extend to the time commitment for online classes, if not increase as online courses take more time than face-to-face courses.

Parents should be treated as an important part of the teaching team, in the K-12 system. They are among the first people I go to for getting background information about a student and their individual situation. When I first introduce myself to parents, via phone or email, I warn them that they will be included in any important information emails about how the student is doing. This helps me to be able to communicate directly with students and just cc parents in the email. It is nice to not have to write a whole separate email. This initial contact with parents is also to be sure that their lines of communication are open. When I call,

I ask that they save my number. If I email, I write out my email for the parents to contact me. (Some email systems block the email or have a reply to another address.)

Communication with parents needs to start as soon as instructors can so that they are helping set up the learning space for the students. The largest factor that students struggle with is making sure they have enough time allotted for the course and the work. Whether it is a course done at school or in addition to the regular school day, the online course will need some extra time. I like to tell my K-12 students that they should plan on five to ten hours a week to work online. The range may increase if mathematics is a difficult subject or if the student has not had much online experience before. Parents help students regulate their time. Many students need some help as they develop the life skill of time management. Parents can be the biggest help in getting students working and dedicating enough time to be successful. Students who are pressed for time generally do not perform well. Students rarely win the catch-up game.

An outline of the course, due dates, and the syllabus are tools that parents need if they are to help. Outlining the course for the parents helps them know what the students are doing, how it connects to previous work, and how it connects to future work. This outline should also include due dates so that parents can ask about specific assignments done by certain times. Specific due dates in an online format need to fulfill a specific purpose. The reason that students take an online course is for the flexibility of work time. When required to log in daily to stay on an instructor's schedule, that can be off-putting. The due dates need to serve a purpose, and the purpose needs to be clear to parents so that they can reinforce it with students. Due dates are a big deal for any interactive part of the course as the students need things from classmates to continue their progress. This makes sense to parents as many of them have jobs that require them to depend on others or others to depend on them.

The syllabus needs to be very clear about expectations and grading. Grading is always the largest focus as it is the result of the course that will continue with them after the course. A clear description of the weighting of different parts needs to be in the syllabus. It always amazes me how much work students will put into figuring out their grade but refuse to do a percentage assignment to improve their grade. The different weights help to show the importance of different parts of a course to students and parents. Consider what information is sent when the tests are most of the grade. Test-heavy courses are possibly why a student failed a face-to-face course. If an instructor wants the parents on their side, they should include some justification for anything that may be different from the parents' expectations. A syllabus should include how the course is set up for student success. I do this in my college courses as a secret video. It is not hard to find, but by labeling it secret it has some excitement in it. In this video, I lay out how each of the parts of the course helps the students and how the many parts are connected and build meaning and purpose for later parts. In mathematics, it is very easy to help bring out how one part leads to the learning in the next and how they all come together into a new skill that they will need in real life or a future math class.

Aside from time allotment, parents are the next closest people to the students to get help from. The initial communication with parents often includes some feedback from parents about their feelings on mathematics. Knowing this is a key for gauging how much time and attention they will provide to the students. If they had success, then they will work with their student. Also, parents will often see the struggle that requires time and attention from the instructor. Ensuring that parents are ready to call on the instructor for help or refer their student to the instructor is key. I really have appreciated when I got a text or email from a parent asking that I reach out to their child on a particular assignment. It gives me a place to start the conversation rather than the general "can I help you?" I can get specific to the assignment and the content to help the student.

The effectiveness of an instructor's online help is also tough to gauge from the student. Following up with the parents after interventions with the student is a nice third-party perspective. I have found out that my well-written example was way off-base and that the students are now more frustrated and confused. Occasionally, the students or parents need a face-to-face third party too. Face-to-face tutors may be necessary for students to be successful. Online instructors should not take this as a failure. It is important to recognize the reality that not every student can be successful in an online course. Tutors would need to be set up with parents and should be considered if the student's needs are not being met.

9.8 USING ONLINE TOOLS

There is some interesting research about the use of digital manipulatives that should also be considered when thinking about the use of online tools. Discussion about the levels of abstraction should be taken into consideration when selecting tools for teaching mathematics (Lee & Tan, 2014). Initial learning in many cases needs to have a physical representation that students use to begin their understanding. Even the use of fingers to count fulfills this purpose. The students' fingers are a tangible object that students can see and touch. The earliest concepts need to have this physical representation. For more mature students who have moved into a later stage of abstraction, digital manipulatives can be used for representations. The later stages of abstraction are the pictorial representation of a physical object, the symbolic representation of the physical object, and finally the abstraction that has no visual reference but uses the idea of the physical object. For more information about the levels of abstraction see Lee and Tan (2014).

Now the use of these tools can serve a great purpose for the teaching of mathematics. Many students do not begin their mathematics education online; therefore much more of the teaching of mathematics online can use these tools without the precautions of students' developmental level of abstraction. Knowing that any use of digital manipulatives requires a level of abstraction in the thinking of students should inform the instructors that a digital tool may need some supplemental instruction for students to understand. If students have never seen the actual physical thing that is being represented digitally, then the students will have a gap in understanding. Keeping this in consideration, I often have my students purchase some physical items. This could be the course cost as many times an online course will not require a textbook.

The use of online textbooks is not the focus of this section but is an important consideration for the online tools that are used. The reading material that can be provided in an online course is limited only by the access of the student. This brings up the important consideration of necessary limits to students and their access. While instructors may consider using the openness of access to the internet for research and learning a fountain of truth that they will send students to, it is more often a hazardous cascade of misinformation and uncreditable ideas. Students often provide blogs and wikis as a source unless specific guidelines are put in place. These guidelines need to be put up with the idea that they are guardrails for success rather than "No Trespassing" signs meant to keep students away from real-world truths. Guardrails serve the purpose of limiting exposure to danger. The danger in the online learning of mathematics appears as shortcuts or misconceptions. Many credible sources discuss errors and mistakes in great detail to help readers know how to avoid or fix such problems. Unless students will put in the dedication to read these sources in full, they can easily short-circuit their own learning and fall for the very traps that the sources are seeking to help them avoid.

Providing several structured readings and activities to be completed to learn a mathematics concept is a great use of the online environment. An important part of setting up an online course will be the decisions on what is necessary and sufficient to learn the mathematics concept and what is supplemental. The supplemental is not required of all students and should not be directly tied to a grade in the course. Experienced instructors see the developing problem in this balance. Many students will not do anything that is not required to pass the class. They are minimalists. They will do the least that is required of them. Often these students are the ones who need the most exposure and interaction with the mathematics content. This quandary leads to a great online tool.

Differentiation in an online classroom can be utilized by restrictions on the content or pathways required. These pathways can be developed by having specific activities and assignments that must be completed at certain levels. In my own online course, I require students to pass a mathematics content quiz at a specific level. This quiz can be taken multiple times and has a large question bank to pull from so the quiz is never the same. I do this so that students must show that they have mastered enough of the content that they are ready for the next mathematical concept. By creating these gateways my learning management system will help to differentiate for me. Students with the necessary level of understanding can move on in the course. Students who need more time on a concept are not allowed to work on anything else until the current concept is mastered. This kind of setup may be new to students and should be explained. Most people understand the progressive nature of mathematics and that a foundation must be in place to understand the next topics.

Additionally, this same gateway setup can be used for students who have mastered certain topics and can be expedited to topics that will provide them an academic challenge. The use of early assessment to gather these data should not directly affect grades. This gateway can then directly affect if a student is required to complete other activities or assignments in the course. Using these gateway tools, the instructor needs to ensure that even the very best student who may test out of many activities or assignments is still exposed to all that is necessary and sufficient for the course. On the other end of the spectrum, there must be enough content and direction that students are not going to be stopped in their

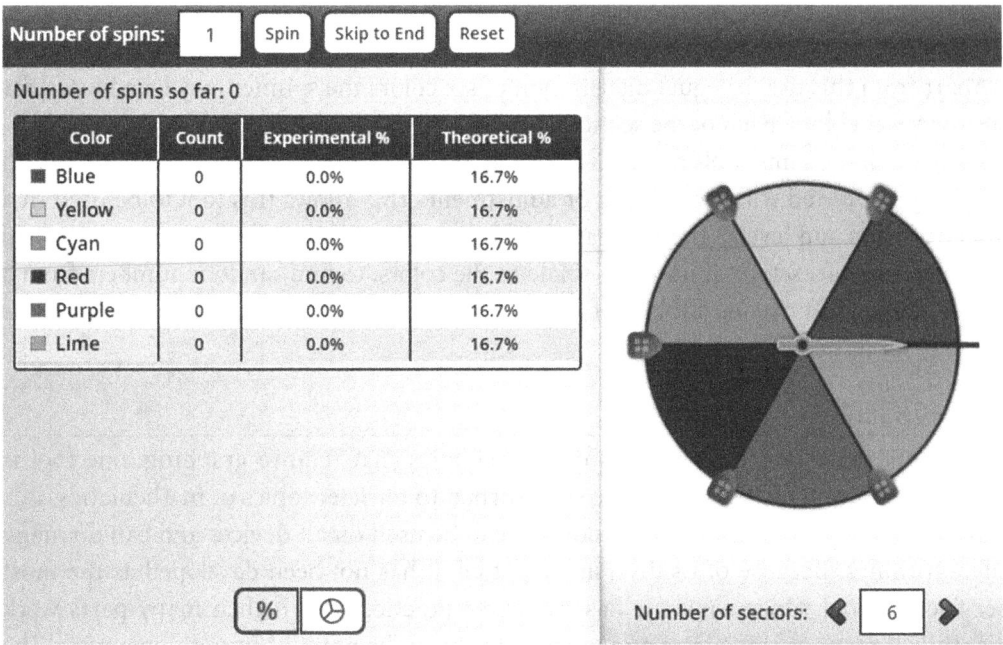

FIGURE 9.1   Screenshot of NCTM's adjustable spinner with default settings.

progression through the course materials. A default screen after several failed attempts to have the student contact their instructor is a nice guardrail to prevent students from going too far looking for answers or to quit the course without some chance for the instructor to intervene.

Apart from the course setup and content that would be housed in a course shell, other tools that can help students in their learning of mathematics should also pique the interest of instructors. One online tool that I prefer over any similar real-world tool is the use of the NCTM's adjustable spinner. This tool is ready to use from the first click to accomplish many things and can be found at NCTM's Illuminations page.[*] The default tool has a spinner with six sections that have different colors and are equally distributed (Figure 9.1). For use as an instructor, this tool can be displayed through a screen share or screen recording. The user may select the number of spins that will be done. The user then clicks spin, and each spin is done in real time on the spinner. There is an option to skip to the end of the number of spins that were put in. This allows instructors to let the spinner go while they explain more about the use of this spinner or to jump to a specific number of spins more quickly. The adjustable spinner can do thousands of spins in a short time. Calculations of the outcomes are tracked within this tool and are displayed alongside the theoretical probability based on how the spinner is setup. This tool does a side-by-side comparison of the theoretical probability with the experimental probability. This type of information helps to expose students to the need for a larger sample size when gathering statistical data. Many of the calculations are done within this tool. By having the tool do much of the calculation

---

[*] www.nctm.org/Classroom-Resources/Illuminations/Interactives/Adjustable-Spinner/.

work for students, the course can then focus on deeper levels of conversation about how to interpret the calculations and data.

Apart from this use for equal distribution of six colors the spinner may be adjusted for more or fewer colors. Each of the sections can be adjusted in size to represent unequal distributions. All the same tools for spins function in the same way. Also, calculations within the tool are adjusted with every drag or adjustment. This allows this tool to be used in a variety of ways and levels. The tool can even be used in a much simpler way to just help pick colors out of a set, or, if numbers relate to the colors, to find random numbers from a specific sized set. It may be a tool that is shown and used but not explained until students are ready for it.

9.9 GEOGEBRA

As students move into higher levels of mathematics that require graphing, one tool is especially helpful to use that can then be returned to for later topics of mathematics. It is GeoGebra. GeoGebra is a universal tool that can be used on all devices and can do many different things. Because of its diversity as a tool, it has not been developed as the most user-friendly tool. Once someone has had some practice with it then many parts work smoothly. It requires some time and practice to get to the point of smooth operation. The use of this tool for online learning is very diverse. It can be used as a simple graphing calculator, to understand what is graphed by manipulating a graph, to draw any pictures or representations by the instructor, for constructions, to graph 3D objects, to work spreadsheets, for data representations, and many other things.

The graphing capability of GeoGebra is greater than that of a graphing calculator. Other online calculators, like DESMOS, function strictly like a hand-held graphing calculator. GeoGebra has a graphing capability in the traditional sense. There is a spot to input equations as a y equals as would be done on a handheld graphing calculator; see Figure 9.2. In addition to this function, equations can be put into GeoGebra in different forms. Linear equations can also be put into GeoGebra in a slope-intercept form and can be changed to standard form or vice versa. This function can help students learn the different versions of how an equation can look. The input of equations and what GeoGebra uses is the only limitation that I have found. GeoGebra does not keep fractions in the equations. Though fractions can be put into GeoGebra, it will convert the fraction to a decimal and round it to the nearest hundredth as a default; see Figure 9.3. The program is more accurate than to the nearest hundredth in the graphing that it does, but the accuracy does have its limits.

This limitation of GeoGebra is a nice opportunity to stress the importance of accuracy and exactness in the tools that are used. This is a mathematical practice of the common core standards that should be discussed some at each level of mathematics (Center for Best Practices, National Governors Association, Officers, & Council of Chief State School Officers, 2010). Graphing accuracy may be the first place this conversation begins as an online course may require the students to print a graph and to hand-draw a line. The accuracy of that line and extending it through the whole graph are both common issues that come up during these early attempts of graphing. As the hand graphing translates to using a program, like GeoGebra, the idea that the graph that we see is only a window of

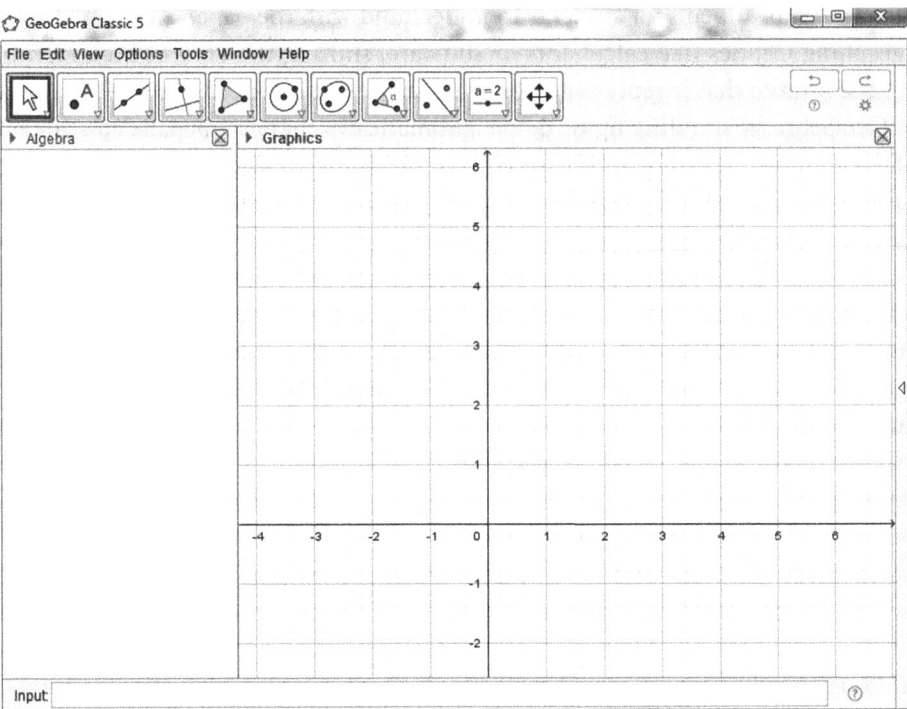

FIGURE 9.2   Screenshot of GeoGebra, highlighting the input aspect.

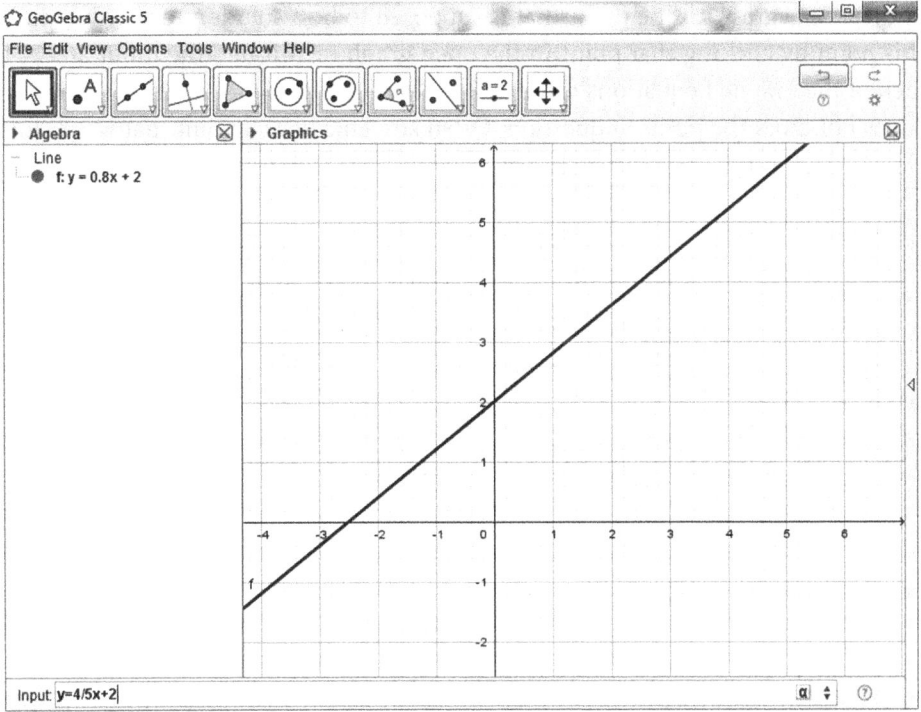

FIGURE 9.3   GeoGebra using decimals even with the input of fractions.

the whole graph is difficult for students to understand with the paper-and-pencil method. With graphing utilities like calculators or software, shifting the view of the graph to see more of the infinite that is represented on the graph can be done. The simple zoom function of GeoGebra by scrolling up or down automatically changes the labeled scale on the axes of the graph so that students see the change and realize how small of a distance or how large of a distance is being represented in the window of the screen.

An additional aspect of using a graphing utility on a computer or more expensive calculator is the ability to see multiple colors for the lines and even shading. This helps students to track things more easily from the paper to the electronic representation. While other graphing utilities, like DESMOS, allow this same capability, GeoGebra features an additional aspect that helps students who struggle with the multiple things on a graph. Each of the equations in the Algebra window has a dot that, when filled, means that the equation is being shown on the graph. By clicking that filled dot it becomes empty and the representation on the graph also is removed; see Figure 9.4. Doing this allows students to focus on just the part of the graph they want for that part of the problem. This has allowed me as an instructor to have the same graph with many different equations graphed, but I only show the ones that I want at the time. It is as easy as the click of the dot to show or not show an equation.

Additionally, as I have used GeoGebra to teach about systems of equations, another feature comes in quite handy. The ability to click on the graph itself and manipulate objects there has been very useful. The ability to add additional objects that are freely drawn onto the graph helps to test points on a graphed equation. My particular method for showing a solution is to begin with the point that is supposed to be a solution to the system. I can plot this point by clicking that point in the open graph or through the input area. I then add each of the graphed equations to the graph. This allows me to see which of the equations does not cross the point in question. Often students have a point that works for one equation, but not both; see Figure 9.5. This way we can see which equation may be giving them trouble.

This ability to manipulate items on the graph is also helpful in the early stages of learning what each part of an equation does. My personal approach to this learning again starts with placing an object on the open graph. This time I do it with two points and then connect them with a line. GeoGebra's Algebra window lists the information of each point and the equation of the line that connects them; see Figure 9.5. By placing objects in the open graph area, they can continue to be moved around in the graph, and the input area information will change based on where objects are placed. This ability to see the equations or coordinates move as I move things on the graph is a nice start for the early introduction

Line

f: $y = 0.8x + 2$

g: $y = -0.67x - 2$

FIGURE 9.4 The Algebra window of GeoGebra with the display dots for two lines.

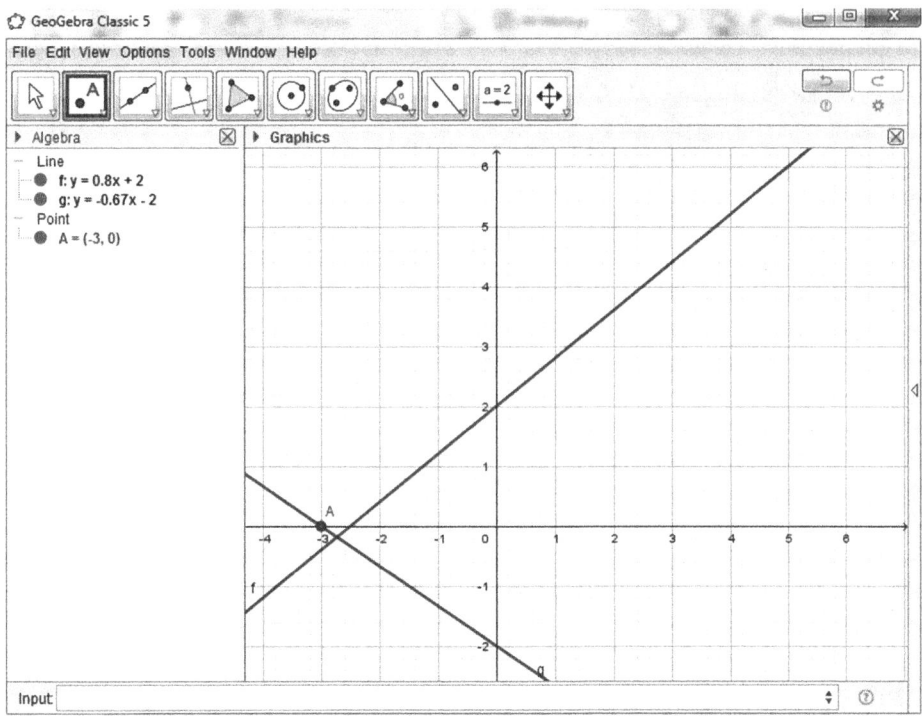

FIGURE 9.5   Screenshot of testing a solution to a system of equations with a point and each line as separate objects.

of coordinates. I can move a point left and right and students will see that the x part of the coordinates is what changes while the y stays the same. I can then move a point up and down and show the vertical placement is tied to the y part of the coordinate. This also helps students to see where coordinates move into the negatives. I've used this to help students determine the types of coordinates in each of the quadrants; see Figure 9.6.

The use of the points on a line is very helpful as I can then move the points around and the line equation changes automatically. At the early introduction, I only use lattice points, points that have integer coordinates. These lattice points help students to be comfortable as they need to do the calculations to start to determine slope. I do not call it slope but reference the number that represents slope in the equation that is automatically created in the input part of GeoGebra. By doing this I can ask the students where the program is coming up with the slope and the y-intercept parts of the equations. Having students build these connections with the graphical representation of the equations has also helped them to realize that any point on the graphical representation is a solution for the x and y of the equation. GeoGebra also can free write over the graphing window. This allows slope calculations within the program rather than in a separate drawing program. Even college students miss this connection between equations and graphs. I can manipulate the points and line to create new equations and test each of the points as a solution to the equation. As students test different ideas about how the slope and y-intercept are determined for each of the equations, the multiple methods reduce the calculations of the slope and the point

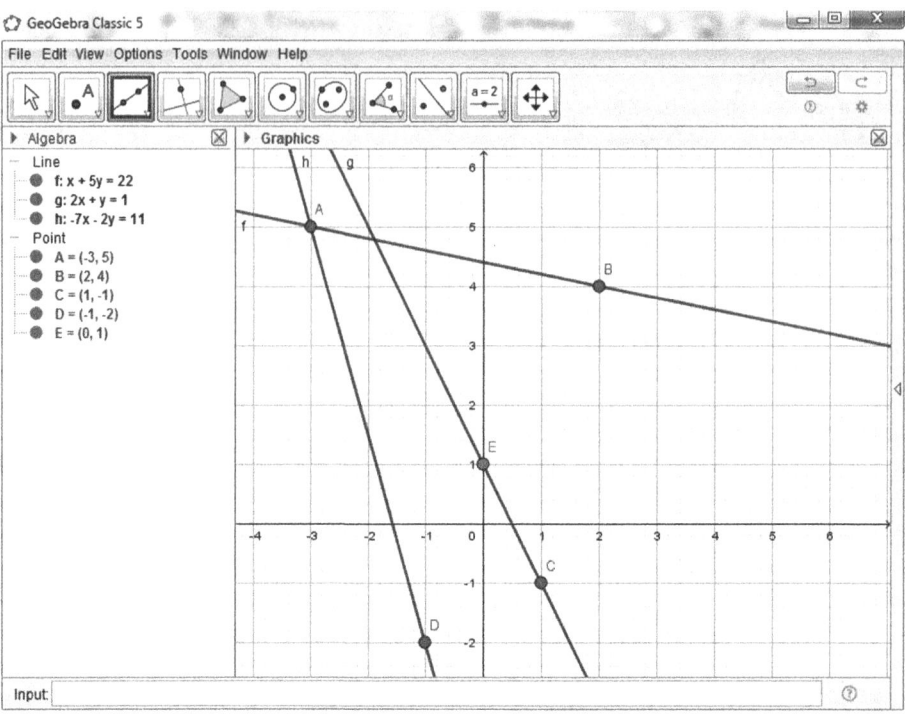

FIGURE 9.6 Equations created by connecting free points on the graph.

where the graphical representation crosses the y-axis. I've used a similar methodology for the graphing of non-linear equations. Having students graph multiple similar equations or draw them and watch what happens to the equations helps them to see what the a, b, and c represent in a standard form of a quadratic equation, ax2+bx+c=0.

The ability to freely draw objects in GeoGebra is a nice tool to help meet some of the Common Core State Standards for Mathematics (Center for Best Practices et al., 2010). This latest integration of mathematics standards to sweep the nation has a few standards that require constructions to be included in geometry content. These standards also specify that students use technology to complete the constructions, as well as other methods. The use of technology in these standards can be especially helpful for instructors. As students are already at a computer, they can enlist the help of a screen recording program, like screencastomatic, to record their work in GeoGebra to perform these constructions. Screen-recording programs can be another useful tool for students to utilize as they need to explain their work or justify a solution. Each student can record their individual work and submit it to the LMS for the instructor to review on their own time. This allows the instructor some individualized time with each student's work. The difficulty to consider is that it takes more time to do this then just checking correct answers. However, the recorded aspect does provide some reprieve as an instructor can skip around in the recording to find the key points of the explanation. This handy tool, like all others, has its advantages and disadvantages.

Constructing in GeoGebra can be done in multiple ways. GeoGebra provides many tools that are a simple clicking of the mouse but would require many steps using a compass and straightedge approach. To help students appreciate the amazing abilities that have been built into GeoGebra, limiting their tools to the compass and line tools of GeoGebra would serve this same purpose. Many constructions can grow in complexity very quickly, but one of the previous attributes of GeoGebra helps in this area too. Being able to hide some of the lines or circles that are drawn can help students to focus on the next step of the equation rather than getting lost in the complexity of the whole construction; see Figure 9.7. GeoGebra uses free-moving objects and fixed objects that are based on the construction and intersection of other objects in the plane. When constructions are done correctly the dragging of the freemoving points can be a nice way to check the accuracy of the construction. This can be a nice alternative to watching all the construction and explanation videos. As an instructor, if students submit their completed constructions, checking just the submitted file can reduce the time load. Then if a student has done it by a different or incorrect method, the video can be used to help clarify the error for better feedback or understand the different method used.

Apart from constructions to meet the standards, the free drawing of shapes can also be useful to see all the different variations of a given situation. I have often used this ability in my face-to-face classrooms and online videos to show that many different types of triangles all have an angle sum of 180 degrees. Objects can be animated to move within any restrictions that have been created. Totally free-moving objects will move in random patterns and directions. I like to at least fix a point on a line or line segment. This gives the point a direction, and using a segment gives it an endpoint so that it will reverse directions

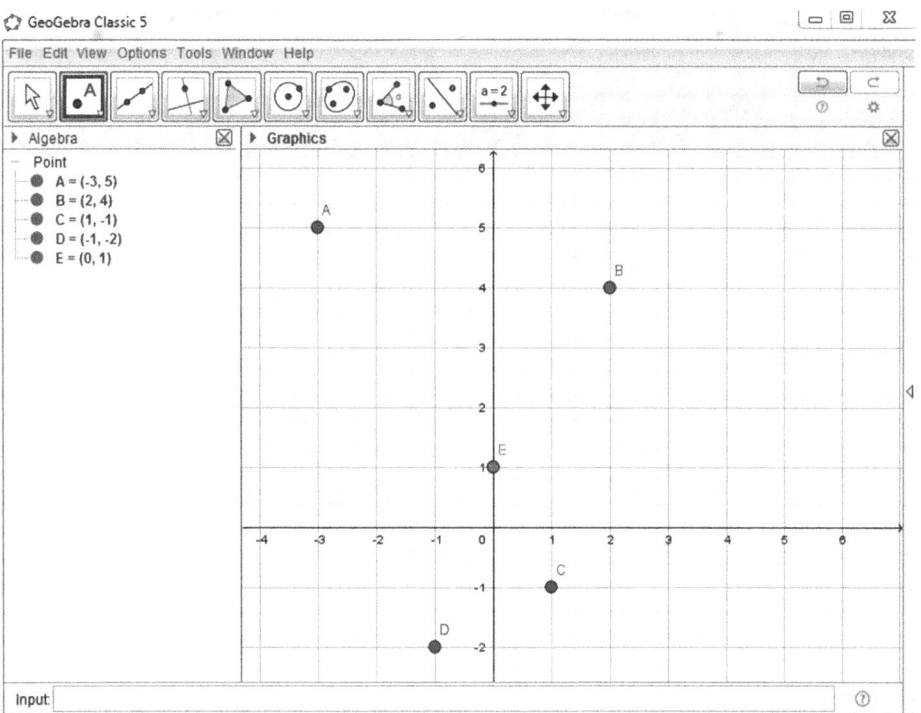

FIGURE 9.7   Free-placed points on the graph and the coordinates listed in the Algebra window.

when it reaches the end. Points can be paused and moved within the restrictions to create each of the different types of triangles. Then I'll allow the points to animate freely so that students can see that the few examples that I provide are not the only ones that work. The showing of calculations on the graph is also a great utility of GeoGebra. This does take some practice as the measurements are assigned various symbols that then need to be used in the equation for the calculation; see Figure 9.8.

As mentioned above, GeoGebra can cut out many additional steps for constructions. One part of this is that it will do all the measurements of a shape and list them all out.

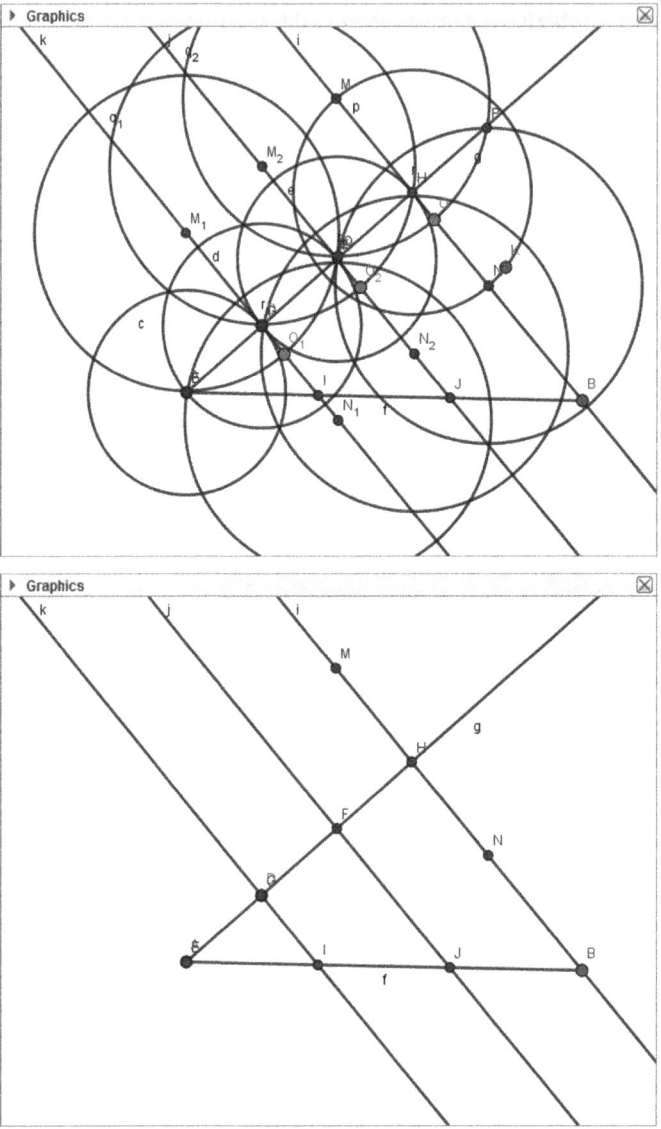

FIGURE 9.8 Comparison of the same construction (trisection of a segment), first with all objects seen required to be completed with the compass and straightedge and the second with many of the objects hidden so that the result is shown.

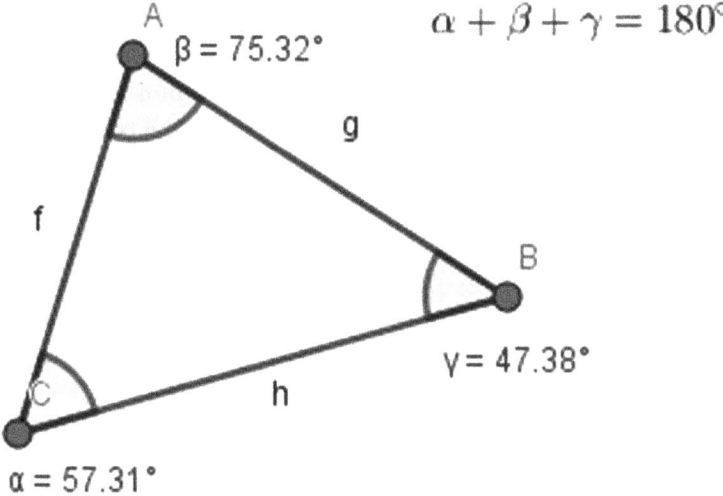

$$\alpha + \beta + \gamma = 180°$$

FIGURE 9.9   A triangle with measured angles and showing the relationship of the angles to sum to 180 degrees.

GeoGebra also has options for drawing a free-form polygon or a regular polygon. The free-form polygon tool can be useful for finding general attributes of polygons and pointing out the difference between concave and convex types. This tool is also helpful in creating a unique shape that can be used to identify different transformations taking place. The regular polygon tool has the user specify a length and then the number of sides. This can be great for seeing what each regular shape looks like and gathering data about angle sums for interior and exterior angles (Figure 9.9).

9.10 A STANDARDIZED METHOD OF DELIVERY

One aspect that has been helpful in the creation of courses and the use of other instructors is the ability to standardize the course content. The online course format allows the creation of assessments and tools that can be distributed to all instructors and that will affect students in the same way. "The main purpose of using Learning Management Systems is to provide a consistent schema for online courses to facilitate the monitoring of students as their learning progresses" (Akdemir, 2010, p. 53). The preloading of a gradebook will standardize how grades are calculated between sections of the same course. Some instructors are only at the level of providing the online course content to students. Advising students about courses to take includes discussion of the method, face-to-face or online, and who to take the course from.

Online courses that have been standardized remove the variation of the value of a course changing based on the instructor for each of the standardized parts. This is also helpful for courses across multiple campuses and multiple schools. Acceptance of a course from another school is much more likely if the courses are standardized within a larger system. The more instructors who teach a standardized course, the larger the pool of people who can help improve the course and develop tools that can be used by all. The spreading of work to develop content and resources helps to lighten the workload.

9.11 CONCLUSION

The teaching of mathematics through the course and communication and the learning by the students are greatly influenced by the instructor's comfort level with the tools in an online environment. The course and communication are the two major tools that an instructor has for helping students learn the content required. Taking the time to set up the course and using all the tools available within the course to cater the course to students' needs and instructor preferences is what separates experienced and preferred instructors from the novice. Using the many methods of communication is helpful for meeting the needs of students in a non-traditional environment as well as making communication clearer. Building relationships with students depends on a foundation being built by the instructor before the start of the course and maintained throughout.

REFERENCES

Aboujaoude, E. (2011). *Virtually you: The Dangerous Powers of the E-Personality.* New York: Norton.

Akdemir, O. (2010). Teaching math online: Current practices in Turkey. *Journal of Educational Technology Systems, 39*(1), 47–64. doi:10.2190/et.39.1.e.

Ashlock, R. B. (2006). *Error Patterns in Computation: Using Error Patterns to Improve Instruction* (9th ed.). Columbus, OH: Pearson.

Bambara, C. S., Harbour, C. P., Davies, T. G., & Athey, S. (2009). The lived experience of community college students enrolled in high-risk online courses. *Community College Review, 36*(3), 219–238.

Center for Best Practices, National Governors Association, Officers, & Council of Chief State School Officers. (2010). Common core state standards for mathematics. *Development, 31.* Retrieved from http://www.corestandards.org/.

Deal III, W. F. (2002). Distance learning: Teaching technology online. *The Technology Teacher, May/June*, 21–26.

Eichhorn, M. S., DiMauro, P. J., Lacson, C., & Dennie, B. (2019). Building the optimal learning environment for mathematics. *Mathematics Teacher, 112*(4), 262–267.

Engelbrecht, J., & Harding, A. (2005). Teaching undergraduate mathematics on the internet. *Educational Studies in Mathematics, 58*(2), 253–276. doi:10.1007/s10649-005-6457-2.

Lee, N. H., & Tan, B. L. J. (2014). The role of virtual manipulatives on the Concrete-Pictorial-Abstract approach in teaching primary mathematics. *Electronic Journal of Mathematics and Technology, 8*(2), 3597–3598.

Li, C.-S., & Irby, B. (2008). An overview of online education : Attractiveness, benefits, challenges, concerns and recommendations. *College Student Journal, 42*(2), 449–458.

Shulman, L. S. (1986). Those who understand: Knowledge growth in teaching. *Educational Researcher, 15*(2), 4–14.

Skemp, R. R. (1987). *The Psychology of Learning Mathematics* (Expanded). Hillsdale, NJ: Lawrence Erlbaum Associates, Inc.

Wiederhold, C. (1995). The question matrix. In *Cooperative Learning & Critical Thinking* (Vol. 1, pp. 15–18). Clemente, CA: Kegan Cooperative Learning.

Wiederhold, C. (1998). *The Q-matrix/Cooperative learning and higher level thinking.* San Clemente, CA: Kegan Cooperative Learning.

CHAPTER 10

Managing Students' Mathematics Anxiety in the Context of Online Learning Environments[*]

Michael A. Tallman and Rosaura Uscanga

CONTENTS

---

[*] We are indebted to Kevin Moore for his thoughtful feedback on previous versions of this manuscript.

10.1 ANXIETY IN THE TEACHING AND LEARNING OF MATHEMATICS

Mathematics anxiety affects a wide variety of learners regardless of age, gender, race, or mathematical ability, although with inconsistent regularity and effect (Betz, 1978; Hembree, 1990; Ho et al., 2000; Meece, Wigfield, & Eccles, 1990; Tobias, 1993; Wigfield & Meece, 1988). While it is often assumed that mathematics anxiety is most commonly experienced by students in the early grades, it is at least equally as prevalent at the secondary (Wigfield & Meece, 1988) and undergraduate levels, even among students who self-select into STEM disciplines (Betz, 1978). Managing students' mathematics anxiety is therefore not the exclusive concern of primary school teachers; all mathematics instructors have the responsibility to understand what mathematics anxiety is, recognize what causes it, and develop strategies to help students manage it. Moreover, because mathematics anxiety is more frequent among females, Hispanics,* and pre-service elementary school teachers (Hembree, 1990), attending to students' anxiety is essential for ensuring equitable access to mathematics and for enhancing future elementary school teachers' interest in, and knowledge of, the subject.

Although the field lacks a normative definition of mathematics anxiety (MA), it is widely acknowledged that MA involves a complex interaction of physiological, emotional, cognitive, and behavioral entailments that manifest from one's appraisal of a mathematical experience as having potentially negative consequences for the self. Being elicited through an appraisal process, MA is a situational experience, not a psychological trait. Some individuals, however, possess a heightened tendency to perceive situations as dangerous or threatening, and thus maintain a greater degree of anxiety-proneness (Zeidner, 2014). For this reason, Cattell (1950) proposed the distinction between *trait anxiety* and *state anxiety*—a contrast later popularized by Spielberger (1972). The former refers to an individual's relatively stable disposition to appraise situations in a way that results in *state anxiety*—a temporary reaction involving somatic sensations of tension accompanied by worrying and self-deprecatory cogitations (Ashcraft, 2002; Epstein, 1972; Ho et al., 2000; Öhman, 2008; Richardson & Suinn, 1972).

Students' mathematics anxiety has been a major topic of concern for educational researchers and practitioners for decades. Research into MA has prioritized the identification of biological, environmental, and personal factors that contribute to its emergence, as well as the documentation of its psychological and behavioral consequences for students' mathematical activity (Ashcraft, 2002; Batchelor, Gilmore, & Inglis, 2017; Eden, Heine, & Jacobs, 2013). Regarding its biological origins, Öhman (2008) emphasizes that the automatic and unconscious perceptual processes that precede the conscious experience of anxiety were naturally selected in our evolutionary history. Consistent with its roots in natural selection, Zeidner (2014) notes that "about 50% of the observed variance in trait anxiety can be accounted for by genetic factors" (p. 271). Scholars have also documented various environmental determinants of anxiety, including early childhood experiences and the

---

* Hembree (1990) explains that there was no observed difference in levels of anxiety between white and black students, but in two studies researchers reported that Hispanics were more anxious than these other two groups. The participants of these studies were limited to college students.

influence of teachers and parents (Drake & Kearney, 2008; Krohne, 1992; Mcleod, Wood, & Weisz, 2007). Personal causes of MA include beliefs in myths regarding mathematical ability, fear of appearing too intellectual or ignorant, self-distrust, an inability to handle frustration, and maintaining a predominantly performance-goal orientation (Bong, 2009; Hembree, 1990; Meece, Wigfield, & Eccles, 1990; Tobias, 1993). Poor cognitive abilities are also associated with MA (Ashcraft & Kirk, 2001; Faust, Ashcraft, & Fleck, 1996; Hopko et al., 2003; Miller & Bichsel, 2004; Maloney, Ansari, & Fugelsang, 2011; Morsanyi, Busdraghi, & Primi, 2014), as are perceptions of underdeveloped mathematics capability (Maloney et al., 2010; Maloney, Ansari, & Fugelsang, 2011; Ma & Xu, 2004; Rubinsten & Tannock, 2010).

The vast majority of research into MA has investigated its consequences for students' mathematics learning and performance. Meece, Wigfield, and Eccles (1990), for example, suggest that lower levels of MA "may facilitate achievement striving, whereas more extreme levels appear to be more disruptive of cognitive and attentional processes, especially on tasks involving higher order thinking skills" (p. 68). Similarly, Wigfield and Meece (1988) claim that while worry can motivate students to invest effort, if the worry becomes too strong it can interfere with performance. Other studies have reported that students' MA is negatively related to their mathematics achievement (Betz, 1978; Hembree, 1990; Ho et al., 2000; Fennema & Sherman, 1977; Richardson & Suinn, 1972; Wigfield & Meece, 1988). Specifically, MA impairs problem-solving ability through a reduction of working memory capacity (Ashcraft, 2002; Ashcraft & Kirk, 2001; Trezise & Reeve, 2017). Generally, the adverse cognitive consequences of anxiety are many, and include compromises to information encoding, storage, processing, and retrieval (Zeidner, 2014). Researchers have also revealed that MA is negatively correlated with students' plans to enroll in future mathematics courses (Betz, 1978; Hembree, 1990), their mathematical self-efficacy (Hackett, 1985), and their selection of and perseverance in STEM majors (Hackett, 1985).

These and other effects of MA make it prudent for instructors and curriculum designers to consider how they might manage students' anxiety, particularly in the context of online learning, which is becoming an increasingly common instructional medium for mathematics at institutions of higher education. While research on the causes and consequences of anxiety is extensive, far fewer studies have investigated the effectiveness of interventions intended to reduce students' MA. Moreover, the majority of such studies employ quantitative and correlational analyses that enable instructors to adopt the particular treatment under consideration, if demonstrated effective. The theoretical contribution of such research with respect to clarifying the phenomenology of MA is limited, and thus does not enable teachers to enact interventions informed by an understanding of the psychosomatic mechanisms that contribute to experiencing MA (Eden, Heine, & Jacobs, 2013). Common treatment programs attempt to mitigate the symptoms of anxiety by reducing levels of physiological arousal and/or equipping the anxious subject with strategies for coping with worry, negative expectations, and task-irrelevant thoughts. These interventions, which seek to moderate the emotional and cognitive experience of anxiety, are of little utility to mathematics instructors who hope to intervene prior to students becoming anxious. Such proactive interventions require a thorough understanding of the cognitive entailments of

MA and how they relate to students' assimilatory schemes, goal structures, and identities as learners of mathematics.

The purpose of this chapter is to offer such a genetic account of MA that makes explicit the cognitive processes from which it originates. Our objective is to elaborate a conceptualization of the eliciting conditions of MA to inform instructional and curricular innovations that seek to decrease the likelihood that students will experience unproductive levels of MA in online learning environments. In addition to being useful for mathematics teachers, our genetic account of MA provides a theoretical foundation for empirical research on MA interventions. We rely heavily on the general theory for the cognitive basis of emotional experience proposed by Ortony, Clore, and Collins (1988)—commonly referred to as the "OCC model"—and add nuance as it relates to anxiety by incorporating a number of related lines of inquiry in the fields of genetic epistemology, identity, and goal theory. We then leverage our conceptualization of the cognitive origins of MA to identify concrete curricular and instructional design principles for online learning environments—grounded in radical constructivism (von Glasersfeld, 1995)—that not only dissuade the cognitive appraisals and constructions that contribute to students feeling anxious, but which promote the conceptual activity that enables students to experience mathematics positively. A key argument emerging from our theoretical analysis is that constructivist curricula and pedagogy, although traditionally justified with reference to their potential affordances for students' mathematical cognition, can also be supportive of students' positive affect.*

10.2 A GENETIC ACCOUNT OF MATHEMATICS ANXIETY

10.2.1 Emotions and the Role of Cognition in Their Elicitation

Our intention to specify the cognitive antecedents of MA, and to infer instructional and curricular recommendations from them, suggests a temporal order of psychological and somatic influences in the experience of an emotion. So as not to give the impression that we categorically dismiss the William Jamesian view that reflexive visceral reactions precede conscious emotional experience, let us say a few words about the nuanced role of cognition in the process of emotion elicitation. We share the perspective expressed by Clore and Ortony (2008) that emotions often originate from physiological reflexes, but that their conscious experience is shaped by cognitive appraisals of both environmental stimuli and somatic sensations. These appraisals have the potential to then intensify, reduce, or alter the feelings associated with the initial physiological reflex, and to transform general affective reactions into a differentiated emotional experience (Cunningham & Zelazo, 2007). The contribution of cognition and feeling in the experience of an emotion is therefore neither additive nor sequential; emotions emerge from the complex co-occurrence of cognitive activity and arousal of the autonomic nervous system. In this model, one can think of emotions as emergent constructions, progressively elaborated and refined through iterative cognitive appraisals of environmental stimuli and somatic states (Clore & Ortony, 2008).

---

* We refer the reader who might consider the phrases "constructivist pedagogy" and "constructivist curricula" paradoxical, or at least nebulous, to Thompson (1991, p. 287).

Privileging a definition of emotion that emphasizes the structure of cognitive appraisals from which they arise—as opposed to the *process* of such appraisals—Ortony, Clore, and Collins (1988) described emotion as a valenced affective reaction to an individual's interpretation and evaluation of the realized or anticipated consequences of events, actions of agents, or aspects of objects. The focus of the appraisal, whether an event, agent, or object, constitutes the primary criterion that distinguishes emotions of qualitatively different types. The desirability of an event, Ortony et al. (1988) explain, is appraised with reference to an individual's *goal structures*, the praiseworthiness of an agent's actions is appraised with reference to an individual's *standards*, and the appealingness of an object is appraised with reference to an individual's *attitudes*. Emotion-inducing stimuli achieve their value through appraisals made with reference to these interwoven elements of an individual's affect. Specific emotional experiences are therefore conditioned by the particular goal structure, standard, or attitude that respectively informs an individual's appraisal of the consequence of an event, action of an agent, or aspect of an object. Since goal structures, standards, and attitudes are subjective cognitive constructions, emotions are an adaptive response to an individual's *interpretation* and *evaluation* of her experiences; they do not issue directly from a reality independent of one's construction.

10.2.2 Mathematics Anxiety as a Prospect-Based Emotion

Figure 10.1 displays the macrostructure of the OCC model. The boxes represent distinct emotion classes. The location of these classes within the structure is determined by the specific cognitive appraisals that contribute to experiencing the emotions within them. The objects of these appraisals and the potential considerations they involve are indicated by the unboxed upper-case text. The lower-case emotion words within each box represent subclasses of emotions that vary in intensity, but which emerge from structurally isomorphic appraisal processes. For instance, well-being emotions within the subclass "joy" might include "elation" and "contentment"—respectively high- and low-intensity forms of joy. The words representing these subclasses of emotions are themselves not important, and were chosen because they serve as neutral examples of qualitatively distinct emotion types.

Anxiety is a prospect-based emotion, and as such results from one's recognition of the prospect of experiencing the personal consequences of an undesirable event. In particular, anxiety is a variant of fear where the emphasis of the appraisal is on the potential psychological (rather than physical) consequences of an event (Ortony, Clore, & Collins, 1988, p. 15). Being an emotion within a class that extends from the left-most branch of Figure 10.1, the appraisals that contribute to experiencing anxiety are made with reference to an individual's goal structures. The intensity of an individual's anxiety—as with any prospect-based emotion—varies directly with the extent to which the individual appraises the potential consequence of an event as undesirable. Bearing principally upon this appraisal is the value of the *focal goal* the event inhibits the individual from achieving (ibid., p. 50). A focal goal is situated within an organic goal structure and is the specific goal against which an individual assesses the desirability of a prospective event. Its value depends on the position of the highest-level goal in the structure whose attainment is facilitated by achieving the focal goal; the higher in the structure it resides, the greater the assessment of (un)

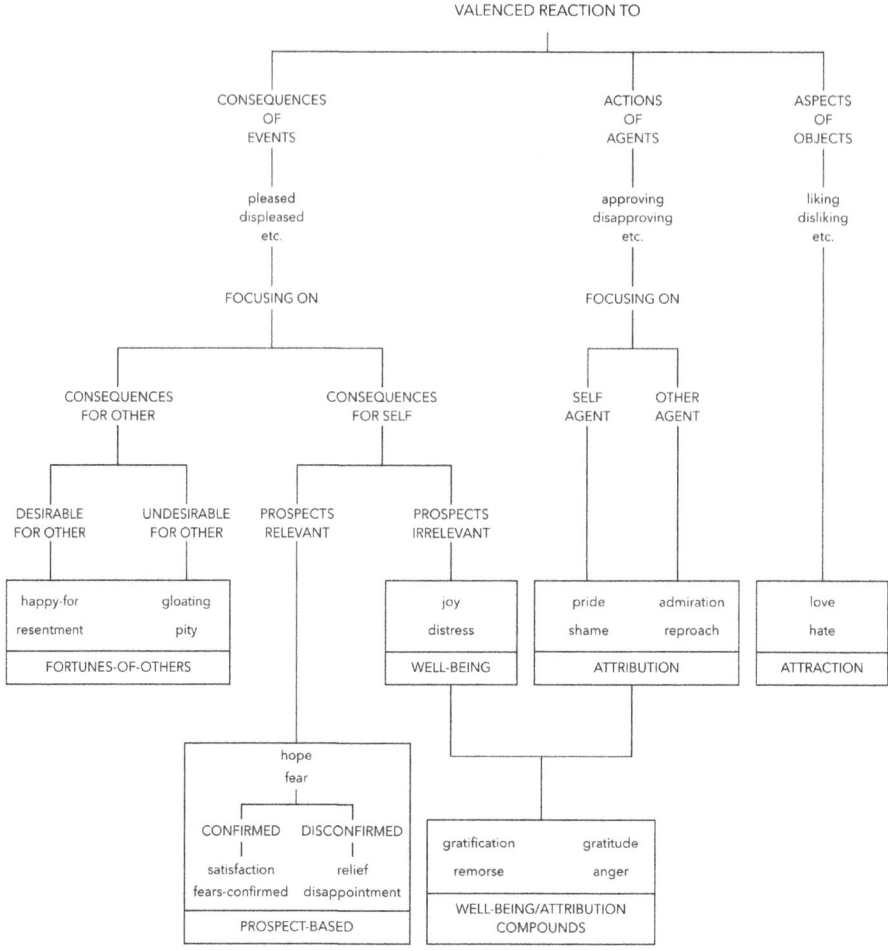

FIGURE 10.1 Structure of qualitatively distinct emotion types (Ortony, Clore, & Collins, 1988, p. 19).

desirability. The number of sub-goals upon which its achievement is contingent determines the position of this higher-level goal in the structure.

10.2.2.1 Goal Structures

Inferring instructional and curricular recommendations from the influence of goals on one's experience of anxiety requires a discussion of goal structures in the context of mathematics learning. Generally, the goals one establishes for oneself define a state of being one desires to enter—they are a projection of an anticipated future (Middleton, Tallman, Davis, & Hatfield, 2015). Goals are always situated within a structure, a kind of fluid hierarchy where the accomplishment of subordinate goals facilitates the achievement of superordinate ones. Fulfilling a goal is rarely a final act; doing so merely serves as a precondition for achieving a higher-level goal in the hierarchy. Any mathematical experience that is goal-directed involves an end state that is both cognitive and affective (Middleton et al., 2015). For instance, a student who seeks to solve a problem, and does so successfully, will not only have a new or modified understanding of some mathematical idea, but will also experience satisfaction as a result of having achieved his or her goal.

The nature of the goals one pursues depends on the perceived challenge of the task at hand. For example, as Middleton et al. (2015) argue, if a student has a poor view of his or her mathematical ability, the goal of obtaining an "A" on a particular assignment might seem unattainable, whereas this might not be the case for a student with strong mathematical self-efficacy. Such perceptions can influence a student's behavior and as a result, a student who does not consider him- or herself proficient in mathematics might avoid trying to earn a high grade in a mathematics class and instead focus on simply passing the course. Goals also differ in the extent to which they address immediate versus future needs. A *proximal goal* is one that can be fulfilled in a short amount of time, whereas *distal goals* take longer since they require achieving a sequence of pre-requisite proximal goals (Middleton, Jansen, & Goldin, 2017; Middleton et al., 2015). The highest-level distal goals are often (implicitly) defined in terms of desired identity states.

The literature on goal orientations identifies two distinct types of goals: *learning/mastery goals* and *performance/ego goals* (Ames & Archer, 1988; Ames, 1992; Dweck & Leggett, 1988; Middleton, Jansen, & Goldin, 2017; Middleton et al., 2015). Learning/mastery goals are consistent with a growth mindset and are defined in terms of developing understanding, increasing competence, and mastering course material (Ames, 1992; Dweck, 1986; Dweck & Leggett, 1988; Middleton, Jansen, & Goldin, 2017; Middleton et al., 2015). Performance/ego goals, in contrast, are compatible with a fixed mindset and prioritize the demonstration of innate ability and the promotion of self-worth through the display of competence to others (Ames, 1992; Ames & Archer, 1988; Dweck, 1986; Dweck & Leggett, 1988; Middleton, Jansen, & Goldin, 2017; Middleton et al., 2015).

The difference between *approach* and *avoidance* orientations is an additionally relevant distinction that explains meaningful differences in the behavior of students who adopt learning/mastery versus performance/ego goals (see Table 10.1). A student who has an approach orientation and establishes learning/mastery goals will attempt to construct understanding and achieve mastery of challenging course material. A student with the same orientation who seeks to achieve performance goals will attempt to be perceived as superior in relation to his or her peers (Middleton et al., 2015). With an avoidance orientation, a student who sets mastery goals might attempt to avoid non-normative understandings, whereas a student who pursues performance goals will try to avoid appearing incompetent (Middleton, Jansen, & Goldin, 2017).

TABLE 10.1   Behaviors Associated with Combinations of Goal Types and Orientations

	Approach	Avoidance
Learning/mastery goals	Construct understanding and achieve mastery of course material	Avoid non-normative understandings
Performance/ego goals	Convey perceptions of superiority in relation to peers	Avoid perceptions of incompetence

To summarize, goals vary along four dimensions: *specificity* (precise versus vague), *proximity* (proximal versus distal), *type* (learning/mastery versus performance/ego), and *orientation* (approach versus avoidance) (Middleton, Jansen, & Goldin, 2017; Middleton et al., 2015). Managing students' MA involves attending to these four interrelated dimensions. With respect to goal specificity and type, reducing students' anxiety requires fostering their construction of precise learning goals by engaging them in mathematical tasks that promote or require understanding. Doing so will enable students to be better equipped to make appropriate evaluations of their mathematical abilities and to address deficiencies therein. With respect to goal type and orientation, several researchers have demonstrated that students who seek to achieve performance goals tend to link failure and lack of ability, which often results in avoidance tendencies and in extreme cases learned helplessness (Ames, 1992; Ames & Archer, 1988; Dweck, 1986; Middleton et al., 2015). In contrast, students pursuing learning goals tend to link effort and progress, thus resulting in approach tendencies, which reduces anxiety and promotes perseverance (Ames, 1992; Ames & Archer, 1988; Dweck, 1986; Middleton et al., 2015). We later return to the importance of instructors promoting students' construction of learning goals and offer specific strategies to this end.

10.2.2.2 Identity

Students' current and desired identities as learners of mathematics are particularly consequential for the degree to which they experience MA. In the case of anxiety, the highest-level distal goal whose achievement is contingent upon the focal goal (which influences the extent to which an individual assesses the potential consequences of an event as undesirable) is a desired identity state. Favored identities are therefore high-level goals that are facilitated by the achievement of the focal goal. We follow Blumer's (1986) definition of identity from his theory of symbolic interactionism, in which he characterizes identity as the kind of object one is to oneself. In contrast to other conceptions of identity commonly cited in the mathematics education research literature (e.g., Cobb, Gresalfi, & Hodge, 2009; Sfard & Prusak, 2005), this simple definition places the construction of identity within the individual to whom it is an identity. Moreover, Blumer's conception underscores the notion that individuals interact with their identities, and that such interaction is consequential in the formation of his or her behavior. Indeed, to be conscious is to engage in a process of indicating objects to oneself and acting towards these objects on the basis of their meaning for the actor. The objects an individual acts upon are not limited to those that exist external to him or her, but include the self-object (or identity), which is treated as if it exists independently of one's construction. Referencing Piaget (1937), von Glasersfeld (1995) explains,

> just as we construct a model of a world, externalize it, and then treat it as though its existence were independent of our doing, so we construct a model of the entity that we call our *self*, and externalize it so that it ends up as 'a thing among other things'.

(p. 123, emphasis in original)

This reified self emerges from the process of interacting with the world and internalizing what one's experiences reveal about oneself. Specifically, as Blumer (1986) observes, the development and evolution of one's identity is a product of his or her interaction with others:

> Like other objects, the self-object emerges from the process of social interaction in which other people are defining a person to himself … [I]n order to become an object to himself a person has to see himself from the outside. One can do this only by placing himself in the position of others and viewing himself or acting toward himself from that position … We form our objects of ourselves through such a process of role-taking.
>
> *(p. 12–13)*

von Glasersfeld (1995) articulated a similar perspective regarding the generation and refinement of one's self concept. He argued that constructing knowledge of the self is a meta-cognitive process in which an individual becomes aware of what they are doing or experiencing, primarily through interaction with others.

This constructivist conceptualization of identity has implications for students' experience of MA and for instructional and curricular innovations that seek to minimize its negative influence. Students are often anxious while engaged in mathematical experiences because they recognize the possibility that their genuine mathematical activity will reveal characteristics of their mathematical competence in particular, and intellectual potential in general, that are antithetical to their desired identities as students and/or future professionals. Mathematical symbols, expressions, and problems, when assimilated by students, frequently represent potential threats to the kind of individual they aspire to be, rather than objects to be interpreted meaningfully and acted upon on the basis of their meaning. As Ortony et al. (1988) explain, the intensity of the prospect-based emotions (which include anxiety) depends on the appraised value of the highest-level distal goal whose fulfillment is contingent upon achieving the focal goal, not the proximal goals that would need to be accomplished en route to achieving the focal goal, such as solving a particular problem, completing an assignment, or even earning a desired grade in a course. Because a sought-after identity constitutes the most significant type of distal goal, threats to its achievement can result in cognitively debilitating levels of anxiety.

10.2.2.3 *Intensity of Emotional Experiences*

In addition to the value of the focal goal being impeded by the potential of an unfavorable event, Ortony et al. (1988) identify several "global" and "local" intensity variables that affect the severity of emotional experiences. Among the global intensity variables most relevant to MA is the *existing level of arousal* variable. Global variables affect the intensity of all types of emotions, and their effects influence whether or not there is an emotional experience at all (Ortony et al., 1988). The *existing level of arousal* variable refers to physiological arousal, which is a product of the cognitive appraisal processes (Ortony et al., 1988, p. 65).

Although arousal is not cognitive in nature, it does influence cognitive processes and products as we previously described. It is important to note that since arousal decays slowly over time, it can carry over to a later emotional experience and thus increase the intensity of reactions to a subsequent situation.

Likelihood is among the local intensity variables most relevant to MA. Local intensity variables only affect the intensity of particular emotion classes (Ortony et al., 1988). The *likelihood* variable refers to someone's belief regarding an event's plausibility (ibid., p. 70). Since anxiety is based on uncertainty, it is intensified when an individual believes that possible outcomes of an event are equiprobable. For example, a student who considers it feasible that he or she will fail at a particular task will experience greater levels of MA than a student who has a high degree of confidence in the outcome of the event, whether success or failure. The experience of MA has an effect on other local intensity variables—namely effort—that influence the intensity of other prospect-based emotions. Effort is sometimes influenced by an individual's assessment of the likelihood of particular outcomes of an event (Ortony et al., 1988). That is, individuals will often expend more effort if they believe doing so can change the likelihood of a specific outcome. In such cases, the effort is referred to as *instrumental effort,* which is expended (or not) for the purpose of changing the likelihood of a(n) (un)desirable event (Ortony et al., 1988, p. 73). In the case of MA, the more effort expended, the greater the undesirability of the prospective outcome, which makes the anxiety experience more severe. It is for this reason that anxiety tends to immobilize effort. This is to be expected if one considers that the desirability of a prospective event is appraised with reference to its implications for one's identity. Because characteristics of an individual's identity are constructed through reflection upon his or her experiences (von Glasersfeld, 1995), fully investing oneself in mathematical tasks more clearly reveals to the individual his or her mathematical aptitude and intellectual capabilities. Thus, students who experience MA often expend little effort to decrease the likelihood of an undesirable outcome. On the surface this seems contradictory, but it is essential to bear in mind that the distal goal that the desirability of a potential outcome is appraised with reference to is a desired identity state, not the more proximal goals of completing a mathematics assignment, achieving a high mark on a test, or even earning a particular grade in a course.

As previously stated, anxiety is based on a degree of ambiguity regarding the outcome of an event (i.e., a mathematical experience), and in the context of mathematics learning this ambiguity is often reflective of how a student appraises his or her psychological resources (i.e., mathematical self-efficacy). Zeidner and Matthews (2011) note, "worry develops when a person perceives his or her ability to cope with a task as unsatisfactory and is uncertain about the consequences of inadequate coping" (p. 15–16). It is important to note that an anxious student, to some degree, recognizes that a positive outcome is at least possible (or else the student would not be anxious but would rather be disappointed or relieved). Hence the reassuring implication that anxious students possess motivation capable of being cultivated by instructors through their encouragement, thoughtful task design, and purposeful orchestration of learning experiences. Reducing a student's perception that their mathematical activity might have negative personal outcomes requires engaging the student in experiences that enhance his or her self-image as a learner and doer of mathematics.

10.2.2.4 Somatic Markers

The OCC model we have discussed is one among many appraisal theories for the cognitive entailments of emotional experience. However, we have not yet clarified the cognitive processes that constitute the appraisals students make that contribute to the onset of MA. We conceptualize the cognitive appraisals that initiate and sustain the experience of an emotion in terms of Piaget's concept of assimilation to a scheme. Piaget and Inhelder (1969) defined a scheme as "the structure or organization of actions as they are transferred or generalized by repetition in similar or analogous circumstances" (p. 4). Piaget broadly defined action to encompass all movement, thought, or emotion that responds to a need (Piaget, 1967, p. 6). Jonckheere, Mandelbrot, and Piaget (1958), as quoted in Montangero and Maurice-Naville (1997, p. 72), explained, "Assimilating an object to a scheme involves *giving* one or several meanings to this object" (p. 59, our emphasis). One's meaning for an object or experience is hence the scheme with which one assimilates it (Thompson et al., 2014). Reflecting Piaget's view that actions (the contents of schemes) maintain behavioral, psychological, and emotive dimensions, Damasio (1994) posed what he called the *somatic marker hypothesis*, which states that individuals construct learned associations between particular feelings and specific classes of actions or stimuli, which represent a potential outcome of an event. These associations have affordances for automating decision-making processes and were naturally selected in our evolutionary history as a result of the survival advantages such automaticity affords. Damasio's somatic marker hypothesis implies that one's assimilation of a stimulus to a scheme activates somatic sensations that indicate to the individual the value of the stimulus, and which mobilizes action patterns independent of conscious cognitive processing. It is in this sense that somatic sensations (i.e., feelings) "mark" the stimuli that are assimilated. Ultimately the somatic sensation becomes a learned association that is reflexive, and is thus an element of an individual's assimilatory scheme. For students who experience MA, mathematical objects (e.g., tasks, expressions, symbols) are "marked" with somatic sensations that co-occur with their assimilation. These somatic markers, abstracted from prior experience, increase students' existing level of physiological arousal, which intensifies their anxiety as these somatic sensations register in consciousness.

10.3 STRATEGIES FOR MANAGING STUDENTS' MATHEMATICS ANXIETY IN ONLINE LEARNING ENVIRONMENTS

Since MA emerges from a variety of subjective appraisals and cognitive constructions in the ways we have described, mathematics teachers can manage students' MA by structuring various features of the learning environment—including the curricular artifacts we design for students—to reduce the likelihood that they will engage in the cognitive activity that results in their feeling anxious. In this section, we leverage our theoretical analysis of the cognitive antecedents and entailments of MA to propose principles of instructional and curricular design for online learning environments that might lessen or even prevent students' MA while simultaneously supporting their construction of productive mathematical conceptions. We justify our proposals by clarifying how they influence the specific cognitive constructions and appraisals from which students' experience of MA manifests.

10.3.1 Assist Students in Establishing Meaningful Learning Goals

A performance-goal orientation invites students to leverage unproductive coping mecha-nisms* that often undermine an instructor's or curriculum designer's intentions for stu-dents' learning. This is because performance goals are fulfilled through accomplishing behavioral tasks; the means by which one does so is irrelevant. The positive affect a student experiences upon having achieved a performance goal is often tempered by the realization that he or she has failed to develop transferable competencies that might facilitate subse-quent successes. There is little to no transfer afforded by leveraging coping mechanisms to satisfy the immediate need of solving a problem or completing an assignment into other, more genuine, mathematical experiences. It is for this reason that a performance-goal orientation is often associated with low mathematical self-efficacy. While pursuing per-formance goals, students regularly experience the limited contexts to which their coping mechanisms apply, while also accepting that they have no other recourse but to persist in attempting to utilize them. This contributes to students' perception of low mathematical ability. In such situations, it is possible that a student recognizes that he or she has not developed particular mathematical sensibilities, understandings, or ways of reasoning that will enable him or her to flexibly engage with novel mathematical tasks. The restricted range of applicability of coping mechanisms makes students uncertain as to whether they will be effective in attempting to solve a particular problem, which increases their uncertainty regarding the outcome of their mathematical engagement. This uncertainty intensifies students' anxiety in a way consistent with the *likelihood* local intensity variable discussed previously.

Managing students' MA therefore involves instructors supporting students' develop-ment of a learning-goal orientation. One can accomplish this by encouraging students to recognize that genuine mathematical proficiency entails the cognitive characteristics of their desired identities as learners, which might include creativity as well as logical, flex-ible, innovative, and critical thinking. Students tend not to associate these competencies with mathematical ability, privileging instead the capacity to efficiently associate rehearsed problem-solving procedures to particular tasks by identifying their surface-level features. The regularity, even alacrity, with which individuals profess incompetence in mathemat-ics, evidently without disappointment or embarrassment, demonstrates their lack of regard for the intellectual faculties they attribute to mathematical aptitude; it is only marginally self-defeating to proclaim that one has failed to master the skill of determining which of the rehearsed problem-solving procedures residing within a vast mental library applies to an esoteric class of procedural exercises. Such proclamations might even be gratify-ing or self-promotional if the individual making them maintains a zero-sum conception of cognitive ability,† and so interprets the essential skill set required for mathematical

---

* An example of such a coping mechanism might be, "A function's derivative at a point is the slope of the tangent line." We consider this a coping mechanism because it enables students to reason about derivatives in a limited range of contexts, but does not allow them to flexibly apply their conception of derivative functions to novel problem-solving situations, nor does it support a conceptual understanding of the relationship between rate of change and accumulation that is essential to understanding the Fundamental Theorem of Calculus (Thompson, 1994).

† Such zero-sum conceptions are not uncommon among those persuaded by the left- versus right-brain fallacies, for example.

proficiency as being developed at the expense of more desirable psychological competencies. Consequently, it is an instructor's responsibility to teach mathematics in a way that supports students' acknowledgement that valued conceptual tendencies such as creative, logical, flexible, and critical thinking are essential to mathematical proficiency, as well as being enhanced through its pursuit. Consistent with Blumer's (1986) interactionist conception of identity and von Glasersfeld's (1995) perspective on the construction of the self, for students to define learning goals in terms of cognitive (rather than observable) states, they must experience in their mathematical activity the utility of employing habits of mind that constitute the cognitive characteristics of their desired identities as learners of mathematics. An instructor can accomplish this by engaging students in accessible tasks for which they have not developed coping mechanisms, and which require mathematical reasoning and sense-making, though on a limited scale. Such tasks are often stated in terms of an applied context that enables students to engage in mathematical reasoning without having to interpret and manipulate symbols for which they do not have meaning (a point to which we later return). Once a student develops an orientation to establish meaningful learning goals, her ability to accomplish them is dependent upon her belief that the material and, more importantly, psychological resources at her disposal are sufficient for coping with task demands.

10.3.2 Enhance Students' Appraisal of Their Psychological Resources

The specific nature of students' learning goals depends on their perceived challenge of the present task. Since challenge is always appraised with reference to an individual's abilities, and since individuals pursue only those goals they feel capable of accomplishing, it is important that instructors attempt to influence students' perception of challenge by shaping students' image of their cognitive resources and, by extension, their anticipation of the affective experiences needed to sustain their mathematical activity when they encounter difficulties. Because challenge is a necessary condition for cognitive re-organization, the intention is not to eliminate students' perception of challenge but rather to foster their appraisal of task demands as manageable. A student's perception of challenge is optimal for her conceptual learning when two conditions are satisfied: (1) the student feels that the resources at her disposal (primarily psychological in the case of mathematics learning) are sufficient for accomplishing the task, and (2) successful completion of the task requires the student to leverage and even enhance the psychological qualities of her desired identity as a mathematics learner, including her reasoning and sense-making capacities and domain-specific mathematical conceptions (Middleton et al., 2015).

Students perceive mathematical tasks, and learning experiences generally, as exceptionally challenging when they assume that productive engagement in them requires knowledge of an expansive catalog of disconnected facts and procedures. The assumption that mathematical proficiency is based principally on one's ability to efficiently recall declarative knowledge increases students' uncertainty as to whether they can successfully participate in mathematics. If, on the contrary, a student recognizes that a small number of essential ways of reasoning are sufficient for engaging productively in a variety of tasks, and if the student has experienced her capacity to engage in these ways of reasoning in several mathematical contexts, then she is more likely to appraise task demands as manageable. Reducing the possibility that students will experience anxiety therefore requires an instructor to understand the role of foundational ways of thinking (e.g., *quantitative* and *covariational reasoning*)* on students' construction of particular mathematical ideas, and to be deliberate in promoting students' reflection on their successful application of these ways of thinking so that they might become explicit, thereby affording students a level of control over them (Tallman & Frank, 2020). This confidence in and control over mathematical *ways of thinking* (Harel, 2008) reduces a student's uncertainty about the outcome of their mathematical activity, thus reducing their anxiety.

The essential point is that students develop the perception of their psychological resources as sufficient for satisfying task demands through guided reflection on their learning experiences (Blumer, 1986; von Glasersfeld, 1995). Specifically, an instructor can support students in enhancing their appraisal of their psychological resources by providing repeated opportunities for them to reflect on how the cognitive components of their desired identities as mathematics learners contributed to their achievement of particular cognitive states, or learning goals. Doing this enables students to recognize that they possess the intellectual capabilities necessary to understand mathematical ideas in a meaningful way, and contributes to their development of a disposition to appraise the challenges they encounter as manageable. Instructors should therefore strive to provide students with repeated opportunities to experience in their mathematical activity the affordances of employing particular ways of reasoning in a variety of contexts.

10.3.3 Support Students' Construction of Quantitative Meanings for Mathematical Symbols, Expressions, and Equations

Since somatic sensations activated upon an anxious student's assimilation of mathematical stimuli enhance his or her level of physiological arousal, which then intensifies their anxiety (in a way consistent with our prior discussion of the *existing level of arousal* local intensity variable), it is important for instructors to engage students in experiences that enable them to construct meaning for mathematical objects so they feel confident acting with and performing operations on these objects on the basis of their meaning. Doing so decreases students' uncertainty about the implications of their mathematical engagement for the proximal goals they seek to achieve, and for their identities as learners and doers of mathematics generally. Moreover, if students can assimilate mathematical symbols to schemes of mental images, actions, and operations, then their cognition is less likely to be immobilized by the particular somatic sensations that accompany their assimilation of these symbols in the absence of such meaning.

We can state this third recommendation in terms of Piaget's distinction between *figurative* and *operative* aspects of thought: students' anxiety is reduced if they are equipped to assimilate mathematical stimuli to schemes for which the figurative aspects of cognition are subordinate to the operative aspects. The advantage of this framing is that it positions us to leverage features of Piaget's genetic epistemology—specifically reflecting and

---

* See Smith and Thompson (2007) and Thompson (1990, 2011) for characterizations of *quantitative reasoning*. See Saldanha and Thompson (1998), Carlson et al. (2002), and Thompson and Carlson (2017) for descriptions of *covariational reasoning*.

reflected abstraction—to clarify how one might support students' construction of quantitative meanings for mathematical representations with the goal of diminishing the reflexive somatic sensations that intensify their anxiety.

An affordance of online instruction is that it can support students' construction of operative structures by allowing them to engage with dynamic applets and interact with virtual manipulatives in ways that are not possible during face-to-face instruction. As we stress below, the effective design of these applets and virtual manipulatives would require: (1) an awareness of the mental process involved in conceptualizing mathematical ideas in particular ways, (2) an awareness of productive ways of reasoning (e.g., quantitative and covariational reasoning), (3) a general theory for the development of mathematical concepts, and (4) an explicit learning trajectory.

10.3.3.1 Figurative Versus Operative Modes of Thought

Piaget (1970) described the distinction between figurative and operative aspects of thought as follows:

> The figurative aspect is an imitation of states taken as momentary and static. In the cognitive area the figurative functions are, above all, perception, imitation, and mental imagery, which is in fact interiorized imitation. The operative aspect of thought deals not with states but transformations from one state to another. For instance, it includes actions themselves, which transform objects or states, and it also includes the intellectual operations, which are essentially systems of transformation.
>
> *(p. 14)*

Müller (2009) summarized the contrast between these two aspects of thought in terms of assimilation: "The operative aspect represents the structuring activity of assimilation, the figurative aspect provides the material for this activity" (p. 223). Being responsible for the organizing function of assimilation, the operative mode of thought is comprised of actions and operations* used to interact with and modify the world. These actions can be either overt or internal depending on the cognitive stage at which a child is operating: a child's operative thought takes the form of behavioral actions during the sensorimotor period and mental actions during the stage of concrete operations (Ginsburg & Opper, 1988, pp. 161–162). Although distinct, these two aspects of thought are complementary and interdependent. The operative mode of cognition influences how one experiences reality through the figurative functions of perception, imitation, and mental imagery. Conversely, the perceptual, imitative, and imagistic constituents of figurative thought provide objects of action and transformation by operative schemes.

Thompson (1985) generalized the distinction between the figurative and operative forms of cognition by emphasizing the control exerted by the latter over the former, and in doing

---

* Operations are reversible mental actions that can be applied to a generalized class of objects without regard to an initial state.

so broadened the definition of figurative aspects of thought beyond only those features of cognition based in perception, imitation, and mental imagery:

> When a person's actions of thought remain predominantly within schemata associated with a given level (of control), his or her thinking can be said to be figurative in relation to that level. When the actions of thought move to the level of controlling schemata, then the thinking can be said to be operative in relation to the level of the figurative schemata. That is to say, the relationship between figurative and operative thought is one of figure to ground. Any set of schemata can be characterized as figurative or operative, depending upon whether one is portraying it as background for its controlling schemata or as foreground for the schemata that it controls.
>
> *(p. 195)*

The control over figurative material by operative structures allows one to anticipate the outcome of applying a scheme and to make propitious decisions in relation to prior events and current circumstances (Thompson, 1985, p. 194; von Glasersfeld, 1995, p. 65). In our prior discussion of the *likelihood* local intensity variable, we explained that anxiety is maximized when an individual considers possible outcomes of an event with either positive or negative implications for the self as equiprobable. The expectations and guidance in decision-making enabled by operative thought lessens anxiety by reducing one's uncertainty about the outcome of one's cognitive engagement and, relatedly, decreasing the possibility that somatic markers will initiate unproductive behaviors in response to a student's assimilation of a mathematical stimulus. State anxiety occurs when an individual experiences limited control over the figurative aspects of thought, and trait anxiety develops when such experiences become habituated. Thompson's (1985) generalization of Piaget's notable distinction therefore demonstrates the essential role of operative thought in reducing students' MA by emphasizing the structure it imposes on figurative experience.

We previously stressed the importance of instructors enabling students to construct meaning for mathematical objects so they feel confident acting with and performing operations on these objects on the basis of their meaning. Piaget's (1970) description and Thompson's (1985) generalization of the operative form of cognition clarifies how constructing schemes that foreground the operative mode of thought is essential to acting with and performing operations on mathematical objects on the basis of their meaning for the actor. If the figurative aspects of a student's thought are uncontrolled by operative structures—that is, if a student's conception of a mathematical idea is dominated by her re-presentation of figurative material without this material being the object of action or transformation—then she is unlikely to be able to assimilate novel figurative material and experiences. This inability establishes the conditions for thought-inhibiting visceral reactions, and for attempting to leverage unproductive coping mechanisms, which as we previously emphasized increases students' uncertainty about the outcome of their mathematical activity, thus increasing their anxiety. If, on the contrary, a student has constructed

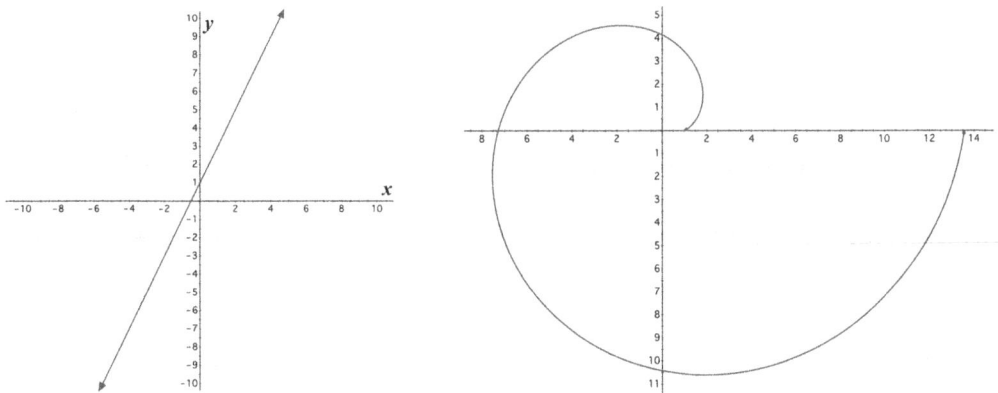

FIGURE 10.2   Linear functions represented in Cartesian and polar coordinates: a source of anxiety for students with a figurative conception of rate of change.

operative structures that control her action on and transformation of figurative material, then she is at least positioned to reason productively about unfamiliar mathematical problems and situations.

To exemplify the affective consequences of students' activity guided by figurative versus operative modes of thought, we consider how two students might respond to the task of approximating the constant rate at which one quantity varies with respect to another by examining two graphs, one in Cartesian coordinates and another in polar coordinates (see Figure 10.2).

The familiar idiom "rise-over-run" is suggestive of a meaning for rate of change based in and constrained to figurative material. Operationally, "rise" refers to the vertical change between two points on the Cartesian graph (often with integer ordinate values) and "run" refers to the corresponding horizontal change (where "change" is measured in units determined by the axes of the coordinate system). This conception is dominated by figurative thought because of its reliance on sensorimotor experience (i.e., moving up and over) and re-presentation of perceptual material (i.e., directed magnitudes that respectively represent "rise" and "run"). Moreover, this meaning is constrained to a particular representation, which is an essential feature of conceptions that foreground figurative aspects of cognition (Moore, Stevens, et al., 2019). A student who operates with this meaning will respond to the task of determining the constant rate at which y varies with respect to x by identifying two points where the graph intersects lattice points of the Cartesian coordinate plane and then computing the ratio of the vertical change (rise) and horizontal change (run) between these points. Although the student might recognize that the selection of points that determine these directed lengths is arbitrary, he does not conceptualize the resulting ratio as a *rate* since, for the student, it has no meaning in terms of the constrained covariational relationship between x and y.[*]

---

[*] A *ratio* is a multiplicative comparison of the measures of two constant (non-varying) quantities while a *rate* defines a proportional relationship between varying quantities' measures (Thompson & Thompson, 1992). Constructing a rate therefore involves images of *smooth continuous variation* (Thompson & Carlson, 2017), as well as the expectation that as two quantities covary, multiplicative comparisons of their measures remain invariant.

Rather than unhesitatingly trying to leverage this "rise-over-run" conception of rate of change to the function represented in polar coordinates, the student will likely notice perceptual novelties that he is not positioned to accommodate (e.g., the gridlines are circles centered at, and lines passing through, the origin; the graph fails the "vertical line test"). Being generally underequipped with operative mathematical structures, over time the student has constructed associations between novel mathematical stimuli and the negative feelings he experienced when expected to reason with or about them. These somatic associations or "markers" enable the student to react reflexively (and often unproductively) to mathematical objects that represent potential threats to the achievement of his learning goals or desired identity state. The immediate visceral reactions that occur in response to the student's recognition of novelties in the polar representation are subsequently refined into a nuanced and differentiated emotional experience. This occurs primarily through his reflection on the extent to which the accomplishment of his learning goals are dependent upon the results of his activity in this context, and whether he considers these results representative of his mathematical ability or intellectual capacity in general. This is how the *existing level of arousal* global intensity variable is operationalized: conscious cognitive processing of reflexive somatic sensations (i.e., somatic markers) influences one's appraisal of an object, agent, or event (experienced or prospective), which results in more intense emotional experiences.

For contrast, consider a student who conceptualizes constant rate of change as a proportional relationship between corresponding changes in the continuously varying measures of a function's input and output quantities. Rather than being constrained by particular perceptual material or sensorimotor experience, these aspects of figurative thought are controlled by operative structures of action and transformation. Although the student who possesses this operative conception might demonstrate similar behaviors as the student whose understanding for rate of change is limited to "rise-over-run," what distinguishes the result of her $\Delta y/\Delta x$ computation is that it represents the constant of proportionality relating the measures of covarying magnitudes Δy and Δx; it is not the direct application of a mnemonic device.

Importantly, the student who possesses such an operative conception will also recognize perceptual novelties of the polar graph, but will nonetheless expect that her meaning is sufficient for coping with them to sensibly determine the constant rate at which the output quantity varies with respect to the input quantity. As a result of this expectation, the task neither represents a threat to the student's ability to accomplish her proximal learning goals nor her identity as a mathematics learner. The student simply has no reason to expect that her engagement with the task will reveal untenable deficiencies in her mathematical knowledge or ability. Confident in the general applicability of her understanding, the student recognizes that she must first conceptualize corresponding changes in the input and output quantities and then multiplicatively compare these changes as they covary. This, of course, requires a robust understanding of the polar coordinate system (Moore, Paoletti, & Musgrave, 2014)—including what it means to measure an angle in radians—and a particular way of conceptualizing the covariation of the input and output quantities in this context (Thompson & Carlson, 2017). Equipped with such an understanding and way of

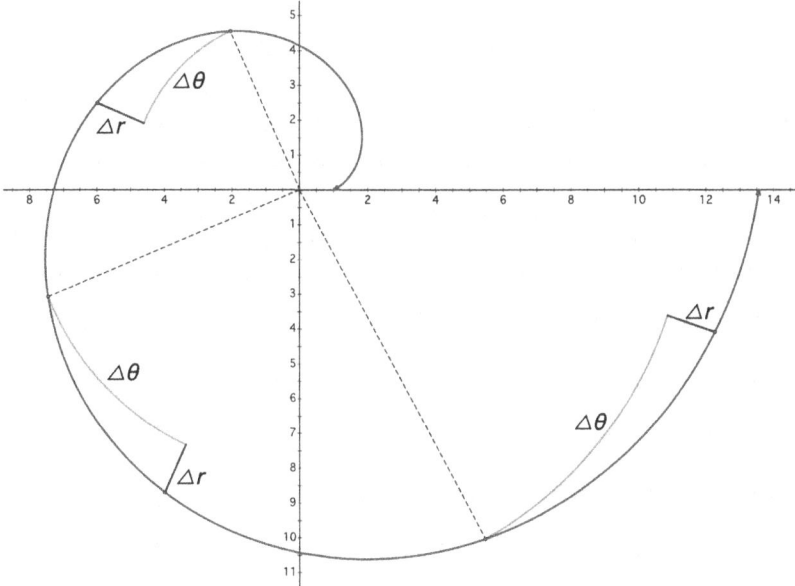

FIGURE 10.3   Constant rate of change in polar coordinates ($\Delta r=2\Delta\theta$). (Note that the length of the green arc representing $\Delta\theta$ is measured in units of the radius r.)

reasoning, the student can coordinate changes in an angle's measure (θ) with corresponding changes in radius (r) to determine the constant rate at which r varies with respect to θ (see Figure 10.3).

In sum, as a result of constructing and repeatedly leveraging mathematical schemes grounded in operative thought, students develop the *expectation* that their meanings are sufficient for effectively coping with novel tasks. This expectation has the potential to reduce or even eliminate the reflexive somatic sensations that "mark" mathematical stimuli as potential threats to one's identity, and which tend to initiate unproductive behavioral reactions (e.g., task avoidance, memorization, the unreasoned employment of coping mechanisms).

Provided the frequency with which mathematics curricula and instruction promote conceptions that are constrained by figurative aspects of thought, it is difficult to overstate the potential for operative structures to reduce students' anxiety. Examples include the "vertical line test" to determine whether a graph represents a functional relationship, conceptualizing the derivative of a function at a point as the slope* of a tangent line, and understanding definite integrals as representing bounded areas, not to mention the predominant focus in mathematics curricula on "the graph," which has the effect of constraining students' understanding of function classes to visual properties of their graphical representations and to the arbitrary conventions of particular representational systems (Moore, Silverman, et al., 2019).

---

* Indeed, "slantiness" might capture with greater accuracy the meaning for derivative at a point promoted in common calculus curricula.

The potential of operative schemes and the problematic nature of students' figurative conceptions are well-documented in the research literature. Moore and Thompson (2015) and Moore (in press), for example, introduce the theoretical construct *static shape thinking* to describe students' meanings for graphical representations grounded in perceptual features of a graph. A student who engages in static shape thinking makes associations between mathematical terms or inscriptions and visual properties of (or actions on) a graph as an object so that these perceptual associations *constitute* one's meaning for such terms and inscriptions (e.g., constant rate of change means "straightness"; exponential growth means "curving up"; quadratic means "U-shaped"; inverse functions mean "flip over the diagonal"). This is in contrast with *emergent shape thinking*, which involves conceiving figurative properties of a graph as having emerged from representing the simultaneous variation of quantities' measures in a coordinate system. A student who engages in emergent shape thinking interprets figurative aspects of a graph as representing properties of constrained covariation (Moore & Thompson, 2015, p. 786). Moore, Stevens, Paoletti, Hobson, and Lang (2019) document the problematic nature of students' static shape thinking and provide evidence of the affordances of emergent shape thinking for enabling students to accommodate novel figurative material and experiences. Moore et al.'s (2019) conclusions support those of other researchers (e.g., Carlson et al., 2002; Moore, Paoletti, & Musgrave, 2014; Moore & Thompson, 2015; Tallman, 2015) who in their different ways have demonstrated the generativity of operative structures or illustrated the limitations of conceptions that foreground figurative aspects of thought.

O'Bryan (2018) and O'Bryan and Carlson (2016) provide other notable examples of the benefits of operative structures and the deficiencies of figurative conceptions. O'Bryan (2018) introduced the theoretical construct *emergent symbol meaning* (a term that intentionally parallels emergent shape thinking) to characterize students' instrumental use of symbols to reason quantitatively. Specifically, a student who engages in *emergent symbolization* uses mathematical symbols representationally, and thus expects that actions on symbols have quantitative significance and that expressions and formulas reflect a *quantitative structure* (Thompson, 1990) that can be discerned through analysis of their composition. O'Bryan (2018) demonstrated that students who lack these expectations "struggle to produce accurate mathematical models and interpret the implications of the mathematical structures in the models they do generate" (p. 234). More encouragingly, O'Bryan and Carlson (2016) document a shift in the rigor and conceptual focus of the mathematical discourse facilitated by a middle school teacher after having engaged in an intervention to support her development of emergent symbol meaning.

Together the deficiencies of figurative conceptions of graphs and mathematical formulas/expressions respectively documented by Moore, Stevens, et al. (2019) and O'Bryan (2018) are relevant to a substantial portion of students' mathematical activity at the secondary and tertiary levels. Although these researchers did not explicitly analyze the emotional experiences of students whose mathematical activity was predominantly informed by figurative aspects of thought, the preceding theoretical discussion suggests that the uncertainty regarding the outcome of one's mathematical engagement and the reflexive somatic sensations that "mark" mathematical stimuli as potential threats to one's identity

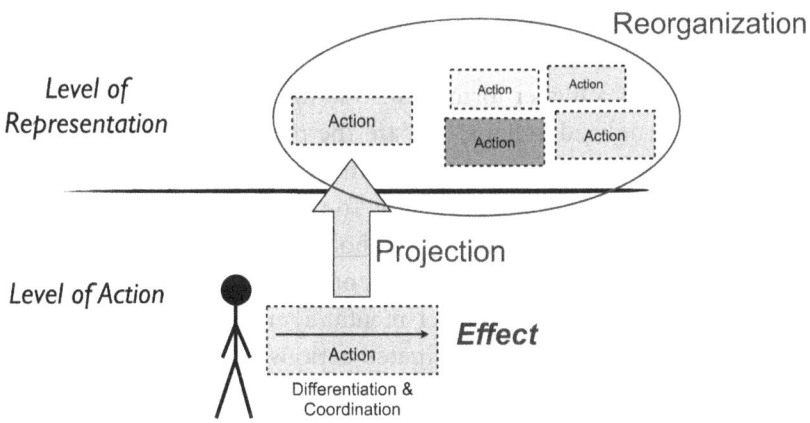

FIGURE 10.4   Reflecting abstraction.

can co-occur with or be products of students' frequent use of figurative meanings. This uncertainty and somatic marking increases the intensity of students' anxiety in a way consistent with our discussion of the *likelihood* and *existing level of arousal* intensity variables (Ortony, Clore, & Collins, 1988). The rationale for supporting students' development of operative structures is therefore not exclusively cognitive; constructing operative mathematical schemes also has the potential to reduce students' anxiety and to promote their positive affect generally. However, the issue of how an instructor or curriculum designer might accomplish this objective remains to be addressed. It is to this issue that we now turn.

10.3.3.2 Piagetian Abstraction as the Mechanism for the Construction of Operative Structures

Piaget proposed reflecting and reflected abstraction as the mechanism for constructing operative structures. As the first author has previously written (Tallman, 2015; Tallman & Frank, 2020), reflecting abstraction involves the subject's reconstruction on a higher cognitive level of the coordination of actions* from a lower level, and results in the development of *logico-mathematical knowledge,* or schemes at the level of operative thought (Chapman, 1988). Reflecting abstraction is thus an abstraction of actions and occurs in three phases: (1) the *differentiation* of a sequence of actions from the effect of employing them, (2) the *projection* of the differentiated action sequence from the level of activity to the level of representation, or the reflected level, and (3) the *reorganization* that occurs on the level of representation of the projected actions (see Figure 10.4) (Piaget, 2001). A subject must differentiate (dissociate) actions from their effects before she can construct an internalized representation of them, a process Piaget referred to as projecting actions to the level of representation (i.e., the level of cognition). Additionally, the subject must coordinate the actions that produced the effect before she can project and represent them on a higher cognitive level. Once a subject differentiates actions from their effect and then coordinates

---

* Piaget (1967) defined action broadly to encompass all movement, thought, or emotion that responds to a need (p. 6).

them, she is prepared to project these coordinated actions to the reflected level where they are organized into cognitive structures, or schemes.

Reflected abstraction is a higher form of abstraction that involves performing mental operations on the internalized actions that are the product of prior reflecting abstractions. Engaging in reflected abstraction results in a coherence of actions and operations accompanied by conscious awareness of them. To consciously operate on actions at the level of representation suggests that one has symbolized coordinated actions at this higher cognitive level. Reflected abstraction thus relies on what Piaget called the *semiotic function*, or the subject's capacity to construct mental symbols to represent aspects of her experience. The subject symbolizes coordinated actions at the level of representation so as to reify the material actions the symbol represents into a form she can use as an object of thought at the level of representation. On this higher cognitive level, the subject can consciously manipulate these symbols independently of re-presenting the coordinated actions they signify, all the while being capable of doing so. The semiotic function is thus the essential mechanism by which reflecting abstraction becomes reflected abstraction. As a result of the conscious awareness of internalized actions that occurs as a byproduct of reflected abstraction, the subject's ability to purposefully assimilate new experiences to the reflected level provides evidence that she has engaged in reflected abstraction. Additionally, performing conceptual operations on the symbols the subject constructs to represent coordinated actions at the level of representation results in increasingly organized cognitive structures. Reflected abstraction is therefore the means by which systems of organized actions at the level of representation become progressively coherent and refined.

Being the mechanisms responsible for the construction of operative structures, engaging students in experiences that engender reflecting and reflected abstractions is essential to reducing or preventing their MA.

10.4 CONCLUDING THOUGHTS

We have suggested that mathematics teachers can design curricular resources and manage online learning environments to minimize students' experience of mathematics anxiety by engaging students in instructional experiences that (1) assist them in establishing meaningful learning goals to pursue en route to achieving a desired identity state, (2) enable them to appraise their psychological resources as sufficient for accomplishing their learning goals, and (3) support their construction of operative schemes for mathematical symbols, expressions, and equations. By repeatedly engaging students in learning experiences that (1) require mathematical reasoning, (2) reveal their possession of the intellectual competencies essential to mathematical proficiency, and (3) provide an experiential basis for mathematical representations, students can begin to have confidence in their abilities and will perceive their mathematical experiences not as a threat to achieving their desired identities as learners and future professionals, but as an opportunity to strengthen valued cognitive capacities that constitute the psychological dimension of these identities. Additionally, such sustained learning experiences will ultimately make the anticipated outcome of students' mathematical engagement predictable, thus decreasing the uncertainty that they will experience negative outcomes from their mathematical engagement.

We hope to have demonstrated that an instructional commitment to fostering students' positive affect does not have to be in opposition to supporting their meaningful mathematical learning. Conversely, a dedication to constructivist pedagogy does not necessarily introduce greater potential for students to experience unproductive levels of anxiety. On the contrary, the instructional and curricular recommendations we have proposed to address the cognitive antecedents of mathematics anxiety can emphasize significant mathematical understandings while also fostering the type of affective engagement essential for their construction. We argue that the recommendations above can be more effectively applied in online learning environments that allow students to interact with virtual manipulatives in ways that are not possible in traditional classroom instruction. Even if a classroom instructor did provide an opportunity for students to interact with virtual manipulatives, time constraints might still interfere; online instruction circumvents several of these issues.

There are many normative practices in mathematics instruction and curriculum design that are antithetical to supporting students' affect, and which actively encourage the cognitive appraisals and constructions that contribute to students experiencing mathematics anxiety. These include but are not limited to a predominant focus on supporting conceptions grounded in figurative aspects of thought, the expectation that conceptual understanding depends on or emerges from procedural fluency (Kieran, 2013), a systemic inattention to the meaning and coherence of mathematical ideas (Thompson, 2013), assessment practices that emphasize performance rather than understanding (Niss, 1993), the relative absence of explicit pedagogical theories guiding mathematics instruction and curriculum development (Simon, 2013), a general failure of mathematics instruction to meaningfully build on viable models of students' mathematical thinking (Jacobs, Lamb, & Phillip, 2010), and the infrequency with which instructors engender in students an affective and intellectual need for learning what they intend to teach them (Harel, 2013).

The pervasiveness of some of these practices partially reflects a lack of opportunity for mathematics instructors to develop the knowledge base required to effectively implement the recommendations we propose. While this knowledge base is extensive, having a clear image of what it means to understand particular mathematical ideas and being aware of a trajectory by which students might construct them are at its foundation (Tallman, 2015). To this point, specifying what one wants students to *do* is different than articulating how one wants students to *understand*. Instruction that is guided by an image of behavioral outcomes instead of a coherent and explicit characterization of the conceptual activity an instructor seeks to engender rarely affords students the opportunity to construct productive mathematical meanings. It is therefore essential to enacting our recommendations that curriculum designers and instructors define learning goals for students in cognitive rather than behavioristic terms, which involves articulating the mental actions and conceptual operations that comprise the understandings the curricula or instruction seek to promote (Tallman, 2015). This entails characterizing what students need to imagine, interpret, visualize, attend to, and conceive to construct the meaning one envisions. Additionally, an instructor must possess a theory for the development of mathematical concepts in general, and an explicit learning trajectory for specific mathematical ideas in particular. Without them, instructors have no recourse but to adopt a "transmission" model of teaching, guided by the implicit assumption that complex mathematical ideas can be transferred to students through clear explanations and/or demonstrations (Simon, 2013). In essence, a particular character of *content* knowledge is essential for designing and facilitating online learning experiences that reduce students' mathematics anxiety; accomplishing this goal is not reducible to a list of pedagogical prescriptions without consideration for the mathematical conceptions they are supposed to support.

REFERENCES

Ames, C. (1992). Classrooms: Goals, structures, and student motivation. *Journal of Educational Psychology, 84*(3), 261–271.

Ames, C. & Archer J. (1988). Achievement goals in the classroom: Students' learning strategies and motivation processes. *Journal of Educational Psychology, 80*(3), 260–267.

Ashcraft, M. H. (2002). Math anxiety: Personal, educational, and cognitive consequences. *Current Directions in Psychological Science, 11*(5), 181–185.

Ashcraft, M. H. & Kirk, E. P. (2001). The relationships among working memory, math anxiety, and performance. *Journal of Experimental Psychology: General, 130*(2), 224–237.

Batchelor, S., Gilmore, C. K. & Inglis, M. (2017). Parents' and children's mathematics anxiety. In U. X. Eligio (Ed.), *Understanding emotions in mathematical thinking and learning* (pp. 315–336). London, England: Academic Press.

Betz, N. E. (1978). Prevalence, distribution, and correlates of math anxiety in college students. *Journal of Counseling Psychology, 25*(5), 441–448.

Blumer, H. (1986). *Symbolic interactionism: Perspective and method.* Berkeley, CA: University of California Press.

Bong, M. (2009). Age-related differences in achievement goal differentiation. *Journal of Educational Psychology, 101*(4), 879–896.

Carlson, M., Jacobs, S., Coe, E., Larsen, S. & Hsu, E. T. (2002). Applying covariational reasoning while modeling dynamic events: A framework and a study. *Journal for Research in Mathematics Education, 33*(5), 352–378.

Cattell, R. B. (1950). *Personality: A systematic theoretical and factual study.* New York, NY: McGraw Hill.

Chapman, M. (1988). *Constructive evolution: Origins anddevelopment of Piaget's thought.* Cambridge University Press.

Clore, G. L. & Ortony, A. (2008). Appraisal theories: How cognition shapes affect into emotion. In M. Lewis, J. M. Haviland-Jones & L. Feldman Barrett (Eds.), *Handbook of emotions* (3rd ed., pp. 628–642). New York, NY: The Guildford Press.

Cobb, P., Gresalfi, M. & Hodge, L. (2009). An interpretative scheme for analyzing the identities that students develop in mathematics classrooms. *Journal for Research in Mathematics Education, 40*(1), 40–68.

Cunningham, W. A. & Zelazo, P. D. (2007). Attitudes and evaluations: A social cognitive neuroscience perspective. *Trends in cognitive sciences, 11*(3), 97–104.

Damasio, A. (1994). *Descartes' error: Emotion, rationality, and the human brain.* New York, NY: Putnam (Grosset Books).

Drake, K. L. & Kearny, C. A. (2008). Child anxiety sensitivity and family environment as mediators of the relationship between parent psychopathology, parent anxiety sensitivity, and child anxiety. *Journal of Psychopathology and Behavioral Assessment, 30*(2), 79–86.

Dweck, C. S. (1986). Motivational processes affecting learning. *American Psychologist, 41*(10), 1040–1048.

Dweck, C. S. & Leggett, E. L. (1988). A social-cognitive approach to motivation and personality. *Psychological Review, 95*(2), 256–273.

Eden, C., Heine, A. & Jacobs, A. M. (2013). Mathematics anxiety and its development in the course of formal schooling—A review. *Psychology, 4*(6), 27–35.

Epstein, S. (1972). The nature of anxiety with emphasis upon its relationship to expectancy. In Spielberger (Ed.), *Anxiety: Current trends in theory and research* (pp. 291–337). New York, NY: Academic Press.

Faust, M. W., Ashcraft, M. H. & Fleck, D. E. (1996). Mathematics anxiety effects in simple and complex addition. *Mathematical Cognition, 2*(1), 25–62.

Fennema, E. & Sherman, J. (1977). Sex-related differences in mathematics achievement, spatial visualization and affective factors. *American Educational Research Journal, 14*(1), 51–71.

Ginsburg, H. & Opper, S. (1988). *Piaget's theory of intellectual development*. Englewood Cliffs, NJ: Prentice-Hall.

Hackett, G. (1985). Role of mathematics self-efficacy in the choice of math-related majors of college women and men: A path analysis. *Journal of Counseling Psychology, 32*(1), 47–56.

Harel, G. (2008). DNR perspective on mathematics curriculum and instruction part I: Focus on proving. *Zentralblatt für Didaktik der Mathematik, 40*, 487–500.

Harel, G. (2013). Intellectual need. In K. Leatham (Ed.), *Vital directions for mathematics education research* (pp. 119–152). New York, NY: Springer.

Hembree, R. (1990). The nature, effects, and relief of mathematics anxiety. *Journal for Research in Mathematics Education, 21*(1), 33–46.

Ho, H. Z., Senturk, D., Lam, A. G., Zimmer, J. M., Hong, S., Okamoto, Y., Chiu, S. Y., Nakazawa, Y. & Wang, C. P. (2000). The affective and cognitive dimensions of math anxiety: A cross-national study. *Journal for Research in Mathematics Education, 31*(3), 362–379.

Hopko, D. R., McNeil, D. W., Lejuez, C. W., Ashcraft, M. H., Eifert, G. H. & Riel, J. (2003). The effects of anxious responding on mental arithmetic and lexical decision task performance. *Journal of Anxiety Disorders, 17*(6), 647–665.

Jacobs, V. R., Lamb, L. L. & Philipp, R. A. (2010). Professional noticing of children's mathematical thinking. *Journal for Research in Mathematics Education, 41*(2), 169–202.

Jonckheere, A., Mandelbrot, B. & Piaget, J. (1958). *La lecture de l'experience*. Paris, France: Presses univ. de France.

Kieran, C. (2013). The false dichotomy in mathematics education between conceptual understanding and procedural skills: An example from algebra. In K. Leatham (Ed.), *Vital directions for mathematics education research* (pp. 153–171). New York, NY: Springer.

Krohne, H. W. (1992). Developmental conditions of anxiety and coping: A two-process model of child-rearing effects. In K. A. Hagtvet & B. T. Johnsen (Eds.), *Advances in test anxiety research* (Vol. 7, pp. 143–155). Lisse, Netherlands: Swets and Zeitlinger.

Ma, X. & Xu, J. (2004). The causal ordering of mathematics anxiety and mathematics achievement: A longitudinal panel analysis. *Journal of Adolescence, 27*(2), 165–179.

Maloney, E. A., Ansari, D. & Fugelsang, J. A. (2011). Rapid communication: The effect of mathematics anxiety on the processing of numerical magnitude. *The Quarterly Journal of Experimental Psychology, 64*(1), 10–16.

Maloney, E. A., Risko, E. F., Ansari, D. & Fugelsang, J. (2010). Mathematics anxiety affects counting but not subitizing during visual enumeration. *Cognition, 114*(2), 293–297.

Mcleod, B. D., Wood, J. J. & Weisz, J. R. (2007). Examining the association between parenting and childhood anxiety: A meta-analysis. *Clinical Psychology Review, 27*(2), 155–172.

Meece, J. L., Wigfield, A. & Eccles, J. S. (1990). Predictors of math anxiety and its influence on young adolescents' course enrollment intentions and performance in mathematics. *Journal of Educational Psychology, 82*(1), 60–70.

Middleton, J. A., Jansen, A. & Goldin, G. A. (2017). The complexities of mathematical engagement: Motivation, affect, and social interactions. In J. Cai (Ed.), *Compendium for Research in Mathematics Education* (pp. 667–699). Reston, VA: National Council of Teachers of Mathematics.

Middleton, J. A., Tallman, M., Davis, O. & Hatfield, N. (2015). Taking the *severe* out of perseverance: Strategies for building mathematical determination. In N. Alpert & C. Kurose (Eds.), *Mathematical instruction for perseverance* (pp. 1–25). Chicago, IL: Spencer Foundation.

Miller, H. & Bichsel, J. (2004). Anxiety, working memory, gender, and math performance. *Personality and Individual Differences, 37*(3), 591–606.

Montangero, J. & Maurice-Naville, D. (1997). *Piaget or the advance of knowledge* (A. Curnu-Wells, Trans.). Mahwah, NJ: Lawrence Erlbaum.

Moore, K. C. (in press). Concept construction, transfer, and graphical shape thinking. In C. Honensee & J. Lobato (Eds.), *Transfer of learning: Progressive perspectives for mathematics education and related fields.* New York, NY: Springer.

Moore, K. C., Paoletti, T. & Musgrave, S. (2014). Complexities in students' construction of the polar coordinate system. *The Journal of Mathematical Behavior, 36,* 135–149.

Moore, K. C., Silverman, J., Paoletti, T., Liss, D. & Musgrave, S. (2019). Conventions, habits, and US teachers' meanings for graphs. *The Journal of Mathematical Behavior, 53,* 179–195.

Moore, K. C., Stevens, I. E., Paoletti, T., Hobson, N. L. & Liang, B. (2019). Pre-service teachers' figurative and operative graphing actions. *The Journal of Mathematical Behavior, 56,* 100692. https://doi.org/10.1016/j.jmathb.2019.01.008.

Moore, K. C. & Thompson, P. W. (2015). Shape thinking and students' graphing activity. In T. Fukawa-Connelly, N. Infante, K. Keene & M. Zandieh (Eds.), *Proceedings of the 18th annual conference on research in undergraduate mathematics education* (pp. 782–789). Pittsburgh, PA.

Morsanyi, K., Busdraghi, C. & Primi, C. (2014). Mathematical anxiety is linked to reduced cognitive reflection: A potential road from discomfort in the mathematics classroom to susceptibility to biases. *Behavioral and Brain Functions, 10*(1), 31.

Müller, U. (2009). Infancy. In U. Müller, J. Carpendale & L. Smith (Eds.), *The Cambridge companion to piaget* (Cambridge Companions to Philosophy, pp. 200–228). Cambridge: Cambridge University Press.

Niss, M. (Ed.) (1993). *Investigations into assessment in mathematics education: An ICME study.* Dordecht, Netherlands: Kluwer Academic.

O'Bryan, A. E. (2018). *Exponential growth, learning trajectories, and online learning environments: Designing for and studying the development of student meanings in online courses.* Unpublished Ph.D. dissertation, School of Mathematical and Statistical Sciences, Arizona State University.

O'Bryan, A. E. & Carlson, M. P. (2016). Fostering teacher change through increased noticing: Creating authentic opportunities for teachers to reflect on student thinking. In T. Fukawa-Connelly, N. Engelke Infante, M. Wawro & S. Brown (Eds.), *Proceedings of the 19th annual conference on research in undergraduate mathematics education* (pp. 1192–1200). Pittsburgh, PA: West Virginia University.

Öhman, A. (2008). Fear and anxiety: Overlaps and dissociations. In M. Lewis, J. M. Haviland-Jones & L. Feldman Barrett (Eds.), *Handbook of emotions* (3rd ed., pp. 709–729). New York, NY: The Guildford Press.

Ortony, A., Clore, G. L. & Collins, A. (1988). *The cognitive structure of emotions.* Cambridge, England: Cambridge University Press.

Piaget, J. (1937). *The construction of reality in the child.* New York, NY: Basic Books.

Piaget, J. (1967). *Six psychological studies.* New York, NY: Random House.

Piaget, J. (1970). *Genetic epistemology.* New York, NY: W.W. Norton.

Piaget, J. (2001). *Studies in reflecting abstraction.* New York, NY: Psychology Press.

Piaget, J. & Inhelder, B. (1969). *The psychology of the child.* New York, NY: Basic Books.

Richardson, F. C. & Suinn, R. M. (1972). The mathematics anxiety rating scale: Psychometric data. *Journal of Counseling Psychology, 19*(6), 551–554.

Rubinsten, O. & Tannock, R. (2010). Mathematics anxiety in children with developmental dyscalculia. *Behavioral and Brain Functions*, 6(1), 46.

Saldanha, L. & Thompson, P. W. (1998). Re-thinking covariation from a quantitative perspective: Simultaneous continuous variation. In S. Berenson & K. Dawkins (Eds.)., *Proceedings of the annual meeting of the international group for the psychology of mathematics education—North America* (Vol. 1, pp. 298–304). Raleigh, NC: North Carolina State University.

Sfard, A. & Prusak, A. (2005). Telling identities: In search of an analytic tool for investigating learning as a culturally shaped activity. *Educational Researcher, 34*(4), 14–22.

Simon, M. A. (2013). The need for theories of conceptual learning and teaching of mathematics. In K. Leatham (Ed.), *Vital directions for mathematics education research* (pp. 95–118). New York, NY: Springer.

Smith, J. & Thompson, P. W. (2007). Quantitative reasoning and the development of algebraic reasoning. In J. J. Kaput, D. W. Carraher & M. L. Blanton (Eds.), *Algebra in the early grades* (pp. 95–132). New York, NY: Erlbaum.

Spielberger, C. D. (1972). *Anxiety: Current trends in theory and research*. New York, NY: Academic Press.

Tallman, M. (2015). *An examination of the effect of a secondary teacher's image of instructional constraints on his enacted subject matter knowledge*. Unpublished Ph.D. dissertation, School of Mathematical and Statistical Sciences, Arizona State University.

Tallman, M. & Frank, K. M. (2020). Angle measure, quantitative reasoning, and instructional coherence: An examination of the role of mathematical ways of thinking as a component of teachers' knowledge base. *Journal of Mathematics Teacher Education, 23*(1), 69-95.

Thompson, P. W. (1985). Experience, problem solving, and learning mathematics: Considerations in developing mathematics curricula. In E. A. Silver (Ed.), *Learning and teaching mathematical problem solving: Multiple research perspectives* (pp. 189–236). Hillsdale, NJ: Erlbaum.

Thompson, P. W. (1990). *A theoretical model of quantity-based reasoning in arithmetic and algebra*. Center for Research in Mathematics & Science Education: San Diego State University.

Thompson, P. W. (1991). To experience is to conceptualize: Discussions of epistemology and experience. In L. P. Steffe (Ed.) *Epistemological foundations of mathematical experience* (pp. 260-281). New York: Springer-Verlag.

Thompson, P. W. (1994). Images of rate and operational understanding of the Fundamental Theorem of Calculus. *Educational Studies in Mathematics*, 26(2–3), 229–274.

Thompson, P. W. (2011). Quantitative reasoning and mathematical modeling. In L. L. Hatfield, S. Chamberlain & S. Belbase (Eds.), *New perspectives and directions for collaborative research in mathematics education* (pp. 33–57). Laramie, WY: University of Wyoming.

Thompson, P. W. (2013). In the absence of meaning. In K. Leatham (Ed.), *Vital directions for mathematics education research* (pp. 57–93). New York, NY: Springer.

Thompson, P. W. & Carlson, M. P. (2017). Variation, covariation, and functions: Foundational ways of thinking mathematically. In J. Cai (Ed.), *Compendium for research in mathematics education* (pp. 421–456). Reston, VA: National Council of Teachers of Mathematics.

Thompson, P. W., Carlson, M. P., Byerley, C. & Hatfield, N. (2014). Schemes for thinking with magnitudes: An hypothesis about foundational reasoning abilities in algebra. In K. C. Moore, L. P. Steffe & L. L. Hatfield (Eds.), *Epistemic algebra students: Emerging models of students' algebraic knowing*, WISDOMe Monographs (pp. 1–24). Laramie, WY: University of Wyoming.

Thompson, P. W. & Thompson, A. G. (1992). *Images of rate*. Paper presented at the Annual Meeting of the American Educational Research Association, San Francisco, CA.

Tobias, S. (1993). *Overcoming math anxiety*. New York, NY: W. W. Norton & Company.

Trezise, K. & Reeve, R. A. (2017). The impact of anxiety and working memory on algebraic reasoning. In U. X. Eligio (Ed.), *Understanding emotions in mathematical thinking and learning* (pp. 133–158). London, England: Academic Press.

von Glasserfeld, E. (1995). *Radical constructivism: A way of knowing and learning.* New York, NY: RoutledgeFalmer.

Wigfield, A. & Meece, J. L. (1988). Math anxiety in elementary and secondary school students. *Journal of Educational Psychology, 80*(2), 210–216.

Zeidner, M. (2014). Anxiety in education. In R. Pekrun & L. Linnenbrink-Garcia (Eds.), *International handbook of emotions in education* (pp. 265–288). New York, NY: Routledge.

Zeidner, M. & Matthews, G. (2011). *Anxiety 101.* New York, NY: Springer.

CHAPTER 11

A Face-to-Face Program of Support for Students in a Hybrid Online Developmental Mathematics Course

Edgar J. Fuller and Jessica Deshler*

CONTENTS

* August 26, 2019.

11.1 INTRODUCTION

Historically, mathematics courses have been viewed as obstacles to student success and to access to science, technology, engineering, and mathematics (STEM) careers. To overcome these obstacles institutions of higher education have provided a variety of ways to meet and support the needs of ever larger, more diverse groups of students (Watkins & Mazur, 2013). STEM-intending students who begin their mathematics study at levels substantially lower than calculus encounter multiple of these potential barriers as they progress through a sequence leading to the courses needed for their majors. As students move through courses, different aspects of their experiences, life events, and internal affective states can affect their studies. Anxiety has been shown by a number of researchers (Ashcraft & Krause, 2007; Hembree, 1990) to play a role in the efficiency and effectiveness of mathematics students. For students entering developmental mathematics courses, the impact of these underlying affective states can be compounded both positively and negatively by the multiple, sequential courses needed to reach the level of mathematics needed in STEM disciplines. As this population grows at major universities the need for better understanding of the interaction of larger, more diverse populations with technology-driven and other support systems as they move through multiple mathematics courses has grown.

At the same time, researchers have shown that peer-mentoring creates a supportive environment that allows students to engage in learning while connecting with role models to whom they can identify more readily (Fox & Stevenson, 2006; Morales, Ambrose-Roman, & Perez-Maldonado, 2016; Rios-Ellis et al., 2015; Dennehy & Dasgupta, 2017). We describe in this paper a face-to-face peer-mentoring program that was implemented to support students in self-paced, computer-based developmental mathematics course at a large university in the Eastern United States and its effect on their levels of mathematics anxiety and their overall success. We specifically wanted to answer the following research questions:

- *How does participating in a face-to-face peer-mentoring program help support student success in an online developmental mathematics course?*

- *How does participating in a face-to-face peer-mentoring program affect the mathematics anxiety of students in an online developmental mathematics course?*

11.2 LITERATURE REVIEW

As reported by Radford, Pearson, Ho, Chambers, and Ferlazzo (2012), approximately 42% of students in the United States enter college needing a mathematics course below the level of college algebra. These students need additional support to gain the mathematical background needed to begin the expected progression of mathematics courses for their chosen majors. At the same time, more than one-third of all students pursuing a STEM degree in the US enroll in mathematics courses such as intermediate algebra (Radford et al., 2012), and, in most cases, these students will need to progress to at least a first semester calculus course. (Intermediate algebra in the US typically covers material that students have been exposed to in secondary schools and is a level below college algebra). This creates a number of chances for students to depart from their STEM program and is one of the reasons that the number of students graduating with a STEM degree remains relatively stagnant. Overall diminishing student retention rates (Hurtado, Eagan, & Chang, 2010; Thompson & Bolin, 2011) were noted in the United States' President's Council of Advisors on Science and Technology (PCAST) report (Olson & Riordan, 2012) that highlighted a need for more than a million additional STEM professionals in the US over the next decade, and so this population of STEM-intending, developmental mathematics students at institutions from across the US is a resource from which we can retain potential members of this future workforce.

11.2.1 Students and Computer-Supported Instruction

In addition, the availability of computer-supported instructional platforms and the changing landscape of higher education has allowed universities to seek more efficient ways of instruction including the use of distance learning and other technologically supported classroom models. Studies have noted consistently increasing offerings of online developmental mathematics courses (Epper & Baker, 2009; Jaggars & Bailey, 2010) as well as the use of computer-assisted instruction to provide low-cost access to large numbers of students at both the two- and four-year college levels (Lei, 2010). These courses pose a specific challenge to higher education in that they are a significant point of access for many students (Zavarella & Ignash, 2009), but the students entering STEM pathways tend to need strong supports if they are to be successful. Several researchers (Bray & Tangney, 2017; Meletiou-Mavrotheris et al., 2017; Senn, 2008) have noted the capacity for change that technology presents in these hybrid/online settings. From online homework systems and computer-based testing (Lei, 2010; Senn, 2008) to the use of online platforms to build learning communities (Swenson & Evans, 2003), researchers and practitioners have utilized different tools to increase access to courses as well as to improve the instructional environment (Young, 2017). These tools especially tend to focus on online assessments ranging from homework to quizzes and exams and may potentially put students in situations where they are combining a high-stakes learning activity with the cognitive load of the content as well as the additional barriers presented by the technology being used (Young, 2017). In particular, serving large numbers of students who need high levels of instructor contact with models that lower the contact level in order to make it more cost-effective can create structures that inhibit student success (Ashby, Sadera, & McNary, 2011; Boylan, Bliss, & Bonham, 1997; Lei, 2010).

One aspect of this potential for inhibition can be traced to the ways in which students respond to the instructional environment (Berland & McNeill, 2010; Kalyuga, 2007). Student attitudes towards coursework (Alexander & Martray, 1989; Catsambis, 1994; Hofer & Pintrich, 1997) and the beliefs they hold regarding their ability to succeed (Bandura, 1997; Hall & Ponton, 2005) can create internal conflict when their preconceived notions do not agree with the feedback they receive from a course. Many students find themselves overwhelmed by the demands of a mathematics classroom, and in courses where students have been placed that contain content that they have already encountered in high school, for example, the added coursework presents multiple opportunities for departure when they feel disillusioned about their academic progress. As noted by Hembree (1990), student interaction with mathematics courses tends to induce an anxious state and this anxiety (Ashcraft & Krause, 2007) in turn tends to inhibit both the ability to reason mathematically as well as the development of mathematical concepts over time. As a result, it is critical that students in early mathematics courses encounter supportive environments so that they can build both the mathematical knowledge needed to be successful and the internal sense of mathematical self-efficacy needed to persist in their courses. This is especially true in courses that incorporate technological supports such as computer-based testing or homework systems (Boylan et al., 1997; Lei, 2010). In these course environments, students encounter both the difficulty of the mathematical material and the potentially negative impacts of the technology involved (Taylor, 2008). Computer-based homework and testing environments present additional failure points to at-risk students by combining the cognitive load of the material being learned with the potential difficulties of the system used to present the material and/or exams.

Researchers have identified other factors impacting anxiety as well. Stoet, Bailey, Moore, and Geary (2016) found that anxiety in mathematics manifests in female students more strongly. In other work, Willig, Harnisch, Hill, and Maehr (1983) found that anxiety levels were predictive of mathematics outcomes for Hispanic students but not all groups. Studies (Jaggars & Bailey, 2010) have also suggested that online courses in particular have complex interactions with first-generation and other underprepared students, and that these students require more substantial supports in these environments to be successful. These complicated interactions of content, course structure, and assessment indicate that efforts to monitor and dispel anxiety may lead not only to increased success but also to more inclusive classrooms.

11.2.2 Peer-Mentoring Supports

As one response to this complexity, studies have shown that peer-mentoring can be an effective way to support students in mathematics. In areas such as chemistry (Wamser, 2006) and physics (Crouch & Mazur, 2001; Watkins & Mazur, 2013), peer instruction has been shown to have a positive impact on student work in mathematical areas and to support higher levels of retention (Weissman et al., 2011). Peer-mentoring has also been specifically shown to reduce anxiety (Rodger & Tremblay, 2003). Peer-mentors provide many of the needed study and time management skills, both aspects of potential struggle (Culler & Holahan, 1980; McKeachie, 1984) for students in hybrid situations where they may lack a high level of instructor interaction.

To develop a program of support, a review of the peer-mentoring literature (Anderson & Boud, 1996; Crisp & Cruz, 2009; Jacobi, 1991; Topping, 1996) was conducted and identified four categories of support for college students: psychological or emotional support, role models, assistance in goal setting and career paths, and subject-specific expertise. Cramer and Prentice-Dunn (2007) further argue that psycho-social support influences identity formation and belonging, but contend that a mentoring program should span multiple areas since these psychological processes work in the context of many other influences. Based on the work in these other settings we implemented a program that intended to

1. Provide a support system for students struggling with the content in developmental mathematics.

2. Increase developmental mathematics students' feelings of campus connection.

3. Help developmental mathematics students navigate curriculum and locate university resources.

4. Increase developmental mathematics students' confidence, involvement in learning, and retention.

5. Cultivate relationships between students who have successfully completed developmental and subsequent mathematics courses and current students in the course.

6. Develop current developmental mathematics students into potential future mentors.

7. Help developmental mathematics students address adjustment issues and improve decision making.

These goals target a number of areas that research indicates are potential failure points for students at a variety of levels.

11.2.3 Theoretical Framework

In order to analyze the ways in which students' attitudes and beliefs in a course environment (Alexander & Martray, 1989; Catsambis, 1994; Hofer & Pintrich, 1997) and sense of self-efficacy in mathematics (Bandura, 1997; Hall & Ponton, 2005) impact their success in a course, we view failure as a departure from their academic plan and adopt Tinto's model of retention and departure (Tinto, 1990). In this framework student persistence is said to depend on both academic integration and social integration. For academic integration, support structures such as online homework systems and tutors provide access to content (Sims & Schuman, 1999). On the other hand, students in courses that are hybrid or online may feel isolated (Murphy & Stewart, 2017), and as a result the social integration aspect also weighs heavily on their retention likelihood. We hypothesize that peer-mentoring in this context bridges the gap between academic and social integration by providing content support at the same time that connections are being made with peers with whom students can strongly identify socially (Rice et al., 2013).

11.3 INSTITUTIONAL SETTING

Our study was conducted at a university established as a land-grant institution—an institution supported through the use of federal land provided to the states to fund education that would include agriculture and mechanics in addition to traditional liberal arts studies. This university is also a comprehensive research university but maintains the original goals of a land-grant institution in that it seeks to facilitate the education of the citizens of the state at all levels. As such, it attracts a significant number of first-generation students—students who are the first in their families to attend college and therefore come from environments that likely lack the understanding of the supports a college student might need. In addition, a large proportion of these students have declared a major within a STEM discipline as indicated by the National Science Foundation categorization of programs and so are considered STEM-intending.

11.3.1 A Course before College Algebra

Although many students at the university will begin their mathematics study in calculus or precalculus, roughly 30% of first-year students admitted to the institution are not prepared for college-level mathematics courses (Fuller, Deshler, Kuhn, & Squire, 2014) at the level of college algebra, defined at this university to be a beginning study of functions and their algebraic and graphical structure. As a result more than 1,000 students per year enroll in a mathematics course offered at the university whose content typically precedes college algebra and would more typically occur in their secondary mathematics courses. Approximately 70% of these students who succeed move on to additional mathematics courses, and we estimate that 48% these successful students enter successor mathematics courses while pursuing STEM degrees.

The course that is the focus of this study is taught as an intermediate algebra course covering foundational topics such as integer and fractional arithmetic, the manipulation of expressions involving unknowns including polynomial, rational, exponential, and logarithmic equations, solving equations, and factoring. The course is offered in a self-paced, hybrid format course based on the completion of online modules while on-campus in a classroom with an instructor who does not lecture and instead offers individualized explanations to students as needed and assists with homework, practice, and exam problems. Students are assigned homework quizzes (HWQs) for seven chapters' worth of material in an intermediate algebra textbook implemented using an online homework system. Students must complete the HWQs with scores of at least 80% that then lead to chapter exams that students must complete with at least 70% mastery. Students are allowed three attempts on an exam before they must redo the HWQs. Once a student successfully completes a chapter exam, he or she may move on to the next chapter of material. Upon completion of all chapter exams, students must pass a comprehensive final exam with at least 70% mastery.

In order to complete material in the course, students use the Pearson My Math Lab online system (Pearson, 2019) to complete homework exercises, HWQs, and exams. The modules are supported by an online text with background content, and students can access that material electronically during class, during homework exercises, and between attempts on HWQs but not during exams. Questions are generally posed as open answer, and students may ask for help during class and HWQs either by clicking a button in the browser or by asking a lab attendant for help. Each student works at a separate computer in a 48-seat or 120-seat laboratory. They may work together at any time other than during exams. They meet four days per week in this lab for one hour, but it is stressed that they must complete material outside the lab as well as during class time.

11.3.2 Peer-Mentoring Program Description

During the fall semester of 2016 we implemented a pilot peer-mentoring program and developed material to support this program using a small group of students and mentors. We selected peer-mentors from past intermediate algebra students who successfully completed the course with a grade of A and subsequently succeeded in a college-level mathematics course (non-remedial) with a grade of A or B. We solicited volunteer mentees for this stage of the program and conducted meetings throughout the term while developing mentor training material, mentor/mentee activities, and establishing assessment protocols. During the pilot we had eight students serve as peer-mentors and only 24 students request to participate in the peer-mentoring program.

In the spring semester of 2017, we implemented the peer-mentoring program by setting up parallel course sections taught during the same time periods each day. We chose two time periods to examine and provided two classes during each time period. After students had enrolled, we chose one section randomly at each time period to be a control section with no intervention and one in which to implement the peer-mentoring. Two instructors were chosen to lead the two parallel sets of classes. That is, each instructor had one mentored and one non-mentored section of the course. We referred to the two mentored sections as "treatment" and each of the 87 students in those sections was required to participate in peer-mentoring activities for the duration of the term. There were a total of 77 students in the two control sections of the course.

Students were assigned to meet with a peer-mentor in groups of eight and were required to attend meetings once per week during the semester. Students were considered to have completed the peer-mentoring if they missed no more than four meetings for the entire 15- week semester. During this phase of the study we had seven of the original eight mentors return as mentors.

11.4 METHODS

In total there were 450 students enrolled in the course during the peer-mentoring study including those in the treatment, control, and other sections of the course. The population consisted of 199 female and 251 male students, and it contained 108 STEM-intending students. The population also contained 136 first-generation students.

In order to measure the impact of the peer-mentoring process we gathered data related to student success in the course along with data collected from survey instruments. We used the Abbreviated-Mathematics Anxiety Rating Scale (AMARS) (Alexander & Martray, 1989), a 25-item shortened version of the Mathematics Anxiety Rating Scale (MARS) instrument of Richardson and Suinn (1972) to measure the anxiety levels of our students. This instrument has been validated and measures anxiety levels in the three areas of exam anxiety, course anxiety, and numerical task anxiety using a collection of Likert scale prompts such as *"Please indicate your level of anxiety when studying for a math test."* Student responses ranged from 1 = Not at all, to 5 = Very much. Three factors have been identified in the survey including exam anxiety (EA—items 1 to 15), numerical task anxiety (NTA—items 16 to 20), and course anxiety (CA—items 21 to 25). A composite score total for each factor was computed with EA ranging from 15 to 75, NTA and CA from 5 to

25. We surveyed students at the beginning of the semester and again at the end to examine changes in anxiety levels. Participation in the peer-mentoring program was required for students, but participation in the research study was voluntary and response rates are reported in the results section. We also collected data on student success in the course to examine the effects of the peer-mentoring program.

To identify difference in outcomes, we compare success data from the treatment and control groups, first using a χ^2 test and then a comparison of mean GPA. For comparison of subgroups such as gender, first-generation, or STEM-intending groups, an analysis of variance of grade point average on a 0 to 4 scale was performed between groups. Finally, to analyze the impact of anxiety, a repeated measures analysis of variance (ANOVA) was performed using data from the pre-survey and post-survey.

11.5 RESULTS

Of the 450 students enrolled in the course, 82 responded (18.2%) to the initial AMARS survey with an average score of 69.29 on a scale of 25 to 125. This group maintained similar proportions with respect to gender (54 female, 28 male), STEM-intending ($N = 25$), and first-generation ($N = 35$) students. We administered a follow-up AMARS survey to measure anxiety levels at the end of the semester, and 43 (9.6%) responded with an average of 65.74 on the same scale. In total, 38 responded to both pre- and post-surveys, and the average decrease was 2.08 points.

11.5.1 Student Success

We found that students in the peer-mentoring program were more successful in the course and demonstrated lower levels of mathematics anxiety at the end of the course than students who did not participate in the program. From the treatment group, 54 of the 87 students completed the peer-mentoring (by missing four or fewer meetings) and, as shown in Table 11.1, 83.3% of these succeeded in the course (defined as a grade of A, B, or C). While the result was not statistically significant ($t(129) = -1.4222$, $p = 0.1574$), it is an indication that students supported in this way tend to succeed at a higher rate. The lack of statistical significance is likely due to the low population of students in the treatment groups relative to the effect size measured. Specifically, we find an effect size (Cohen's d) of $d = 0.2443$ for the difference in the mean success rate of the treatment and control populations. With this effect size, a power analysis indicates that a sample size of $N = 172$ would be necessary to establish a significant difference in mean likelihood of success for an independent sample t-test ($\alpha = 0.05$).

TABLE 11.1 Success Rates for Students Completing the Peer-Mentoring Program

	Success		Fail		Total
	N	%	N	%	N
All non-mentored sections	216	75.2	70	24.8	286
Paired instructor control group	56	74.0	21	26.0	77
Mentoring treatment sections	45	83.3	9	16.7	54

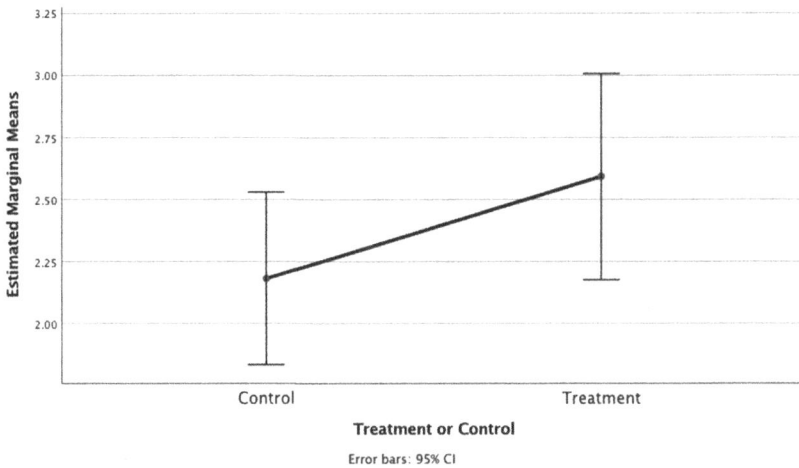

FIGURE 11.1   Means and estimated confidence intervals for grade point average of mentored and non-mentored students.

The performance of both mentored and non-mentored students can be better analyzed in general by comparing the mean GPA of the populations of the treated and control groups using the typical 4.0 scale. Using this scale, the marginal means of the populations can be computed and 95% confidence intervals imputed that provide some insight into the expected variation in this performance. These data are shown in Figure 11.1. In this plot we can see that with the current population size, the ability to predict with confidence the actual mean is constrained, and so while we see a difference in the success rates, we cannot rule out the possibility that this increase is due to random chance. It is, however, a large difference and suggests that a larger implementation of the program would likely provide more conclusive results.

11.5.1.1 Gender Subgroups
As summarized in Table 11.2 and depicted graphically in Figure 11.2, student success results differed by gender for this study. Success rates for male students were similar to the overall trend observed for the entire population in that the treatment group exhibited a higher rate of success. However, female students do not appear to have benefited in the same way as their male counterparts from the program despite the same success rates in the course.

Again, the differences were not observed to be significant ($t(60) = -0.51142$, $p = 0.6109$), but a larger sample size may be able to clarify these effects and their significance. The effect size within the female population is smaller, however ($d = 0.12419$), and a power analysis in this case indicates that a population of at least $N = 693$ would be needed to establish a significant difference in the likelihood of success.

11.5.1.2 STEM-Intending Subgroups
One of the primary concerns of this study was to observe the impact of the course and the peer-mentoring program on students planning to complete STEM majors. The success

TABLE 11.2   Success Rates of Students Completing Peer-Mentoring by Gender

		Success		Fail	
		N	%	N	%
F	Other	94	77.0%	28	23.0%
	Control	25	78.1%	7	21.9%
	Treatment	25	83.3%	5	16.7%
M	Other	122	74.4%	42	25.6%
	Control	31	68.9%	14	31.1%
	Treatment	20	83.3%	4	16.7%

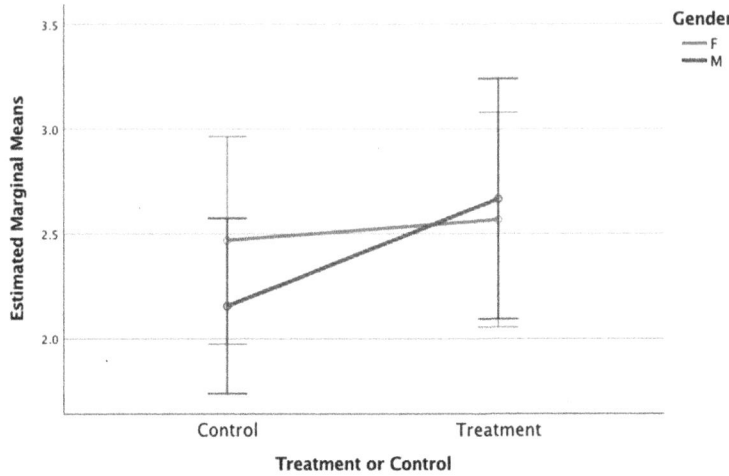

FIGURE 11.2   Means and estimated confidence intervals for GPA of mentored and non-mentored students by gender.

TABLE 11.3   Success Rates of Students Completing Peer-Mentoring by STEM Major Choice

		Success		Fail	
		N	%	N	%
Non-STEM major	Other	165	76.0%	52	24.0%
	Control	42	71.2%	17	28.8%
	Treatment	31	81.6%	7	18.4%
STEM major	Other	51	73.9%	18	26.1%
	Control	14	77.8%	4	22.2%
	Treatment	14	87.5%	2	12.5%

rates of STEM-intending students overall were similar to the non-STEM-intending population (Table 11.3), while the success rate for STEM-intending students in the peer-mentored group was higher. They succeeded at a rate that was 9.7% higher than the control population and 13.6% higher than the remaining population of STEM-intending students. In Figure 11.3 we see that the peer-mentoring program had a more positive effect on STEM majors when compared to the non-STEM majors.

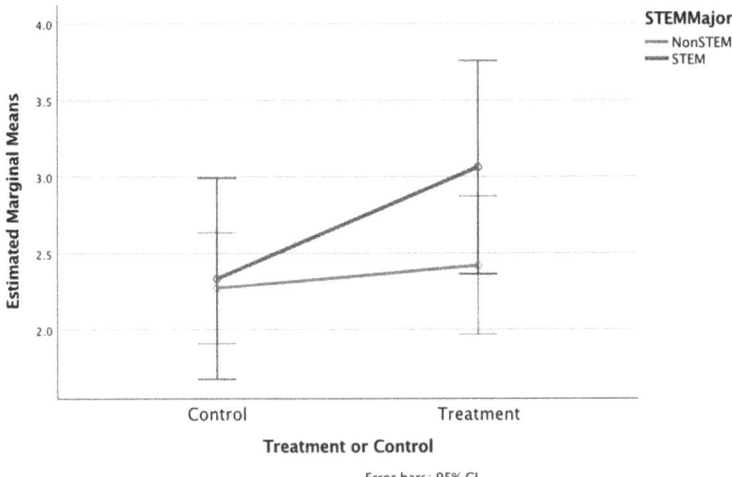

FIGURE 11.3 Means and estimated confidence intervals for GPA of mentored and non- mentored students by STEM major choice.

TABLE 11.4 Success Rates of Students Completing Peer-Mentoring by First-Generation Status

		Success		Fail	
		N	%	N	%
Not first-generation	Other	153	79.3%	40	20.7%
	Control	43	74.1%	15	25.9%
	Treatment	31	79.5%	8	20.5%
First-generation	Other	63	67.7%	30	32.3%
	Control	13	68.4%	6	31.6%
	Treatment	14	93.3%	1	6.7%

The difference was not found to be significant as noted above ($t(32) = -0.72599$, $p = 0.4731$), but the effect size for STEM-intending students is similar to the general population at $d = 0.2294$. To distinguish effects in this case, a sample size of $N = 197$ is indicated.

11.5.1.3 First-Generation Students

In addition to considering the STEM intention of our students, we analyzed outcomes for this population. As shown in Table 11.4 and in Figure 11.4, we see a substantial increase in success compared to other first-generation students not in the peer-mentoring program with more than 90% of these students successful if mentored compared to less than 70% if not. Due to the small number of peer-mentored students, none of these results were found to be statistically significant ($t(32) = -1.8177$, $p = 0.07849$). The effect size in this case was medium, however, at $d = 0.54473$, and so this result would be easily confirmed or invalidated with a larger population.

11.5.2 Measured Anxiety

Student anxiety was measured by the AMARS instrument in the second week and again in the last week of the semester. In Table 11.5 we see the average total anxiety for students

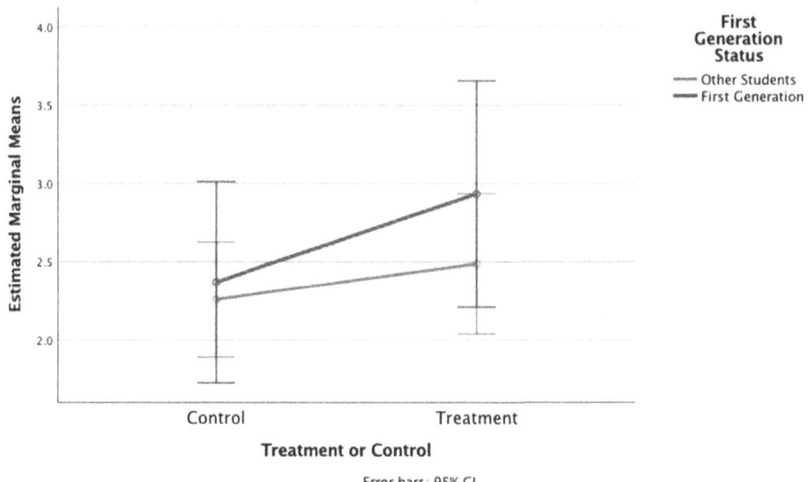

FIGURE 11.4   Means and estimated confidence intervals for GPA of mentored and non- mentored students by first-generation status.

TABLE 11.5   Anxiety Levels of Responding Students

	Total Anxiety (Pre)	Total Anxiety (Post)
Other sections	66.25 (N = 53)	63.87 (N = 30)
Control	78 (N = 12)	76 (N = 4)
Peer-mentoring completed	72.58 (N = 12)	66.57 (N = 7)

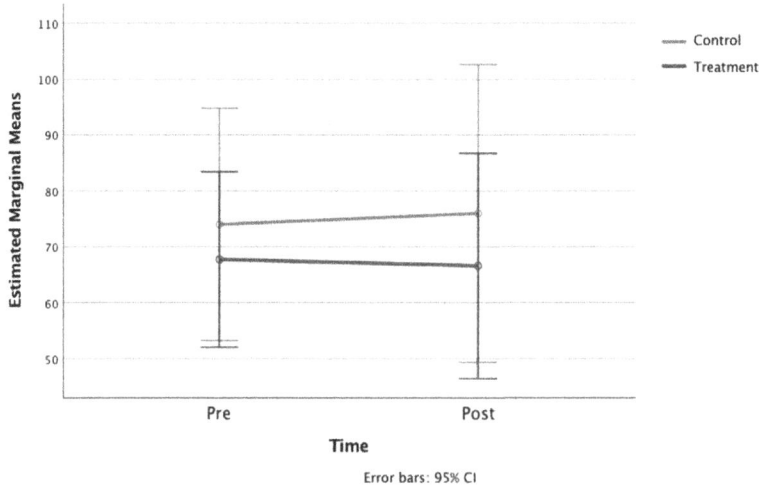

FIGURE 11.5   Measured anxiety in treatment and control groups.

by group (control, treatment, or other). The observed differences are again too small for the size of the sampled population to be able to conclude significance of the outcome overall. The overall trend for all students was a slight drop in anxiety, and this was paralleled within the treatment and control groups. Figure 11.5 shows the estimated variance in mean

TABLE 11.6   Pre- and Post-Survey Anxiety Means by Population Groups

	Pre-survey		Post-survey	
	Mean	N	Mean	N
Non-STEM	70.74	57	64.34	29
STEM	66.00	25	79.00	9
F	70.89	54	64.41	27
M	66.21	28	76.18	11
Non-first-generation	68.68	47	65.56	25
First-generation	70.11	35	72.15	13

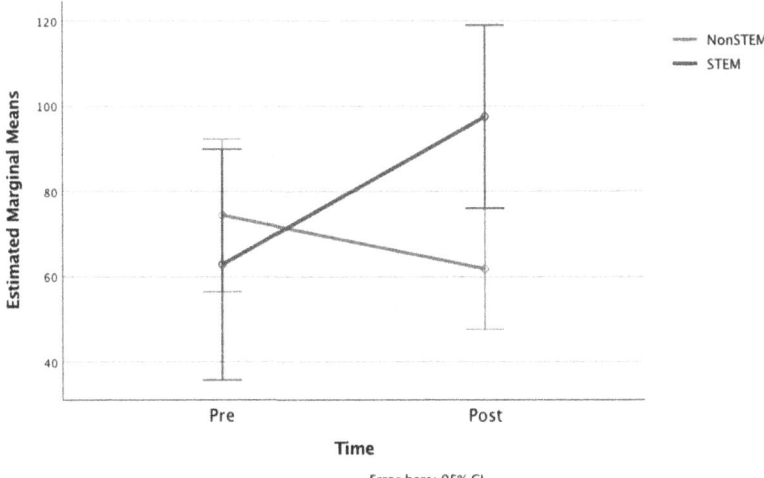

FIGURE 11.6   Measured anxiety controlling for population differences: STEM majors.

anxiety levels for the treatment and control groups for the two time periods. In this plot we see that there is a small decrease in the anxiety levels of the treatment group but that this decrease cannot be considered different from a number of other possible outcomes.

As shown in Table 11.6, however, the anxiety response across population subgroups had interesting patterns. Students in these sub-populations within the control and treatment groups exhibited differences in the change in anxiety response that were significant. A repeated measures ANOVA shows the difference of means to be significant from the pre- to post-survey ($F (1, 7) = 26.339$, $p < 0.001$) for STEM-intending students while the between-subjects variance compared to non-STEM-intending students was not. The comparison of the predicted variance in the means for the STEM- and non-STEM-intending student populations shown in Figure 11.6 illustrates the likelihood that peer-mentoring appears to actually increase levels of anxiety among STEM students while decreasing anxiety in non-STEM-intending participants.

On the other hand, a similar analysis for first-generation students (Figure 11.7) finds a decrease in anxiety during the course that is significantly different from what would be expected from random chance (repeated measures ANOVA $F (1, 7) = 30.640$, $p < 0.001$) over time. Non-first-generation students demonstrated an increase.

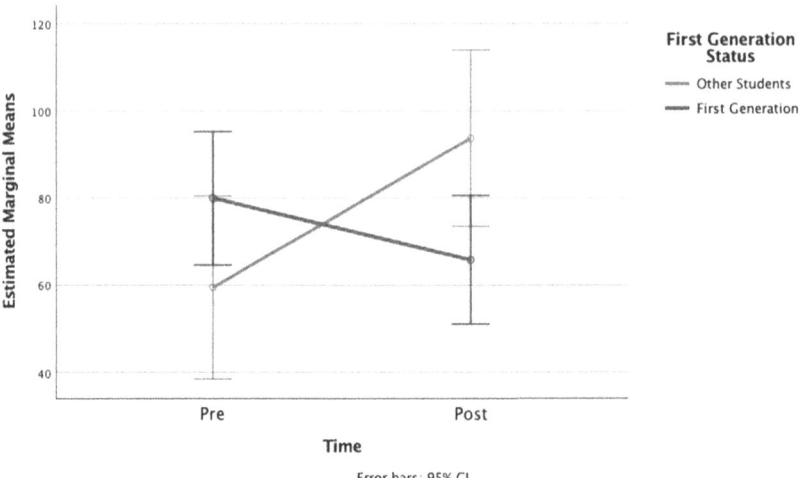

FIGURE 11.7   Measured anxiety controlling for population differences: first-generation students.

Unfortunately, the gender demographics of the post-survey respondent population within both the treatment and control sections differed too much from the pre-survey group, and the results there were encouraging in their support for female students but not enough data points were available for both genders for analysis.

11.6 CONCLUSIONS

In this chapter we report the outcomes from a pilot project that implemented a peer- mentoring support system for students in a hybrid, online developmental mathematics course using computer-based homework and testing systems in the instruction of a large number of students. In this work, we implemented this peer-mentoring program with 87 students from a total population of 450 entering the course in the spring of 2017. From this group, 54 students completed the mentoring activities, and we present outcomes from this group and compare them with a group of 77 students in a comparison group that controlled for time of day and instructor effects. Students were not assigned at random, but the sections were chosen randomly.

We observed increases in student success for the mentored population but could not determine if these differences were related to the peer-mentoring program specifically. These increases were supported for female, STEM-intending, and first-generation students as well, but no statistical significance could be identified between groups in these cases either.

We observed decreases in anxiety as measured by the AMARS instrument for the treated group that were not significant, but underlying effects in the STEM-intending and first-generation groups that were shown to be different than what would be expected from random effects. In particular, peer-mentoring as we have structured it decreased anxiety among first-generation students in a significant way, but STEM-intending students show increased levels of anxiety. It may be that the interaction with mentors made STEM intending students more cognizant of their situation which manifested as reported anxiety, while this same interaction with first-generation student decreased anxiety levels by resolving concerns these students were developing. Encouragingly, the mentoring supported female students as well, although, again, not enough data were obtained for analysis. Tinto (Tinto, 1990) notes that the two aspects of integration, academic and social, must be present for optimal retention of student populations, and anxiety in mathematics is an inhibitor for both mathematical cognition (Hembree, 1990) and social interaction (Degnan & Fox, 2007). A possible conclusion here is that some portion of the increase in success for students, in particular first-generation and students who perceive themselves as underprepared or challenged in some way, is due to a decrease in their level of anxiety in the course from peer-mentoring. This effect may be small enough that much larger numbers of mentored students would need to be studied in order to clarify the nature of any impacts, but the result is encouraging.

In all this, one immediate observation is that the response rates for both the pre-survey (18.2%) and the post-survey (9.6%) were relatively low, and so some of these conclusions need further work in a larger implementation of the program to see if they can be repeated or better understood. It appears that anxiety is lowered for target sub-populations but possibly not overall, and so our peer-mentoring process may need adjustment to better serve students more broadly. Most encouraging is the increase in success, though it is not yet clear from this work if the impact on anxiety is related to the increase in success. Given that the size of the effect of the treatment on rates of success for most populations is in the range of $d = 0.2$, further study with a population roughly twice the size of this study, around $N = 200$, would shed more light on some of these findings.

11.7 RECOMMENDATIONS FOR PRACTITIONERS

Student interaction with technology continues to evolve as technology improves and more advanced tools for instruction are developed. The ubiquity of online and computer-assisted instructional platforms has broadened the scope of student interaction with these systems.

The peer-mentoring program we have described in this chapter is a useful approach for supporting students in situations where student interaction with online or computer-based instruction will be high. Given the trends we observed during the implementation of our program, we also suggest monitoring student anxiety levels. Student anxiety related to their interaction with online systems can be quite high and can increase during the semester as they may become frustrated with systems that seem agnostic about their success. Support should be put in place that considers student progress as well as their attitudes towards the instructional systems to which they are exposed. As indicated by Tinto (Tinto, 1990), social and academic integration are necessary for student persistence, and this integrative process is driven by student attitudes towards their classroom environment. Peer-mentoring can provide a sense of community to students, especially first-generation students, who are at risk of departing their academic studies prematurely.

Additionally, developing some means of engaging with students such as discussion boards or chat groups that allow instructors to measure student affect either formally or informally can enhance the sense of belonging felt by students, and the information developed can be used to move students towards more positive engagement.

REFERENCES

Alexander, L., & Martray, C. R. (1989). The development of an abbreviated version of the mathematics anxiety rating scale. *Measurement and Evaluation in Counseling and Development. 22*(3), 143-150.

Anderson, G., & Boud, D. (1996). Extending the role of peer learning in university courses. *Research and Development in Higher Education, 19*, 15–19.

Ashby, J., Sadera, W. A., & McNary, S. W. (2011). Comparing student success between developmental math courses offered online, blended, and face-to-face. *Journal of Interactive Online Learning, 10*(3).

Ashcraft, M. H., & Krause, J. A. (2007). Working memory, math performance, and math anxiety. *Psychonomic Bulletin & Review, 14*(2), 243–248.

Bandura, A. (1997). *Self-Efficacy: The Exercise of Control.* New York, NY: Freeman.

Berland, L. K., & McNeill, K. L. (2010). A learning progression for scientific argumentation: Understanding student work and designing supportive instructional contexts. *Science Education, 94*(5), 765–793.

Boylan, H. R., Bliss, L. B., & Bonham, B. S. (1997). Program components and their relationship to student performance. *Journal of Developmental Education, 20*, 2–9.

Bray, A., & Tangney, B. (2017). Technology usage in mathematics education research–a systematic review of recent trends. *Computers & Education, 114*, 255–273.

Catsambis, S. (1994). The path to math: Gender and racial-ethnic differences in math- ematics participation from middle school to high school. *Sociology of Education, 67*(3), 199–215.

Cramer, R. J., & Prentice-Dunn, S. (2007). Caring for the whole person: Guidelines for advancing undergraduate mentorship. *College Student Journal, 41*(4), 771–778.

Crisp, G., & Cruz, I. (2009). Mentoring college students: A critical review of the literature between 1990 and 2007. *Research in Higher Education, 50*, 525–545.

Crouch, C. H., & Mazur, E. (2001). Peer instruction: Ten years of experience and results. *American Journal of Physics, 69*(9), 970–977.

Culler, R. E., & Holahan, C. J. (1980). Test anxiety and academic performance: The effects of study-related behaviors. *Journal of Educational Psychology, 72*(1), 16.

Degnan, K. A., & Fox, N. A. (2007). Behavioral inhibition and anxiety disorders: Multiple levels of a resilience process. *Development and Psychopathology, 19*(3), 729–746.

Dennehy, T. C., & Dasgupta, N. (2017). Female peer mentors early in college increase women's positive academic experiences and retention in engineering. *Proceedings of the National Academy of Sciences, 114*(23), 5964–5969.

Epper, R. M., & Baker, E. D. (2009). Technology solutions for developmental math: An overview of current and emerging practices. *Journal of Developmental Education, 26*(2), 4–23.

Fox, A., & Stevenson, L. (2006). Exploring the effectiveness of peer mentoring of accounting and finance students in higher education. *Accounting Education: An International Journal, 15*(2), 189–202.

Fuller, E. J., Deshler, J. M., Kuhn, B., & Squire, D. (2014). Tracking success of pre- college algebra workshop students through subsequent college mathematics classes. *PRIMUS: Problems, Resources, Issues in Mathematics Undergraduate Studies, 24*(1), 46–60.

Hall, J. M., & Ponton, M. K. (2005). Mathematics self-efficacy of college freshman. *Journal of Developmental Education, 28*(3), 26.

Hembree, R. (1990). The nature, effects, and relief of mathematics anxiety. *Journal for Research in Mathematics Education, 21*(1), 33–46.

Hofer, B. K., & Pintrich, P. R. (1997). The development of epistemological theories: Beliefs about knowledge and knowing and their relation to learning. *Review of Educational Research, 67*(1), 88–140.

Hurtado, S., Eagan, K., & Chang, M. (2010). *Degrees of Success: Bachelors Degree Comple- Tion Rates Among Initial Stem Majors.* Higher Education Research Institute at UCLA Research brief.

Jacobi, M. (1991). Mentoring and undergraduate academic success: A literature review. *Review of Educational Research, 61*(4), 503–532.

Jaggars, S. S., & Bailey, T. (2010). *Effectiveness of Fully Online Courses for College Students: Response to a Department of Education Meta-Analysis.* Community College Research Center, Columbia University .

Kalyuga, S. (2007). Enhancing instructional efficiency of interactive e-learning environ- ments: A cognitive load perspective. *Educational Psychology Review, 19*(3), 387–399.

Lei, J. (2010). Quantity versus quality: A new approach to examine the relationship between tech- nology use and student outcomes. *British Journal of Educational Technology, 41*(3), 455–472.

McKeachie, W. (1984). Does anxiety disrupt information processing or does poor infor- mation processing lead to anxiety? *Applied Psychology, 33*(2), 187–203.

Meletiou-Mavrotheris, M., Mavrou, K., Vaz-Rebelo, P., Santos, S., Tenhonen, P., Riska, M., … Pilt, L. (2017). Technology adoption in higher education: A cross-national study of university faculty perceptions, attitudes, and practices. In P. Tripathi, & S. Mukerji (Eds.), *Handbook of Research on Technology-Centric Strategies for Higher Education Administration* (pp. 295–317). IGI Global.

Morales, E. E., Ambrose-Roman, S., & Perez-Maldonado, R. (2016). Transmitting success: Comprehensive peer mentoring for at-risk students in developmental math. *Innovative Higher Education, 41*(2), 121–135.

Murphy, C. A., & Stewart, J. C. (2017). On-campus students taking online courses: Factors associ- ated with unsuccessful course completion. *The Internet and Higher Education, 34*, 1–9.

Olson, S., & Riordan, D. G. (2012). *Engage to Excel: Producing one Million Additional College Graduates with Degrees in Science, Technology, Engineering, and Mathematics* (Tech. Rep.). Washington, DC: Executive Office of the President. (Report to the President).

Pearson. (2019). *Mymathlab Homepage.* Retrieved 2019-06-01, from https://www.pearsonmylaba ndmastering.com/northamerica/mymathlab/.

Radford, A., Pearson, J., Ho, P., Chambers, E., & Ferlazzo, D. (2012). Remedial coursework in post- secondary education: The students, their outcomes, and strate- gies for improvement. berke- ley, ca: Mpr associates. *Inc. Retrieved from ERIC database.(ED537852).*

Rice, L., Barth, J. M., Guadagno, R. E., Smith, G. P., McCallum, D. M., et al. (2013). The role of social support in students perceived abilities and attitudes toward math and science. *Journal of Youth and Adolescence, 42*(7), 1028–1040.

Richardson, F. C., & Suinn, R. M. (1972). The mathematics anxiety rating scale: Psycho- metric data. *Journal of Counseling Psychology, 19*(6), 551.

Rios-Ellis, B., Rascon, M., Galvez, G., Inzunza-Franco, G., Bellamy, L., & Torres, A. (2015). Creating a model of latino peer education: Weaving cultural capital into the fabric of academic services in an urban university setting. *Education and Urban Society, 47*(1), 33–55.

Rodger, S., & Tremblay, P. F. (2003). The effects of a peer mentoring program on academic success among first year university students. *Canadian Journal of Higher Education, 33*(3), 1–17.

Senn, G. J. (2008). Comparison of face-to-face and hybrid delivery of a course that requires technol- ogy skills development. *Journal of Information Technology Education: Research, 7*, 267–283.

Sims, R. L., & Schuman, A. H. (1999). Learning in an online format versus an in-class format: An experimental study. *THE Journal, 26*(11), 54–56.

Stoet, G., Bailey, D. H., Moore, A. M., & Geary, D. C. (2016). Countries with higher levels of gen- der equality show larger national sex differences in mathematics anxiety and relatively lower parental mathematics valuation for girls. *PLOS ONE, 11*(4), 1–24. doi:10.1371/journal. pone.0153857.

Swenson, P. W., & Evans, M. (2003). Hybrid courses as learning communities. In S. Reisman, L. G. Flores, & D. Edge (Eds.) *Electronic Learning Communities Issues and Practices, 27*–72.

Taylor, J. M. (2008). The effects of a computerized-algebra program on mathematics achievement of college and university freshmen enrolled in a developmental mathe- matics course. *Journal of College Reading and Learning, 39*(1), 35–53.

Thompson, R., & Bolin, G. (2011). Indicators of success in stem majors: A cohort study. *Journal of College Admission, 212*, 18–24.

Tinto, V. (1990). Principles of effective retention. *Journal of the First-Year Experience & Students in Transition, 2*(1), 35–48.

Topping, K. J. (1996). The effectiveness of peer tutoring in further and higher education: A typology and review of the literature. *Higher Education, 32*, 321–345.

Wamser, C. C. (2006). Peer-led team learning in organic chemistry: Effects on student performance, success, and persistence in the course. *Journal of Chemical Education, 83*, 1562.

Watkins, J., & Mazur, E. (2013). Retaining students in science, technology, engineering, and mathematics (stem) majors. *Journal of College Science Teaching, 42*(5), 36–41.

Weissman, E., Butcher, K. F., Schneider, E., Teres, J., Collado, H., Greenberg, D., & Welbeck, R. (2011). *Learning Communities for Students in Developmental Math: Im-Pact Studies at Queensborough and Houston Community Colleges*. New York, NY: National Center for Postsecondary Research.

Willig, A. C., Harnisch, D. L., Hill, K. T., & Maehr, M. L. (1983). So- ciocultural and educational correlates of success-failure attributions and eval- uation anxiety in the school setting for black, hispanic, and anglo chil- dren. *American Educational Research Journal, 20*(3), 385–410. doi:10.3102/00028312020003385.

Young, J. (2017). Technology-enhanced mathematics instruction: A second-order meta-analysis of 30 years of research. *Ed-Ucational Research Review, 22*, 19–33. doi:10.1016/j.edurev.2017.07.001.

Zavarella, C. A., & Ignash, J. M. (2009). Instructional delivery in developmental mathe- matics: Impact on retention. *Journal of Developmental Education, 32*(3), 2.

CHAPTER 12

A Practical Guide to Discussions in Online Mathematics Courses

Glenn F. Miller and Kathleen H. Offenholley

CONTENTS

12.1 INTRODUCTION

Online learning can be seen as the historical continuation of distance education that began with the correspondence courses of the 1800s, followed by the radio "school of the air" in the 1920s and 1930s, and the telecourses in the latter half of the 20th century (Kentor, 2015). This history can cloud how people view online learning—as if it is, at its heart, a correspondence course, with its most important component being written lecture notes, perhaps with some videos thrown in. However, to view online learning in this manner is to ignore one of its most important and powerful features—the online discussion board.

In this chapter, we will make the particular case for using discussion boards in online mathematics courses. We will outline our research on the best way to give students the opportunity to engage in online discussions, using the strands of mathematical thinking (Kilpatrick, Swafford & Findell, 2001) as a lens through which to view these discussions.

We will also provide some practical advice for managing and assessing student participation in this component of an online mathematics course.

12.2 THE VALUE OF DISCUSSIONS IN AN ONLINE COURSE

Online discussion boards are a way for students to communicate with both faculty and other students. Communication is typically asynchronous—one student may post to the discussion at 3 in the afternoon, while another student posts at 1 am. This ability to communicate at any time keeps online learning flexible, ideal for working students and parents.

While no experimental research has been conducted that isolates the effect of online discussion boards per se, there is strong evidence that the type of communication that discussion boards facilitate has a direct effect on student learning in online mathematics courses. Jaggars and Xu (2016) studied four aspects of online course design (organization and presentation, objectives and alignment, interpersonal interaction, and technology). Of these, the only factor that significantly affected student outcomes in the course was interpersonal interaction, comprising both student–instructor and student-to-student interaction. They found "that students view an effective online instructor as one who … encourages students to interact with their classmates, their instructor, and the course material" (p. 281). The discussion board is one area of the online course where this interaction can occur. The authors rated the quality of the interaction in each course in their study, where quality was defined as interaction that reinforces course content and objectives. In those online courses rated more highly for the quality of interaction, students' course grades were significantly higher than in courses with less high-quality interaction. In fact, the authors found that either a high level of student-to-student or a high level of faculty-to-student interaction will suffice to improve student learning.

Bliss and Lawrence (2009) also discussed the relative importance of faculty participation in discussion boards. Their study of the use of discussion boards in online college mathematics courses found that there seems to be an optimal level of instructor interaction in discussion boards in such courses: too many faculty posts discourage student-to-student interaction, but too few faculty posts leave students feeling that their problems are not being sufficiently addressed. The main focus of their inquiry was on evaluating the quality of student posts. Using the quality measure developed, they determined that "the measure of quality of posts was correlated with presence of feedback." Offenholley (2012) found similarly that instructors influenced the types of discussions in which their students engaged: when instructor posts were *evaluative* (by rating, clarifying, and expanding what was said), there was a positive correlation with the number of times students posted, and when the instructor's evaluations contained more mathematics, so did the student's.

12.3 MATHEMATICIANS COMMUNICATE AND COLLABORATE

Given the benefits of online discussions, it is alarming to hear people ask, "Do you really need a discussion board in a math class?" This is often followed by, "What do you even talk about?" as if mathematics were a world apart, in which numbers and variables could exist without words to connect or explicate them. It is perhaps forgivable when non-mathematicians express such ideas, but it is most concerning when our fellow mathematics faculty say the same things, because the true work of the mathematician is so often a collaborative, communicative process. The most famous of all collaborative mathematicians is, of course, Paul Erdos, whose prolific work with other mathematicians gave rise to the "Erdos number"—the number of co-authorships away one is from a co-authorship with Erdos. Grossman and Ion (1995), keepers of the Erdos Number Project, write that collaboration among mathematicians in general is on the rise, at least insofar in terms of published works:

> while over 90% of all papers fifty years ago were the work of just one mathematician, today scarcely more than half of them are solo works. In the same period, the fraction of two-author papers has risen from under 10% to about one third. Also, in 1940 there were virtually no papers with three authors, let alone four or more; now about 10% of all papers in the mathematical sciences have three or more authors. (Grossman (2002), p. 8)

Since mathematicians so often work collaboratively, it is natural that students of mathematics ought to work in this way as well. Moreover, the online format lends itself to the development of this mode of working and thinking about mathematics, since discussion forums are part of most course management systems.

In addition, employers that will be hiring our students are increasingly looking for "soft skills," such as the ability to collaborate and communicate, along with ability in the field (Schanzenbach, D. W. et al., 2016). The intent of mathematics educators should be to transmit to students the importance both of doing mathematics and being able to "communicate mathematical ideas clearly and coherently both verbally and in writing to audiences of varying mathematical sophistication" (Zorn, 2015, p. 18). Our courses can and should encourage students to be the best future mathematicians that we could wish them to be, or to be the best at any other career that requires communication of technical knowledge.

12.4 ONLINE DISCUSSIONS AND LEARNING

"I struggle every semester to find proper discussion prompts. Many students simply don't want to complete them."

The instructor above was writing in response to a survey sent out by the authors, which asked online faculty to describe their best and worst discussion prompts. It is clear from this response and from conversations over the years with colleagues that finding ways to encourage online discussion is not always easy. It is the authors' intention in this chapter to provide systematically a framework for creating good online discussion prompts, so that colleagues, both novices and experts in teaching mathematics online, can have tools to find the prompts they need to encourage discussion in their classes.

This study is informed by the results of questions sent out to mathematics faculty who teach online from two-year and four-year colleges across the United States. Faculty were asked to describe their best and worst online discussion prompts, and to write about what they saw as essential in an online mathematics discussion. Surveys were sent via email to addresses obtained from membership lists of committees of mathematics professional organizations and from faculty lists at a large university system in an urban area. In total,

38 faculty responded. We considered each response to be akin to an individual interview in a qualitative study. Faculty wrote at length and provided thoughtful responses. It was clear that most of our respondents had thought about what worked, what did not, and why. These respondents form a self-selected group that does not represent all online college mathematics faculty, but a representative statistical sample was not our intention; rather, we hoped to use these faculty responses to create a framework for what constitutes a good discussion prompt, one that will get students to engage more in the class.

12.5 THEORETICAL MODEL

The Strands of Mathematical Proficiency (Kilpatrick, Swafford & Findell, 2001) describe core competencies that mathematics educators should develop in their students and are evident in the standards promulgated by mathematics education organizations at many levels (Zorn, 2015; Koestler, 2013; and CCSSI, 2010). Their usefulness across so many levels of mathematics education regardless of the branch of mathematics made them ideal for categorizing the respondents' discussion prompts. The five strands, as described by Kilpatrick et al., are as follows:

> conceptual understanding—comprehension of mathematical concepts, operations, and relations
>
> procedural fluency—skill in carrying out procedures flexibly, accurately, efficiently, and appropriately
>
> strategic competence—ability to formulate, represent, and solve mathematical problems
>
> adaptive reasoning—capacity for logical thought, reflection, explanation, and justification
>
> productive disposition—habitual inclination to see mathematics as sensible, useful, and worthwhile, coupled with a belief in diligence and one's own efficacy.
>
> *(p. 116)*

These strands were originally conceptualized as strands woven into rope, to show how each aspect is interwoven with the others. For clarity, Figure 12.1 shows the five strands as circles, so that our mapping of discussion topics onto these strands can be shown more clearly in the subsequent figure.

The five categories of mathematical proficiency relate in a nearly one-to-one correspondence with the categories that emerged from reading the discussion prompts that faculty found successful: prompts that encourage synthesis and analysis, social and emotional prompts, study strategy prompts, prompts that ask students to solve content problems, prompts that ask for applications, and prompts that ask students to explain and justify their reasoning (see Figure 12.2).

Kilpatrick et al. (2001) write, in referring to the strands,

> These strands are not independent; they represent different aspects of a complex whole ... The most important observation we make here, one stressed throughout

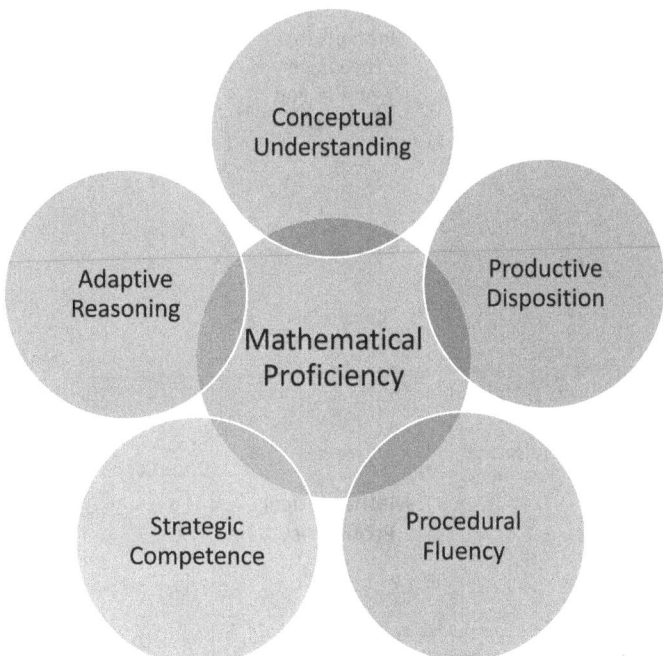

FIGURE 12.1   Strands of mathematical proficiency.

this report, is that the five strands are interwoven and interdependent in the development of proficiency in mathematics … Mathematical proficiency is not a one-
dimensional trait, and it cannot be achieved by focusing on just one or two of these
strands.

(p. 116)

Similarly, the discussion categories that emerged from the respondents are not independent, but overlap and influence each other, with no one discussion type being sufficient to
engage students in all the types of mathematical thinking necessary for a deep understanding of the material. However, as with the mathematical strands, presenting these prompts in
separate categories has the advantage of showing that multiple approaches are necessary to
achieve mathematical understanding and student engagement with the material.

12.6 SYNTHESIS AND ANALYSIS

Synthesis and analysis prompts encourage conceptual understanding. Synthesis and analysis
are two components of the original Bloom's taxonomy often referred to as higher-order
thinking skills. Analysis represents the "breakdown of a communication into its constitu-
ent elements or parts such that the relative hierarchy of ideas is made clear and/or the
relations between ideas expressed are made explicit" (Bloom, 1956, p. 201). Having students compare and contrast topics helps them to discern which technique to apply to a
given problem (sometimes referred to as meta-cognition). Some have posited that online
courses have an advantage over face-to-face courses by giving students the time needed to

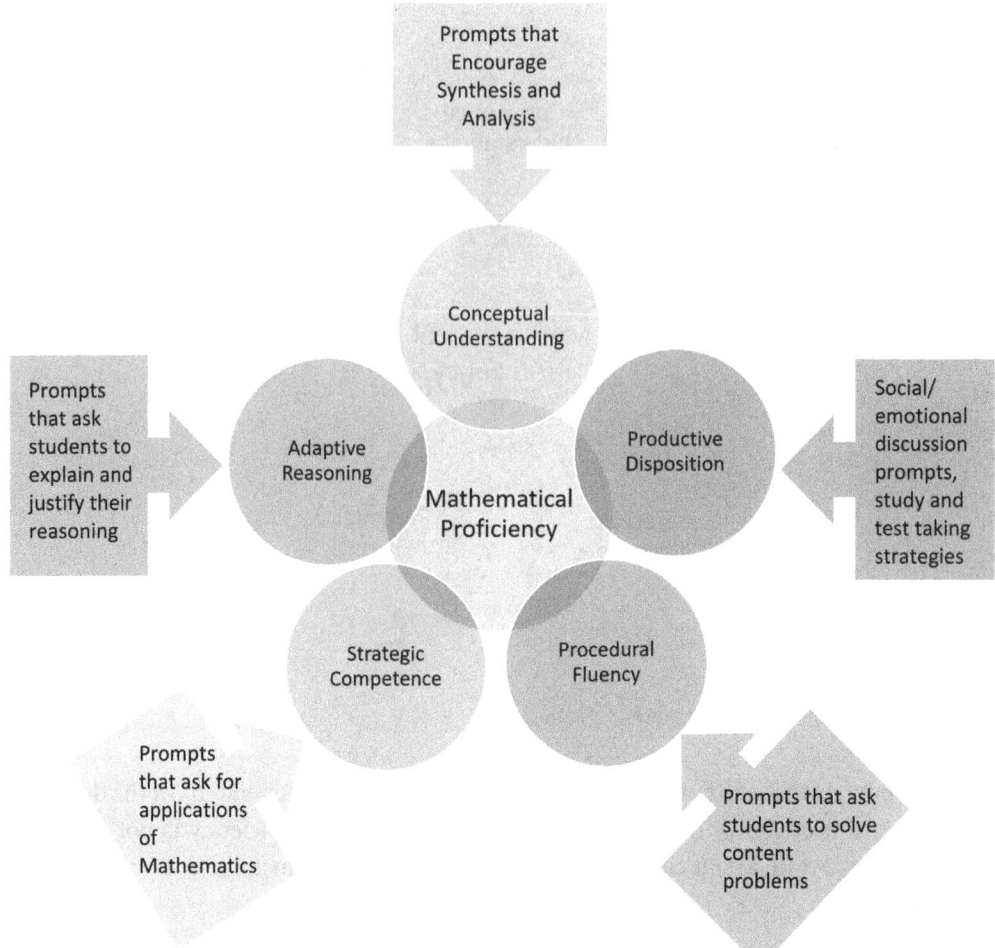

FIGURE 12.2   Mapping discussion prompts.

think more deeply about a question before offering their response (Lee, 2014). This time for reflection can make it possible for questions requiring higher-order thinking skills that require "slow thinking" (Kahneman & Egan, 2011).

It is understood that it is difficult, if not impossible, to address conceptual understanding alone—of course, procedural fluency must also be present in order for students to fully understand the underlying concepts. However, separating conceptual prompts into their own category allows faculty to focus specifically on getting students to think about, and have an appreciation for, the mathematical concepts in the course.

Examples of synthesis and analysis prompts. One instructor has a simple, elegant solution to getting students to think about the concepts in her class. She writes, "My most popular conceptual question is where students are asked to post their muddiest, clearest, and most important points in the chapter. That is always an active discussion." In particular, by having students find the most important points, students can reflect on the overarching mathematical goals and theory.

Ten instructors in the survey pointed out the need to find a way to elicit specific, *different* examples from each student, so that one well-written response comparing and contrasting topics does not prematurely end the discussion. This prompt on the topic of probability shows an example of asking for students to make the relations between ideas explicit, and by asking for an example, gets different responses from different students:

> Describe, in a short paragraph, the relationship between independent events and mutually exclusive events. Are independent events also mutually exclusive? Your answer should include (but should not be limited to) an answer to the following questions:
> Give an example of two events A and B that are mutually exclusive, and explain why they are dependent events. Give an example of two events that are dependent, but are not mutually exclusive. Explain!

Notice that the instructor provides some scaffolding for students in the form of the type of examples a response requires, examples that should lead to the more explicit relationship between the concepts of independence and mutual exclusivity.

Another instructor uses the following prompt:

> How does replacing x by –x affect the domain of any given function? Is the domain always different? Give two examples to justify your answer. Provide the graph of your example equations. (HINT ** Look at any given function and its reflection function and see if the domain changes. This question should be answered after doing this investigation.)

Again, this prompt allows for a variety of different answers, and gives students time to reflect and explore a concept.

Asking students to evaluate, another key verb in Bloom's taxonomy of higher-order thinking (Bloom, 1956), can also help students with conceptual understanding. One instructor had success when he

> shared an article by Keith Devlin on how early work in probability paved the way for mortgages. Students were asked to comment on the article, decide if they thought mortgages were good or bad (or maybe a mix), and why?

The instructor commented that, "I believe the question worked for two reasons: it went beyond what we were focused on in the classroom and it didn't have a right or a wrong answer."

Synthesis involves the "putting together of elements and parts so as to form a whole" (Bloom, p. 203). Discussion board prompts of this type require students to investigate an example from different perspectives and describe the relationship between the equivalent solutions from different approaches. For example, students might be asked to solve a problem graphically, numerically, and analytically and then comment on the relative merits of each of the approaches. This leads students to a synthesis of the various viewpoints and a more profound understanding of what a function is.

Similarly, compare-and-contrast questions, such as completing the square versus factoring for a quadratic equation, combined with the requirement of giving examples (so that multiple responses are possible) will aid students in developing the important mathematical skill of choosing the appropriate, most efficient technique for a particular problem.

12.7 SOLVING CONTENT PROBLEMS

Solving content problems assists in procedural fluency. Students typically gain procedural fluency by solving multiple problems of graded difficulty level for a particular topic. In a traditional course, this is usually accomplished via textbook problems or an online homework system, but incorporating skill building into the discussion board as well can contribute to the needed sense of community and encourage collaboration among students.

Examples of content prompts. Again, the challenge is to create a discussion of the content as opposed to just one problem for which there is essentially one correct solution that, once posted, would effectively end discussion. Faculty respondents to the survey often remarked that their worst discussion prompts were those in which they asked a question with only one answer. Instructors in the survey had versions of doing this by customizing a parameter in the content problem for each student, having students select different problems, or having each student post a troublesome problem and having classmates post solutions. In all cases, it helps to continue the discussion when students are encouraged or required to comment on their classmates' posts.

One instructor's clever idea to vary the parameters of a problem is through what she calls "round-robin" problems. In a round robin, the instructor presents a typical problem and then a numbered list of varied values of key parameters in the problem. The first student respondent solves the problem (with detailed explanation, of course!) The second student to post will post the answer with the second set of values for the parameter, and so forth. In this way, every student gets a chance to work the problem and can add his or her own explanation for their unique version of the problem. But, also, all students, except the first, have the benefit of having a solved example or examples on which to model their work, giving the weaker students a relatively easy way to get that crucial first step towards solving a given type of problem. Asking students to buy a real bag of M&M candies and then do a hypothesis test in several steps on the proportion of the candies of a certain color also provides an applied version of this type of post (since the steps all essentially follow the same form to conduct the hypothesis test, but the numbers and conclusions are different).

A variation of the above is to find problems that can be solved in several ways, for example, a set of algebra problems that can be solved graphically, or by substitution, or by elimination. Each student must join a discussion thread to solve the problem one way, then join a different discussion to solve a different problem using one of the other methods, and finally, join a third discussion to solve another problem using a method they have not tried yet. In each discussion, the answer should be the same using all three methods, which provides verification that the student is doing the work correctly. At a higher level of mathematical sophistication, calculus students might be asked to come up with three different graphs that have the same limit as x approaches a given number, or to find three different ways a derivative could fail to exist as x approaches a given number.

Another elegant way to get students to give many different answers was offered by one instructor who wrote, "I also love non-routine questions, with no one, correct answer. Something that is NOT just another homework question. Like, 'A very steep line goes through the point (3, –5). What could be the equation of the line?'"

Some instructors have reported success in asking students to post exam questions or textbook questions for solutions from other students or the instructor. One instructor writes, "I ask students to post a problem they are having trouble with, and they ask their classmates for help in solving it." She finds this to be a successful prompt for any of the courses she teaches. But other instructors reported that students do not post a sufficient number of solutions or solutions that are too brief. The successful prompt must clearly mandate participation and specify the type of questions and the level of detail and rigor required in posted solutions. For example, one instructor tells students to "explain the steps you used as if you were explaining to a fellow student who does not know how to work this problem."

One calculus professor reported moderate success having students create their own related rates or optimization problem for another student to solve, with the stipulation that, "This problem **must be your own creation** – it cannot come out of a book, from a lecture, from another student, or from somewhere on the web." In the second post, the student must solve another student's posted problem, and in a third, must check another student's solution and either suggest specific corrections or explain why it is correct.

> Be sure to provide **SPECIFIC MATH DETAILS** in all posts, including this one. Also, be sure to respond to any questions or comments that other students make to your work, either by correcting your own work, or explaining the answers to the student questions, or both.

Notice the level of detail required in the prompt in order to get students to answer appropriately.

12.8 APPLICATIONS OF MATHEMATICS

Applications of mathematics encourage strategic competence. Kilpatrick et al. (2001) describe *strategic competence* as "the ability to formulate mathematical problems, represent them, and solve them … similar to what has been called problem solving and problem formulation in the literature of mathematics education and cognitive science" (p. 124). Allowing students to choose their own application can be powerful motivation as students learn how the mathematics topic might be relevant to their own life, including to their personal academic or career interests. As one survey respondent wrote, "The best prompts enable students to rethink the material and see other ways it may applied, to also make the material they have learned personally relevant, and thus to achieve much deeper understanding."

Examples of application prompts. Many respondents had students watch videos, read articles, or do other internet research as a way to find applications. The prompts they then chose varied widely in complexity. For example, one instructor of quantitative literacy simply asked students to find a topic they were studying in the news and post about it. "It was awesome because they had to relate what we were learning to the class." From a different instructor, a slightly more complex prompt for a quantitative literacy class asked students to "…find the highest interest rate you can on a savings account and find the lowest interest rate you can on a credit card (ignore 0% temporary teaser rates—find the rate that kicks in after the temporary time period expires) … It put the textbook problems in a realistic light. You cannot find a savings account that pays 9% compounded quarterly." This prompt could be changed to include more mathematics by having students then use those rates to describe (with detailed computations) an investment strategy that could result in them having a million dollars. Note that we do not make a value judgment as to which type of prompt is better—often a simple prompt can be a quick entry point to get students interested, while the additional problem solving can be useful if students need the practice.

Again, varying from simple to complex, but this time in a statistics course, one instructor had students select and describe a TED Talk of their choice that incorporated the statistics learned in the class. Another instructor had students find an article in which sampling was used; students then had to write out the population, parameter, sample, statistic, and the type of data in the study. Students found articles on an immense variety of non-trivial topics of interest to them. This prompt required that the instructor provide copious feedback, but resulted in the students having over 20 examples for this topic which forms the foundation of a typical introductory statistics course.

In algebra, one of the most inventive prompts was the following: "Given what you have learned so far about functions, create a scenario that is somewhat relevant to an experience you have had or hope to have and represent with all four views (words, equation, table, and graph). In addition to your response, determine an appropriate domain and range for a classmate's scenario." The instructor wrote, "My students have been very creative in their responses and it gives me some ideas about what they are interested in."

For a similar type of question at a higher level of mathematics, calculus students can be asked to find a variable in their life that is, for example, "increasing at a decreasing rate"—the students must then give the value of their function, its derivative, and second derivative at a certain time (citing the appropriate units for each).

Finally, for a very basic, easy-to-use personal prompt for any class, one respondent asked students to give an example of how a particular topic might be used in their major or future occupation. "Makes students think outside their math class box," he wrote.

An application prompt can have also personal significance to a student without being directly about his or her own life. For example, an instructor in a foundations of mathematics course (liberal arts mathematics) asked students to analyze an argument made by a public figure in terms of a formal assessment of the logical validity. The instructor found that "students love discussing this question. It connects to current events, and it gets them to mathematize their thoughts with logic."

"How does probability relate to gambling?" The rather general nature of the prompt allows for varied responses, but clearly asks students to apply course content to an application that most have at least some familiarity. For a more specific version of this prompt, one can ask students to compare the probability of winning the lottery with the probability of other events of interest to them, such as being hit by a car or being struck by lightning.

Finally, one instructor has a way to do an application prompt that could work in any course:

> I assign an objective covered in the course to each student. They must supply an example of an application of that objective … The students really get involved with this discussion post. I see students reply with 'oh, wow. I didn't realize what we were learning in this class could be used to build a rocket' or 'it is interesting to see an application that doesn't involve Peter buying 20 watermelons. This is actually real life!'

12.9 SOCIAL AND EMOTION PROMPTS AND STUDY TIPS

For the *productive disposition* strand, the discussion prompts fall into two categories: those that ask students to address their feelings directly in reaction to prompts about math anxiety and other affective issues, and those that ask students to advise other students, through study tips and test-taking strategies. Both types of prompts can encourage students to feel more positively about mathematics, whether through sharing feelings or through helping others and getting help.

Social and emotional interaction. Prompts of this type aim to explore students' motivations and their confidence to do mathematics (self-efficacy). Often, by having less confident students engage with peers and the instructor, students' perceptions of their ability can change, and that, in turn, increases their achievement (Kim, Park, & Cozart, 2014). Examples of engaging prompts of this type asked students to comment on past math experiences or to read articles about the effects of math anxiety. Sometimes a direct approach is the best way to dispel the fear of mathematics or tendency of students to express the self-defeating mantra "I hate math."

Introductions. In many online classes, it is the introduction that sets the tone for the class, and that begins the processes of collaboration through group discussion. Yet instructors often found that the introduction was their least successful prompt, partly because their directions were too broad or too vague, such as, "Introduce yourself." The instructors that mentioned the introduction as their best prompt had detailed directions for what they wanted their students to post, often accompanied by their own introductory post that modeled what the instructor wanted to see. What follows are two examples of such prompts.

A good and bad math experience:

> I ask my students to write about a bad math experience and a good math experience – as part of our introductions – I model with my own bad experience (trying to memorize times tables) and my good one – thinking about infinity while angling two mirrors to reflect myself infinitely. Using this as part of our intro week airs out

anxieties of the past, allows students to find commonalities of experiences and to sympathize with each other, but also accentuates the positive. Students cheer each other on, and my own example shows that math is not just about numbers.

How statistics impacts your life:

> Please read "Course Overview: The Big Picture" before posting. Please post a couple of paragraphs about yourself. Please include information (as much as you are willing to share) about your academic interests, extracurricular activities, and career goals. **Think of ways that statistics, as it is defined in my essay, might have impacted your life up to now or might impact your life in the future (in light of your academic or career goals).** The main goal of this is an "ice breaker" – a way for you, the students, to start to get to know one another. The secondary goal is to get you thinking about how central mathematics in general and statistics in particular is in today's society. I will react to your posts with any additional applications I can think of to the areas that you express as your academic and non-academic interests. But please notice the bold type faced section above. This assignment is graded and I must be able to determine from your post that you have read my essay and included a response to the bold part of my prompt in your post.

While the first prompt specifically mentions feelings, the second one does not, yet both encourage students to think about themselves in a deeply personal way in connection to mathematics. Both prompts are broadly applicable across a variety of mathematics classes. Although the first prompt was for liberal arts mathematics and the second for a statistics class, faculty who teach calculus or higher-level mathematics classes could certainly ask about career goals and future plans; faculty who teach developmental mathematics classes could ask about good and bad math experiences, and about future jobs in which the students might use mathematics.

It is also worthwhile noting that both prompts are graded. The second prompt mentions this explicitly within the post, while the first professor has this written elsewhere in her grading policies.

For a lighter touch, one of our respondents simply asks students to "Post a trivia fact about yourself," and then to "Look for 2 or 3 'kindred spirits' and respond to their posts."

Social and emotional posts during the semester. Relating mathematics to our students' lives and emotions need not be limited to the introductions at the beginning of the semester. In a math class for future elementary school teachers, a middle-of-the-semester prompt asks students to read an article about teachers and math anxiety, and then discuss the effect that teachers have on their pupils' attitudes about math. Students are then asked to name at least two things they will do to help their students be less anxious about math. Note that students discuss math anxiety but end the post with what they can do about it. Care must be taken to end these types of posts on a positive note of some kind. One respondent found that her *worst* prompt occurred when she asked student why they liked or disliked a particular topic because "inevitably they will say they dislike the topic and that just builds to utter dislike by most if not all of the students in the discussion thread."

Several faculty members had students watch a video on some aspect of self-efficacy or math anxiety. For example, one had students watch "A Math Major Talks About Fear" and then comment on it. The instructor reported that, "Everyone said it was so helpful, and it made them less stressed out about the class."

Study skills and tips. Many instructors found that prompts asking students how they prepared for exams or asking students to identify online resources that are helpful is a way to transfer the skills of successful students to their peers who have not been successful yet. The idea that mathematics is learned, rather than an innate ability, is related to a growth mindset (Dweck, 2014); when weaker students actually hear the work that good students do to become good students, they can be encouraged to do the same. In addition, these discussions contribute to the students' productive disposition for learning the course concepts and content.

Instructors contributed creative approaches to this type of prompt. One asks students, "What more can YOU do to help yourself succeed in this course?" This can be an effective catalyst in changing student behavior. The results might also constitute useful formative assessment for instructors. Flipping this question around to focus on the instructor, another asks: "What more can I do to assist you in reaching your academic goals (other than giving extra credit)?" which makes the feedback even more useful in formative assessment.

Another subtle and creative way to delve into and perhaps improve student study habits is this prompt:

1) Are you keeping up with the reading, assignments, etc.? Do you think this course has the 'right' amount of learning material, homework, etc.?

2) Do you 'believe' in homework as an effective means of increasing mathematics learning?

The first prompt helps students who might not have anticipated the amount of work in an online course to come to terms with their effort vis-à-vis the course requirements. The second could promote a lively discussion about student perceptions of what is required to learn mathematics.

Finally, students can be asked to share websites or other resources they have found. To merit giving credit for students who share learning resources, one could require more explanation and specificity as this instructor does:

Post a link to a video that explains one of the concepts from class. Briefly describe the video. Do you think that the video does a good job explaining the concept? Can you think of any ways that it could have been improved?

This thread actively encourages students to seek help in areas that are difficult for them as well as providing additional resources that all students can use.

Asking students to share study tips and resources can give students a sense of agency and can help build the classroom community. In addition, instructors are spared the time required to make lists of resources, and can chime in to give just a few of their own favorite ideas within the discussion.

12.10 ADAPTIVE REASONING

The final strand, *Adaptive Reasoning*, as described by Kilpatrick et al (2001),

> refers to the capacity to think logically about the relationships among concepts and situations … In mathematics, adaptive reasoning is the glue that holds everything together, the lodestar that guides learning … [it includes] not only informal explanation and justification but also intuitive and inductive reasoning based on pattern, analogy, and metaphor. (p. 129)

In online learning, nearly all types of online discussion prompts should address this strand, because nearly all prompts should ask students to use logic and inference, and then to explain and justify their reasoning.

12.11 CONCLUSION AND RECOMMENDATIONS

Successful communication is just as important in online as in face-to-face mathematics courses, but clearly there are differences, so that instructors may have to spend time learning how to elicit fruitful discussions in an online environment.

For instructors embarking on your first online course, or for those looking to expand current online discussions, hopefully these discussion categories will be useful to you. As you plan, here, again, are the five categories, but now with some ways for you to think about adding your own ideas:

1. Synthesis and analysis—think of key concepts that students should be able to compare and contrast or to communicate in non-technical terms.

2. Solving content problems—think of certain key skills that could be practiced online in a collaborative format, with students using other students' work as a model for their own.

3. Applications of mathematics—think of how the mathematics could be applied and what would personalize the concepts for your students.

4. Social and emotional interaction—think of how students can share their feelings about mathematics with other students, without ending on a negative note—what would encourage a growth mindset around mathematics learning?

5. Study skills and tips—think of how students share their strengths and ideas with other students, and how to help students understand that working hard at mathematics is what helps them become better.

Common characteristics of successful prompts include some effort to personalize the course content, providing an opportunity for students to make the course content relevant to their academic or career interests. Successful prompts tend to be very specific, and expectations for the quantity and quality of student responses should be very explicit. It may help to think about how the responses will be assessed when creating the prompt. For certain problems where students are asked to create their own examples, a sample post is usually beneficial if not essential. Not many survey respondents used the discussions in a group format; that may be a useful subject for future studies. Instructors who put effort into designing creative prompts and signal to students that their participation is valued through consistent feedback should be rewarded with more interaction leading to better student performance in the course.

A common thread among survey respondents was that not only do online math discussions require good prompts, they also require an appropriate structure in the course, in which the grading and expectations for the posts are made clear, including the quantity, the quality, and the frequency (for example, if all students only post right before a deadline, then not much discussion will occur). Then students need formative and summative assessment of their participation to maintain the desired quantity and quality of their posts. Having discussion posts be a graded part of the course will highlight their importance, giving extrinsic motivation while the students come to gradually understand the intrinsic value of classroom collaboration. Successful instructors gave clear signals that the discussion is a valued part of the course.

Now, given a clear structure of expectations as outlined above, we can begin to answer the question, "What do you even talk about?" The answer is as clear as the goals of our courses: that students be able to do more than just solve problems; they should be able to think deeply about these problems and compare and contrast them with others, they should be able to apply the mathematical knowledge, they should be able to communicate about mathematics with clarity and precision, and they should develop as life-long learners acquiring techniques for learning mathematics specifically but also learning skills that can transfer to other disciplines as well.

BIBLIOGRAPHY

Bliss, C. A., & Lawrence, B. (2009). From posts to patterns: A metric to characterize discussion board activity in online courses. *Journal of Asynchronous Learning Networks, 13*(2), 15–32.

Bloom, B. S. (1956). *Taxonomy of Educational Objectives. Vol. 1: Cognitive Domain*. New York: McKay, 20–24.

Common Core State Standards Initiative. (2010). *Common Core State Standards for Mathematics*. Washington, DC: National Governors Association Center for Best Practices and the Council of Chief State School Officers.

Dweck, C. S. (2014). *Mindset: The New Psychology of Success*. New York: Ballantine Books.

Grossman, J. W. (2002). Patterns of collaboration in mathematical research. *SIAM News, 35*(9), 8–9.

Grossman, J. W., & Ion, P. D. (1995). On a portion of the well-known collaboration graph. *Congressus Numerantium, 108*, 129–132.

Jaggars, S. S., & Xu, D. (2016). How do online course design features influence student performance? *Computers & Education, 95*, 270–284.

Kahneman, D., & Egan, P. (2011). *Thinking, Fast and Slow*. New York: Farrar, Straus and Giroux.

Kentor, H. (2015). Distance education and the evolution of online learning in the United States. *Curriculum and Teaching Dialogue, 17,* 21–34.

Kilpatrick, J., Swafford, J. & Findell, B. (2001). *Adding It Up: Helping Children Learn Mathematics.* Washington, DC: National Academy Press.

Kim, C., Park, S. W., & Cozart, J. (2014). Affective and motivational factors of learning in online mathematics courses. *British Journal of Educational Technology, 45*(1), 171–185.

Koestler, Courtney. (2013). *Connecting the NCTM Process Standards and the CCSSM Practices.* Reston, VA: The National Council of Teachers of Mathematics.

Lee, S. M. (2014). The relationships between higher order thinking skills, cognitive density, and social presence in online learning. *The Internet and Higher Education, 21,* 41–52.

Offenholley, Kathleen H. (2012) A Discourse Analysis of the Online Mathematics Classroom, *American Journal of Distance Education,* 26:4, 236-248, DOI: 10.1080/08923647.2012.727727

Schanzenbach, D. W., Nunn, R., Bauer, L., Mumford, M., & Breitwieser, A. (2016). *Seven Facts on Noncognitive Skills from Education to the Labor Market.* Washington: The Hamilton Project.

Zorn, P. (Ed.) (2015). *Undergraduate Programs and Courses in the Mathematical Sciences: CUPM Curriculum Guide, 2015.* Washington, DC: Mathematical Association of America. http://www.maa.org/sites/default/files/pdf/CUPM/pdf/CUPMguide_print.pdf.

PART 3

Using Technology

CHAPTER 13

Cognitive Load Theory and Mathematics Instruction through MOOCs

E. Zimudzi, S. Kesianye, K. G. Garegae, S. Mogotsi,
A. A. Nkhwalume, and M. J. Motswiri

CONTENTS

13.1 INTRODUCTION

Massive Open Online Courses (MOOCs) in higher education have enjoyed a phenomenal rise in popularity in recent years. Esposito (2012) described MOOCs as a distance education environment providing open content that any individual anywhere on earth can freely register to and take courses; they are a flexible and open form of self-directed, online learning designed for mass participation. Harding (2012) emphasizes international student participation systems that provide the opportunity to take courses from respected universities by students in developing countries through online tools such as videos, discussion forums, and even peer-marked assignments.

The nature of the traditional courses contrasts with online courses because anyone can sign up for the MOOC. The learners do not have to be enrolled at a university. The traditional institution restricts the type of learner based on economics, geographical distance, pre-requisites, and attendance limits (Liyanagunawardena, Adams, & Williams, 2012). MOOC developers break these barriers by providing free and equal access to the courses for all. The MOOCs are typically paced around a weekly structure, and the enrolled students access relevant sources during their own time. Due to the ubiquitous nature of the Internet, it can be argued that these MOOCs are vital for low-income students and countries of the Third World, in which most of the African countries belong. In addition, many universities and schools do not have qualified teachers for some advanced courses, and the MOOCs can provide such services since the leading universities are typically the developers and instructors for the MOOCs (Aditomo, 2009).

Some researchers, e.g. Atkeson (2014), suggest that it would be better to use MOOCs as a supplement in blended learning where there is instructor presence for feedback and following an anticipated pacing of activities. Yet, others have argued for the heutagogical approach, e.g. Blaschke (2012), to teaching and learning where learners are highly autonomous and self-determined and where emphasis is placed on the development of learner capacity and capability with the goal of producing learners who are well-prepared for the complexities of today's workplace, partially due to the ubiquity of Web 2.0, and the affordances provided by the developments in Internet and multimedia technology. They argue that with its learner-centered design, Web 2.0 offers an environment that supports a heutagogical approach, and that MOOCs can richly provide such experiences.

Learning theory suggests that learning happens best under conditions that are aligned with the human cognitive architecture. The structure of human cognitive architecture, discernible through the results of experimental research, indicate that short-term memory is limited in the number of elements it can contain simultaneously. Sweller (1988) builds a theory that treats schemas as the cognitive structures that make up an individual's knowledge base. These schemas, that may contain other schemas acquired over time, allow learners to treat multiple elements as a single element. The contents of long-term memory are sophisticated structures that permit us to perceive, think, and solve problems (Sweller, 1994). Thus, the difference between an expert and a novice is that a novice has not acquired the schemas of an expert. Learning requires a change in these schematic structures of long-term memory seen as infinite (Figure 13.1).

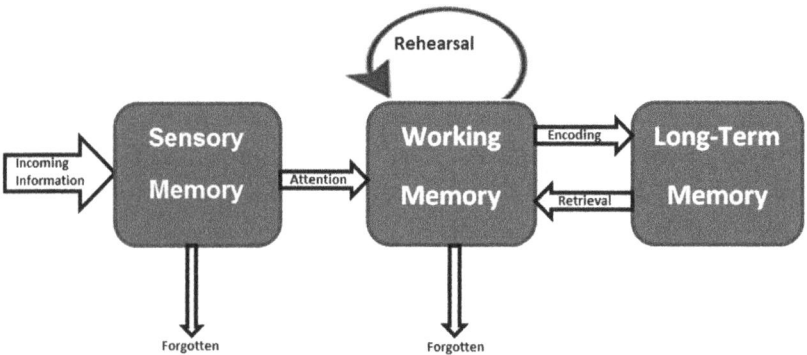

FIGURE 13.1   Information processing model. (Adapted from Atkinson, R. C. and Shiffrin, R. M. (1968): *Human Memory: A Proposed System and its Control Processes.*)

Cognitive load theory is concerned with techniques for reducing the working memory load which has limited capacity to process information from the environment in order to facilitate the changes in the long-term memory associated with schema acquisition. It can, however, access extremely large amounts of previously processed and organized information from long-term memory (Brünken, Plass, & Leutner, 2003; Chandler & Sweller, 1991; Chen, Woolcott, & Sweller, 2017; Sweller, 1988). From an instructional perspective, the information contained in instructional material must first be processed by working memory. For schema acquisition to occur, instruction should be designed to reduce working memory load (Mayer, Griffith, Jurkowitz, & Rothman, 2008; Mayer & Moreno, 2003).

There is not much that we can do about the intrinsic cognitive load because some tasks are generally more complex than others, and the level of learning material will have different levels of intrinsic cognitive load depending on the experience of the learner. Extraneous load occupies working memory but is irrelevant to the intended objectives of content material. It requires extra working memory for learning and problem-solving, and instructional designers have dedicated much research to the reduction of this load (Gillmor, Poggio, & Embretson, 2015). The third, called germane, is the mental effort that learners dedicate to learning and solving a problem. It increases with student motivation to participate in the learning process. The three are additive to comprise the total working memory. Content needs to be transferred from the learner's working memory to their long-term memory in such a way that it reduces the extraneous cognitive load, and if possible present content in a way that increases the germane cognitive load (Gillmor et al., 2015; Mayer et al., 2008; Mayer & Moreno, 2003).

This chapter examines higher education mathematics content in MOOCs' impact on the cognitive load of the students. Our aim is to analyze how the contributions from the cognitive load theory in particular and learning theories are being reflected in the design of courses in MOOCs. A qualitative focus group strategy is used from university students enrolled for a MOOC. Interview questions were mainly based on strategies that have been suggested from research literature, e.g. by Mayer and Moreno (2003) and by Chen et al. (2017).

13.2 MOOCs IN MATHEMATICS

Current researchers agree that the current MOOCs emerged from the work of Canadian scholars Stephen Downes and George Siemens (Anders, 2015) at the University of Manitoba. Since then, the success of open online course in artificial intelligence created by Stanford University and MIT (Chapman, Goodman, Jawitz, & Deacon, 2016), which attracted over 160,000 learners from 190 countries, led to the creation of online learning platforms that offer MOOCs such as Coursera, Udacity, and edX (Chapman et al., 2016) to service the production and delivery of MOOCs. These courses provide quizzes, peer-graded assignments, projects, and/or exams. Courses are provided on demand, and users do not necessarily have to complete a course during a defined time. All materials are provided including videos, lecture notes, and assessment materials. Critics of the MOOC movement cited low completion rates, high development costs (Fischer, 2014), and pedagogical issues (Blaschke, 2012; Doug, 2013).

Two distinct types of MOOCs may be distinguished; cMOOCs and xMOOCs, the former referred to as connectivist MOOCs, the reason being that Canadian researchers George Siemens, Stephen Downes, and Dave Cormier prepared the concept based on the principles of the theory of connectivism (Kesim & Alt, 2015). cMOOCs are designed to generate network effects for learning, facilitating self-organized patterns of collaborative learning. Learners are allowed to participate using their own blog sites and social media accounts. In contrast, xMOOCs platforms predominantly employ a cognitive-behaviorist pedagogical approach (Anders, 2015). It is exemplified by content-based training delivered at scale through a one-to-many distribution model, e.g. as professionally produced short video lecture series, typically delivered by one instructor, and integrated quizzes, readings, practice problems, and testing to help the student with maintaining focus and material retention. These possibilities have been brought about due to the recent developments in software and hardware, specifically developments in Internet technologies and multimedia.

Results from various studies indicate that students highly rated MOOCs in which tutors were willing to interact with students (Hew, 2016), where there was social interaction, i.e. students could share the knowledge and ideas among themselves (Chu, Chen, & Tsai, 2017; Hew, 2016; Zheng, Rosson, Shih, & Carroll, 2015), and the availability of online learning resources and activities that engaged students depending on their diverse learning preferences. However, some researchers, e.g. Fischer (2014) and Bali (2014), have emphasized the importance of learning design and environmental factors. It is clear that there is a gap for studies in pedagogical studies in MOOCs. The future design of MOOCs needs to be grounded in human cognition for effective instruction.

13.3 COGNITIVE LOAD THEORY AND MATHEMATICS LEARNING

Studies by Gillmor et al. (2015) and by Mayer et al. (2008) indicate that extraneous cognitive load does not contribute to the learning process. Learning is slowed when cognitive load exceeds working memory capacity. These studies encouraged maximizing the signal, i.e. the message to communicate, to noise, i.e. the extraneous information that detracts from learning, ratio, and argued for the elimination of extraneous visuals and text information, an approach they called *weeding*. Mathematical cognitive load is mainly exacerbated by problems in symbolic decoding, computational fluency, and conceptual understanding.

Chen and Wu (2015) investigated the use of videos in sustained attention, emotion, cognitive load, and learning performance, and found video lecture types enhance performance, and that sustained attention induced by the voice-over presentation type is markedly higher than that in picture-in-picture type.

Many studies recommend breaking down mathematics content into smaller segments, and also allowing the learner to control the pace of learning (Chen & Wu, 2015; Inventado & Scupelli, 2016; Kirschner, Paas, & Kirschner, 2009; Shadiev, Wu, & Huang, 2017). Also, content designers should opt to present information using various media, i.e. combining video, pictures, and animations (Chen & Wu, 2015; Yung & Paas, 2015). This assists in sharing the content between the visual and the verbal channel. Incidental processing (Mayer & Moreno, 2003) should be removed by removing non-essential music, or/and decorative graphics from the content to decrease extraneous load (Ayres, 2006; Chen & Wu, 2015;

TABLE 13.1   Desirable Features to Reduce Learners' Cognitive Load

Features	Explanation
Breaking down subject content into manageable chunks	Divide content into manageable units to allow students to focus their attention on the key concepts and enhance the power of retention and recall.
Sequencing content delivery from the easier to the more difficult	Simple-to-complex sequencing allows for the thoughtful release of content that encourages motivation and retention of learning.
Inclusion of clear objectives/goals	Well-formulated goals tell the learners what is expected of them, and the direction in which teaching leads.
Clear instructions	Too much time is spent on problem-solving the instructions as opposed to new schema formation due to lack of clarity in instructions.
Relate learning material to assessment	Assessment determines whether the objectives have been achieved or not, and provides data for tutors and learners to measure success/failure, and provides the necessary feedback and self-reflection.
Worked examples	Learning in mathematics can be seen as a process of generalizing from specific examples, and to reduce the "noise" during concept formation to reduce the working memory load.

Shadiev et al., 2017). Thus instructional designers have to be aware of the cognitive requirements that designs impose (Table 13.1).

As instructional designers, we need to be aware of the cognitive requirements our designs impose and ensure that our learners can meet those requirements. Thus, reducing extraneous cognitive load alleviates student stress and anxiety which are correlated to the learner's performance (Gillmor et al., 2015; Renkl & Atkinson, 2003; van Merrienboer, Kirschner, & Kester, 2003).

13.4 METHODOLOGY

A systematic review of the literature was conducted to identify seminal papers about MOOCs and the relevant educational literature, mainly on the cognitive load theory. Various MOOCs on Coursera, edEx, and Udacity were visited to access common core mathematics content courses accessed by students at the University of Botswana. One course was identified for use with the students taking a mathematics course, and studying for a mathematics education program. The researchers used the course to collect the data and analyze how the course had been designed, if the designers took into account the learning theories and in particular the cognitive load of the course content and materials when designing, how the course has impacted on student learning, and how the course influences learning outcomes.

The following questions guided the study:

1. Do MOOCs follow a sound pedagogy and organizational approach to online learning that leads to quality outcomes and experiences for students in mathematics?

2. What new pedagogies and organizational mechanisms might be required if MOOCs are to deliver high-quality learning in mathematics?

After perusing mathematics courses from the three online MOOC platforms, a popu-lar MOOC from edX was identified. We selected a calculus course, Calculus BC link www.edx.org/course/ap-calculus-bc-0, specifically designed for students who had taken a pre-calculus course, and were currently enrolled for a calculus course at the University of Botswana Mathematics Department during the first semester of the academic year 2017/18, running from August 2017 to December 2017. The course generally satisfies the general education requirements at the university, and is self-paced, allowing the enrolled students to achieve mastery in one concept before moving to the next. The course had content in differentiation, series, and integration techniques. It was scheduled for about 11 weeks, almost the same period a similar course should take at our university. The course was also chosen due to the reason that basic calculus skills continue to be a prob-lem for a significant proportion of higher education students (Loch, Jordan, Lowe, & Mestel, 2014).

We employed a focus group strategy, identifying themes and looking for associated data fitting under the categories of intrinsic and extraneous cognitive loads. The partici-pants were all from the Department of Mathematics and Science Education undertaking a program in mathematics education. They were initially conducted using their university emails. The purpose of the study was established, and a plan of study was developed. We coded the same data that we had transcribed onto a word processor in pairs as researchers, and discussed our findings to assess coding similarities and differences.

13.5 RESULTS AND DISCUSSION

Four asynchronous virtual focus group (Saldana, 2008) discussions comprising three to five in-service teachers taking the MOOC course in mathematics were conducted. A total of 17, six males and 11 females, participated in the study. All were studying for a Bachelor of Education specializing in Mathematics Education.

The participants largely agreed that the MOOC selected for assisting in the calculus course helped to enrich their knowledge in calculus. All the students have vast experience using YouTube videos for learning and some Internet resources in other mathematics and education courses.

In terms of our findings, we can see examples and evidence of a reduction of cognitive load by the use of MOOCs in mathematics teaching and learning from the responses of the focus groups. In the excerpts below from focus group interviews, we can note that the designers of the MOOC in calculus factored the cognitive load into the content of the course. Responses indicate that the tutors have broken down course content into manage-able chunks that is digestible by the learners, and the content is sequenced in chapters that are easy to follow. The instructions were clear and easy to follow, with some exercises and worked examples that were easy to follow but a little more basic than challenging mate-rial, thereby lacking in the ability to provoke thinking and their arousal of interest. The objectives for each of the sections were clearly written with what all that the students were expected to do to achieve them. The assessment material was available, though there were responses indicating that they could have more of the assessments. The increased use of multimedia made it easier to understand much of the content. All these contributed to lessening the cognitive load, and helped in the learning of calculus content, though more could be done to reduce the load (Table 13.2).

One learner participant said:

"I was able to choose my own path within each lesson, and could jump between lessons to review some material done earlier that I could understand." This indicates that the MOOC was broken down and sequenced into some chunks that the students could tackle individually.

Also, another one said:

"Each topic was broken down into small chunks, with short instructional videos, interactive graphs," and providing cognitive aids one student said *"the course provided many practice problems."*

This indicates that the course designers attempted to factor in the signal-to-noise ratio by *"weeding"* (Mayer & Moreno, 2003) the extraneous content and avoiding incidental processing not relevant to the task. The practice problems provided, the multimedia use, including graphics and text, worked as cognitive aids, assisting in offloading some of the cognitive demands of working memory.

TABLE 13.2   Summary of Perceptions and Action Ideas Expressed by Learners Enrolled for the Calculus MOOC

Perceptions Expressed by Learners Enrolled for the Calculus MOOC	
Positive Perceptions of Learners	**Concerns Raised**
Increased use of multimedia in teaching	There is minimal interaction between the tutors and the students.
Hands-on experience	No feedback for contributions/some assignments except for some quizzes.
Inclusion of many worked examples in calculus	Some students who enrolled did not have any background, making it difficult to interact.
Breaking down subject content into manageable chunks	The course was graded on a pass/fail basis, not a percentage.
Sequencing content delivery	The work-as-you-please environment does not motivate the learner to complete all tasks.
Inclusion of clear objectives/goals	Technical problems with accessing the MOOC sometimes.
Clear instructions	Lack of scaffolding during attempts to solve calculus problems.
Relate learning material to assessment	
Move beyond traditional/established classroom	
Action Ideas to Make Improvements to the MOOC	
Enroll only learners with enough background, not by interest.	No feedback for contributions/some assignments except for some quizzes.
Assessment should be carried out and tasks graded frequently and by the tutors if a MOOC has to be taken seriously.	Need to include more practice examples to assist the learners.
There is need for frequent interaction between the lecturers and the students which is not possible due to the large numbers.	More support from peers doing the same course.

And another one said:

> *"Engage in video instruction, exam style questions and interactive videos"* and also *"The tutors made sure that the learning objectives were explained,"* and that *"I liked most the flexibility of the course. I could always find time to study the content material. I could also get assistance sometimes from the other students enrolled for the same course."*

Another one said:

> *"The course tutors concentrate only on the calculus content of the course, no other distracting content. The objectives were always clear."*

The tutors were consistent in how they developed their videos, starting always with some introduction, followed by the content and some examples, and then some summary, and practice problems to consolidate the content taught. For example, one student said:

> *"The tutors were consistent with their videos, starting with the introduction stuff, and then some developed content material, followed by some examples, and the conclusion. Then they gave some exercises to work on."*

One participant added:

> *"The tutors always provided feedback for all the questions I asked, including some feedback and some collaborative ideas from some students who enrolled for the same course."*

According to the cognitive load theory, individual learning becomes less efficient as the difficulty of content increases. Therefore opportunities for collaboration must always be provided to divide the cognitive processes among the learners, and thereafter re-integrate the information and coordinate the learning. In our case, the students reported that the tutors provided feedback in acceptable times, although some said they never received feedback on some tasks they undertook. And the students on the focus group got some feedback from the other students who were enrolled for the same calculus course. At the same time, there was a lot of scaffolding also provided by some student and tutor feedback throughout the course if they noted some concern from the students.

There were various concerns that the learners raised about the course, including that the interaction between the learners and the tutors was minimal. Although there was some feedback from the tutors, and mainly from the peers, learners felt that their understanding of the course material, of the instructions, and of the examples could be improved if there was frequent interaction between the learners and the tutors, thereby increasing the learners' cognitive load. The tutors cannot individualize the learning because of the numbers involved in most MOOCs. This desirable frequent interaction could provide the necessary feedback for assignments and scaffolding whenever the learners encountered the problems. And related to this was the concern that some the students did not contribute much to the betterment of the others probably because they did not have the background to enter the course which also led to a very high dropout among the enrolled learners. And the environment did not motivate the learners to work, or to complete any tasks that were assigned.

Collaboration with other students in other environments was a positive among the learners, though, sometimes, learners had to resort to their tutor teaching a similar course at the university for informal assessments. The MOOC tutor was not available for assessing enrolled learners' work except for a few multiple-choice quizzes. There was no feedback from the tutors. If the MOOCs are to be successful, there is need for one-on-one interaction with every learner. The sheer number of the learners makes it really difficult for the teacher to have a one-on-one relationship with the learners, making the tutor unable to assist the learners adequately. Some responses indicated that the practice examples and worked examples on the MOOC were below standard, and were easy to follow for the university level. This problem has been identified with most of the MOOCs that enroll any learner interested without the basic knowledge requirements of the course, and eventually dwell much on the basics of the course.

13.6 DISCUSSION

The study sought to examine to what extent the design of MOOCs in mathematics follows a sound pedagogy and organizational approach, taking cognizance of the cognitive load on the students in higher education, and to analyze the user reflections on new organizational mechanisms that might be required for effective learning and teaching in the design of MOOCs. Guidelines, e.g. Alraimi et al., (2015); Leppink (2017); Margaryan et al., (2015); Sweller (2018), revolve around minimizing extraneous cognitive load so that only a minimum of working memory resources is required for cognitive processes that do not contribute to learning, always working around specific learning goals and appreciating the multifaceted relation between learning and assessment. Data collected from the focus group discussion indicate that the MOOC that was selected employed some strategies to reduce the extraneous cognitive load and optimize germane cognitive load. The learners highly rated the MOOC, indicating that the MOOC designers broke down the content into manageable chunks and sequenced the content, thereby reducing the cognitive load. Objectives were clear and included in the sections of the MOOC. Some assessment was included though below what was expected by the learners. Their reported level of assessment meant that the MOOC catered for the basic learner in calculus, and did not have strict pre-requisites for entry into the course. Although there was sharing of content and examples by the learners, the content shared was of much lower level than what the students expected. The tutors did not interact much with the learners; there was sharing of knowledge and resources by peers. The participants agreed, however, that the content helped to enrich their knowledge in calculus.

Lack of provision of assistance is one of the most frustrating elements in learning mathematics. Educators use the metaphor of scaffolding to refer to the strategy of a temporary support for learning how to support the learning of complex tasks (Brahimi & Sarirete,

2015; Margaryan et al., 2015; Phan, Ngu, & Yeung, 2016). In MOOCs, this support can be provided by the tutors, and by peers, thus, the importance of provision of possibilities for collaboration when designing MOOCs. Our findings indicate that the tutors are generally not present, and they could not provide the necessary scaffolding whenever required. No feedback was provided for the tasks. The respondents reported that the tutors only provided the original material, and no follow-up. The enrolled numbers for the MOOC could hinder the capability of the tutors to interact with individual learners. The peers could provide some assistance but generally came late, or never came. The MOOC also provided the students with opportunities to practice by providing many worked problems as examples and many exercises for the consolidation of concepts, though most of the examples were mainly for the beginner level.

The findings indicate that there is more to do before mathematics MOOCs follow a sound pedagogical approach. The organizational approach would be similar to other online courses, but they lack the necessary support that the learners need to reduce their cognitive load. The recommendation of using a simple-to-complex approach sequencing (van Merrienboer et al., 2003), chunking, inclusion of learning goals and objectives, relating learning to assessment, and the use of worked examples that work with conventional teaching, is not as effective because the learners have no immediate scaffolding (Renkl & Atkinson, 2003) when required. The MOOC used partially followed some organizational approach to online learning with some experiences for learners in mathematics learning. The use of multimedia, including short videos, text, and voice (Chen & Wu, 2015; Chen et al., 2017; Mayer & Moreno, 2003), or a combination would be seen to reduce the cognitive load of the learners though it would not be enough on its own.

13.7 CONCLUSIONS

Cognitive load theory builds upon established models of human memory that include working memory and long-term memory. Working memory can only process a limited number of information elements at any given time. When the cognitive load exceeds the learner's working memory capacity, performance and learning are impaired. Education researchers have developed and tested a number of instructional techniques that decrease extraneous load and optimize germane load so that learners are not overwhelmed. In this chapter, we argue for the need to rethink and develop MOOCs, taking the conditions for learning into account if we are to understand learning on MOOCs. Instructional design in MOOCs has to incorporate the cognitive architecture when designing media for use with learners. And the learners have to receive the necessary support during the learning period if the MOOCs have to remain relevant for some time to come.

The literature shows fairly consistent views on pedagogical considerations that MOOCs should involve and aim for. These include the constant interaction between the tutors and the learners, peer collaboration, providing scaffolding during the learning, making available support of online resources and activities, and a problem-oriented course with clear explanations (Hew, 2016). The effects of the use of different multimedia should also be considered (Chen & Wu, 2015). This actually assists in reducing the cognitive load of mathematics content in MOOCs.

MOOC tutors need to be mindful about how they approach designs for teaching in complex mathematics environments. Activities that are germane in nature lead to rapid schema development in learners. Extraneous load not directly related to learning needs to be minimized or eliminated if possible. Many worked examples and practice questions should be used within a course to make efficient use of the learners' working memory and develop learner schemas. These MOOCs could provide the needed assistance as heutagogical approaches take center stage with adult learners.

13.8 RECOMMENDATIONS

The study sought to examine if the designers in mathematics courses on MOOCs consider the cognitive load of the materials they develop, and to encourage the use of the human cognition architecture. Designs that factor in the cognitive load reduce the effort that the learner has to put in in order to understand the content. Solving problems requires a great deal of mental effort directed at understanding the problem. Studying an appropriately structured problem involves less extraneous load. And there are various strategies that can be used, including removing unnecessary complexity and distraction, providing constant scaffolding, and providing learners with opportunities for collaborative learning in MOOCs. MOOCs in mathematics have to break down content and sequence it for teaching and learning. The objectives should be clear, with clear instructions on how to do tasks. The inclusion of many, varied examples and practice questions is also crucial. All this is necessary during the design of instruction. During instruction, the learners have to have frequent interaction with the tutors, and scaffolding and feedback have to be provided, with deadlines given to assignments.

13.9 LIMITATIONS OF THE STUDY

Focus group strategy results cannot be generalized; however they could provide a baseline to future researcher and mathematics content designers and teachers. And conducting research on the Internet is challenging because of "Internet time" (Karpf, 2012), i.e. the rapidly changing context of content modifications that need to be taken seriously. The sample of the study was also small.

REFERENCES

Aditomo, A. (2009). Cognitive Load Theory and Mathematics Learning: A Systematic Review. *Anima Indonesian Psychological Journal*, 24(3), 207–217.

Alraimi, K. M., Zo, H., & Ciganek, A. P. (2015). Understanding the MOOCs Continuance. The Role of Openness and Reputation. *Computers & Education*, 80, 28–38. doi:10.1016/j.compedu.2014.08.006.

Anders, A. (2015). Theories and Applications of Massive Online Open Courses (MOOCs): The Case for Hybrid Design. *International Review of Research in Open and Distributed Learning*, 16(6), 39–61.

Atkeson, S. (2014). Havard-MIT partnership unveils new MOOCs for k-12. Education Week, 34, p. 8. Retrieved 05 February 2018 from http://goo.gl/s1Q049.

Ayres, P. (2006). Impact of Reducing Intrinsic Cognitive Load on Learning in a Mathematical Domain. *Applied Cognitive Psychology*, 20(3), 287–298. doi:10.1002/acp.1245.

Bali, M. (2014). MOOC Pedagogy: Gleaning Good Practice from Existing MOOCs. *MERLOT Journal of Online Learning and Teaching, 10*(1), 44–56.

Blaschke, L. M. (2012). Heutagogy and Lifelong Learning: A Review of Heutagogical Practice and Self-Determined Learning. *The International Review of Research in Open and Distance Learning, 13*(1), 56–71.

Brahimi, T., & Sarirete, A. (2015). Learning Outside the Classroom Through MOOCs. *Computers in Human Behavior, 51,* 604–609. doi:10.1016/j.chb.2015.03.013.

Brünken, R., Plass, J. L., & Leutner, D. (2003). Direct Measurement of Cognitive Load in Multimedia Learning. *Educational Psychologist, 38*(1), 53–61.

Chandler, P., & Sweller, J. (1991). Cognitive Load Theory and the Format of Instruction. *Cognition and Instruction, 8*(4), 293–332.

Chapman, S. A., Goodman, S., Jawitz, J., & Deacon, A. (2016). A Strategy for Monitoring and Evaluating Massive Open Online Courses. *Evaluation and Program Planning, 57,* 55–63. doi:10.1016/j.evalprogplan.2016.04.006.

Chen, C. M., & Wu, C. H. (2015). Effects of Different Video Lecture Types on Sustained Attention, Emotion, Cognitive Load, and Learning Performance. *Computers & Education, 80,* 108–121. doi:10.1016/j.compedu.2014.08.015.

Chen, O., Woolcott, G., & Sweller, J. (2017). Using Cognitive Load Theory to Structure Computer-Based Learning Including MOOCs. *Journal of Computer Assisted Learning,* 293–305. doi:10.1111/jcal.12188.

Chu, H. C., Chen, J. M., & Tsai, C. L. (2017). Effects of an Online Formative Peer-Tutoring Approach on Students' Learning Behaviors, Performance and Cognitive Load in Mathematics. *Interactive Learning Environments, 25*(2), 203–219. doi:10.1080/10494820.2016.1276085.

Doug, C. (2013). MOOCs and the Funnel of Participation. In *Third Conference on Learning Analytics and Knowledge (LAK 2013)*, 8-12 April 2013, Leuven, Belgium, pp. 185–189. doi:10.1145/2460296.2460332.

Esposito, A. (2012). Research Ethicsin Emerging Forms of Online Issues Arising From a Hypothetical Study on a MOOC. Retrieved 28 September 2017, from http://www.eric.ed.gov/PDFS/EJ985433.pdf.

Fischer, G. (2014). Beyond Hype and Underestimation. Identifying Research Challenges For the Future of MOOCs. *Distance Education, 35*(2), 149–158. doi:10.1080/01587919.2014.920752.

Gillmor, S. C., Poggio, J., & Embretson, S. (2015). Effects of Reducing the Cognitive Load of Mathematics Test Items on Student Performance. *Learning and Instruction, 22*(2), 12–18. doi:10.1016/j.learninstruc.2014.02.004.

Harding, N. (2012). The Massive Open Online Course Revolution Hits the UK. Retrieved 02 February 2018, from http://thepositive.com/mooc-massive-online-courses-uk.

Hew, K. F. (2016). Promoting Engagement in Online Courses: What Strategies Can We Learn from Three Highly Rated MOOCS. *British Journal of Educational Technology, 47*(2), 320–341. doi:10.1111/bjet.12235.

Inventado, P. S., & Scupelli, P. (2016). Design Patterns for Math Problems and Learning Support in Online Learning Systems. *Proceedings of the 10th Travelling Conference on Pattern Languages of Programs - VikingPLoP, s,* 1–16. doi:10.1145/3022636.3022644.

Karpf, D. (2012). SOCIAL SCIENCE RESEARCH METHODS IN INTERNET TIME. *Information, Communication and Society, 15(5),* 639-661. doi: 10.1080/1369118X.2012.665468.

Kesim, M., & Alt, H. (2015). A Theoretical Analysis of Moocs Types From A Perspective of Learning Theories. In *Procedia - Social and Behavioral Sciences* (Vol. 186, pp. 15–19). Elsevier B.V. doi:10.1016/j.sbspro.2015.04.056.

Kirschner, F., Paas, F., & Kirschner, P. A. (2009). A Cognitive Load Approach to Collaborative Learning: United Brains for Complex Tasks. *Educational Psychology Review, 21*(1), 31–42. doi:10.1007/s10648-008-9095-2.

Leppink, J. (2017). Cognitive Load Theory Practical Implications and an Important Challenge. *Journal of Taibah University Medical Sciences*, *12*(5), 385–391. doi:10.1016/j.jtumed.2017.05.003.

Liyanagunawardena, T. R., Adams, A. A., & Williams, S. A. (2012). MOOCs : A Systematic Study of the Published Literature 2008–2012 ln. *The International Review of Research in Open and Distance Learning*, *14*(3), 202–227.

Loch, B., Jordan, C. R., Lowe, T. W., & Mestel, B. D. (2014). Do Screencasts Help to Revise Prerequisite Mathematics? An Investigation of Student Performance and Perception. *International Journal of Mathematical Education in Science and Technology*, *45*(2), 256–268. doi:10.1080/0020739X.2013.822581.

Margaryan, A., Bianco, M., & Littlejohn, A. (2015). Instructional Quality of Massive Open Online Courses (MOOCs). *Computers & Education*, *80*, 77–83. doi:10.1016/j.compedu.2014.08.005.

Mayer, R. E., Griffith, E., Jurkowitz, I. T. N., & Rothman, D. (2008). Increased Interestingness of Extraneous Details in a Multimedia Science Presentation Leads to Decreased Learning. *Journal of Experimental Psychology: Applied*, *14*(4), 329–339. doi:10.1037/a0013835.

Mayer, R. E., & Moreno, R. (2003). Nine Ways to Reduce Cognitive Load in Multimedia Learning. *Educational Psychologist*, *1520*(38), 43–52. doi:10.1207/S15326985EP3801.

Phan, H. P., Ngu, B. H., & Yeung, A. S. (2016). Achieving Optimal Best : Instructional Efficiency and the Use of Cognitive Load Theory in Mathematical Problem Solving. *Educational Psychology Review*. doi:10.1007/s10648-016-9373-3.

Renkl, A., & Atkinson, R. K. (2003). Structuring the Transition From Example Study to Problem Solving in Cognitive Skill Acquisition : A Cognitive Load Perspective. *Educational Psychologist*, *38*(1), 15–22.

Saldana, J. (2008). An Introduction to Codes and Coding. In *The Coding Manual for Qualitative Researchers*. Edited by Uwe Flick, London. Sage Publications Limited. pp. 1–31.

Shadiev, R., Wu, T. T., & Huang, Y. M. (2017). Enhancing Learning Performance, Attention, and Meditation Using a Speech-to-Text Recognition Application: Evidence from Multiple Data Sources. *Interactive Learning Environments*, *25*(2), 249–261. doi:10.1080/10494820.2016.1276079.

Sweller, J. (1988). Cognitive Load During Problem Solving : Effects on Learning. *Cognitive Science*, *12*, 257–285.

Sweller, J. (1994). Cognitive Load Theory, Learning Difficulty, And Instructional Design. *Learning and Instruction*, *4*, 295–312.

Sweller, J. (2018). Measuring Cognitive Load. *Perspectives on Medical Education*, *7*, 1–2. doi:10.1007/s40037-017-0395-4.

van Merrienboer, J. J. G., Kirschner, P. A., & Kester, L. (2003). Taking the Load Off a Learner's Mind : Instructional Design for Complex Learning. *Educational Psychologist*, *38*(1), 5–13. doi:10.1207/S15326985EP3801.

Yung, I. H., & Paas, F. (2015). Effects of Computer-Based Visual Representation on Mathematics Learning and Cognitive Load. *Educational Technology & Society*, *18*(4), 70–77.

Zheng, S., Rosson, M. B., Shih, P. C., & Carroll, J. M. (2015). Understanding Student Motivation, Behaviors and Perceptions in MOOCs. *Proceedings of the 18th ACM Conference on Computer Supported Cooperative Work & Social Computing - CSCW '15*, (March), 1882–1895. doi:10.1145/2675133.2675217.

Technological Pedagogical Content Knowledge for Meaningful Learning and Instrumental Orchestrations

A Case Study of a Cross Product Exploration Using CalcPlot3D

Monica VanDieren, Deborah Moore-Russo, and Paul Seeburger*

CONTENTS

* This research is supported in part by the National Science Foundation under Grant Numbers 1524968, 1523786, and 155216.

14.1   INTRODUCTION

This chapter begins to examine the design, appropriation, and transformation of an online, dynamic, visualization exploration activity for multivariable calculus. Because both how a digital resource is actually used in the classroom and the teacher's evolving knowledge of how a resource can aid in student understanding are intertwined (Borys & Choppin, 2017; Gueudet & Trouche, 2009; Lawless & Pellagrino, 2007), in this study we focus on a single digital resource (the CalcPlot3D cross product exploration) through multiple lenses.

The development of an online learning resource, such as a mathematical applet, should take into account not only the target audience, but also the mathematical content itself, related pedagogical issues, and technological implementation challenges. For example, the ways that content can be represented, where the content fits into the curriculum (including the prior knowledge expected of students), common student misconceptions, system requirements (related to memory and compatibility), and types of digital devices that might be used are just a few of the issues that a developer must consider. Many, but not all, must also be contemplated by instructors along with some additional considerations that are particular to their local settings. For instructors, introducing an applet in the classroom may require attention to pedagogical factors, such as activity types appropriate for teaching specific content with particular technology (Grandgenett, Harris, & Hofer, 2011; Harris, Mishra, & Koehler, 2009) and social arrangements for classroom participation. However, applet adoption should also entail self-reflection regarding the instructor's understanding of the content (especially when the online tool allows for exploration), preferences for pedagogical strategies, and knowledge of the technological platforms (including hardware and browsers) to be used to support the online tool. Furthermore, many of the mathematical, pedagogical, and technological issues to consider overlap with one another (Harris et al., 2009). The developer and individual instructors using the resource may have significant variations in their knowledge of the mathematics, pedagogy, and technology (and the connections between the three) that could limit implementation (Drijvers, Tacoma, Besamusca, Doorman, & Boon, 2013; Harris et al., 2009), but these variations can also be leveraged by the developer to improve and transform the online tool (Hansen, Mavrikis, & Geraniou, 2016).

Instructors' varied experiences and knowledge, along with other factors such as their working environment and the needs of their students, impact how an online resource is actually implemented in the classroom (Gueudet & Trouche, 2009). Teachers ultimately choose how to use a resource based on their preferences and needs, and not on the designer's intent or teaching guides (Drijvers, Doorman, Boon, Reed, & Gravemeijer, 2010). Therefore, a designer must be sensitive to these complexities surrounding the appropriation of an online resource into heterogeneous classrooms and must understand instructors' agency in the design process as they make choices on how to adapt the resource to fit their circumstances and needs (Borys & Chopin, 2017).

This self-study focuses primarily on the developer (the third author) and one early-adopting instructor (the first author) over ten years of development, implementation, and revision of the CalcPlot3D cross product exploration for multivariable calculus. The second author is an educational researcher who facilitates this self-study. Identifying how teachers structure a lesson using an artifact in the classroom (e.g., group work, teacher demonstrations, etc.) can help designers in revision stages of the design cycle. Therefore, in this study we will examine the variety of ways in which the CalcPlot3D cross product exploration was used by the developer and an early-adopting instructor who has used the tool in her classroom for eight years. We aim to answer two questions which are made explicit in the section of this paper titled Theoretical Framework and Research Questions.

While this self-study examines one exploration in a particular dynamic visualization applet for multivariable calculus, our results are more universally applicable since they demonstrate fundamental knowledge that affects the design cycle and rollout of online educational manipulatives and underlying models of appropriation of curricular resources (Trgalová & Rousson, 2017). This research could be used to inform the design of new online learning tools, to help support teachers as they shift to implementing new technologies, and to prepare pre-service teachers and graduate teaching assistants for their professions. This study also adds to the emerging, but limited, body of research exploring the collaborative design of a virtual manipulative within a community of practice and the connection between the design and the professional knowledge of the instructors (Hansen et al., 2016). Finally, on a theoretical level, this study tests the synergy of two frameworks (i.e., instrumental orchestration and Technological Pedagogical Content Knowledge (TPACK) for meaningful learning in information and communication technologies) to research questions in undergraduate mathematics education.

14.2 CROSS PRODUCT

We begin with a general review of the literature surrounding the mathematical content, pedagogy, and technology related to teaching the cross product. We take the social constructivist view of Olive and colleagues (2009) that the integration of digital technologies into the mathematics classroom has changed both the nature and construction of mathematical knowledge. "Technology is likely to change not only the content of school mathematics but also the processes of school mathematics and the nature of mathematical understandings" (Heid, 2005, p. 357). Through technology, students are likely to develop multi-representational views of mathematics and a nearly kinematic understanding of concepts (Heid, 2005). Therefore, as we describe the challenges in instruction and implementation of online tools in this section, we will underscore issues related to multiple representations and visualizations of the content.

Critical features of the cross product in mathematics. One way in which the nature of mathematical knowledge has changed is that technology allows students to manipulate representations of mathematical objects as tangible objects and observe invariant relationships (Olive et al., 2009). According to variation theory, a student learns a concept when they perceive the invariant relationships between its critical features. Critical features are aspects or conditions of a topic that are necessary for understanding (Runesson, 2006). The critical features of vectors and of the cross product include: magnitude, direction, angle between two vectors, location of the vectors, and relative orientation of two vectors (VanDieren, Moore-Russo, Wilsey, & Seeburger, 2017).

Technology has shifted the emphasis of mathematics, allowing students to play with ideas before mastering algebraic manipulations, to visualize dynamic relationships, and to connect formal and informal mathematics (Olive et al., 2009). By manipulating a mathematical object to observe properties of invariance, students can transition from conjecturing to formalizing about a mathematical concept (Olive et al., 2009). What this means in terms of the cross product is that students need not know how to compute the cross product, find its length, nor decipher the symbolic representations in order to gain some understanding about the concept such as articulating the effect of changing the angle between vectors **u** and **v** of fixed length on the magnitude of **u** × **v**, or predicting and observing the direction of **u** × **v** based on the directions of **u** and **v**.

Pedagogical challenges related to student understanding of the cross product. Although vectors are regularly presented in both high school and preliminary college courses, students often will not be exposed to the vector cross product until they undertake college-level calculus coursework. Both students (Kustusch, 2016; Scaife & Heckler, 2010; Van Deventer, 2006) and pre-service physics teachers (Govender & Gashe, 2016) tend to struggle with the concept of the cross product. Students can become confused with choosing one of the many mechanisms for applying the right-hand rule, and this may lead to incorrect execution of the rule (Kustusch, 2016; Scaife & Heckler, 2010; Van Deventer, 2006). One source of difficulty with the right-hand rule may be that students do not recognize the differences between relative orientations of two vectors (Govender & Gashe, 2016; VanDieren et al., 2017). In other words students may struggle to distinguish between the orientation of the pair of vectors **u** and **v** and the pair **u** and **w** in Figure 14.1.

To help alleviate some of the difficulties students have with the right-hand rule (Kustusch, 2016), several mnemonics (Greenslade, 1980), paper manipulatives (Van Domelen, 1999;

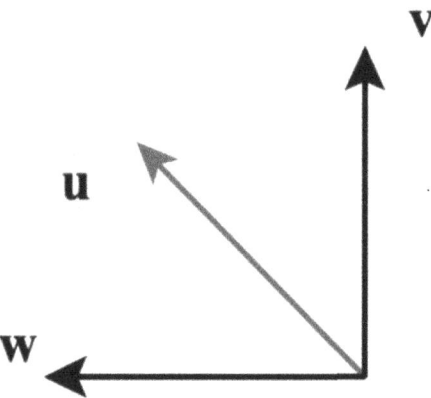

FIGURE 14.1 Two pairs of vectors (**u** and **v**, **u** and **w**) with the same interior angle, but different orientations.

Nguyen & Meltzer, 2003), and online resources (e.g., the vector and cross product tools listed on the repository site www.merlot.org/merlot/index.htm) are available for teachers to implement in their instruction.

Technological limitations related to student understanding of the cross product. Beyond the pedagogical challenges that arise from the content, technologically based challenges add to the complexity of teaching the cross product. These stem from students' relationships with the technology, their spatial reasoning and visualization abilities, and any hardware and software limitations.

Students tend not to question the output of a computer. However, even with digital mathematical resources that have undergone the strictest testing, there are still limitations in terms of floating point computations, round-off errors, graphical glitches around singularities, etc., that may, at best, confuse, or at worst, mislead a student (Olive et al., 2009). Additionally, students may develop unintended and unwanted approaches (such as shortcuts) through their use of technology (Kosheleva & Giron, 2006). It is, therefore, important for instructors to have the technological, pedagogical, and mathematical understanding to manage these situations (Olive et al., 2009).

While research is mixed on how well multi-representational environments support student learning, one common finding is that learners have difficulties translating between different representations (Ainsworth, 2006; Kozma, 2003). In particular, students struggle with visualization (Miller-Young, 2013) and spatial reasoning (Kustusch, 2016) of vectors in three dimensions. Furthermore, the complex learning tasks associated with multiple representations stand in the way of the potential advantages of using them. For instance, while online visualization tools provide students with a similar reference frame as an instructor's use of a whiteboard when executing the right-hand rule, online visualization may thwart the transfer of knowledge and skills between different media, in particular between the computer screen and paper. This presents a concern since exams are usually in pen-and-paper format (Kustusch, 2016).

To help better design online visualization tools, Ainsworth (2006; 2008) identifies five areas of understanding necessary for learning with multiple representations. These are understanding: (a) the form of the representation, (b) which operations to apply to a representation to retrieve the relevant domain information, (c) how to select an appropriate representation, (d) how to construct an appropriate representation, and (e) how to relate representations. Research on student understanding of vectors indicates that students have difficulties with each of these areas:

a. Students often confuse the syntax and semantics of $v \times u, u \times v,$ and $\|u \times v\|$ (VanDieren et al., 2017; Zavala & Barniol, 2010) and do not distinguish the graphical relationships of the pair of vectors u and v and the pair u and w in Figure 14.1 (Govender & Gashe, 2016; Scaife & Heckler, 2010; VanDieren et al., 2017).

b. Students tend to inappropriately apply operations and ideas from trigonometry and measure the angle between two vectors as a negative value or as an angle greater than 180 degrees (Kustusch, 2016; VanDieren et al., 2017). They may also interpret the dot product as a vector rather than a scalar value (Van Deventer, 2006), or the cross product as a scalar rather than a vector (VanDieren et al., 2017; Zavala & Barnial, 2010).

c. Students do not use real-life contextual clues to aid in interpreting vector diagrams and they tend to rely on formulas in their responses even if a formula is unnecessary or insufficient to answer the problem (Miller-Young, 2013).

d. Students often do not realize the need to relocate vectors via parallel transport to construct graphically the sum of two vectors (Nguyen & Meltzer, 2003) or to visualize a cross product (Kustusch, 2016).

e. Students also may have difficulty consistently interpreting vectors when they are presented in multiple representations (e.g., vectors presented graphically, symbolically, numerically, or situated in a physical context) (Nieminen, Savinainen, & Viiri, 2013).

Finally, inherent in any digital resource are technological issues such as portability, accessibility, interconnectibility, and interface that may affect teaching and learning (Heid, 2005). We will not address all of these issues here; instead we will focus on a major change that took place in 2015 when most mainstream browsers began to phase out their support of Java. When this occurred, instructors and students were required to change the way they accessed the digital resource, CalcPlot3D. First, error and warning messages had to be overridden, then only certain browsers would load CalcPlot3D, and finally a new version of CalcPlot3D running in JavaScript had to be accessed. Instructors and students unfamiliar with the shift to JavaScript from Java may have been deterred from continuing to use the resource during this time period when they experienced problems loading the CalcPlot3D Java applet in their preferred browser.

14.3 THEORETICAL FRAMEWORK AND RESEARCH QUESTIONS

This study is framed by two related theories: instrumental orchestration and Technological Pedagogical Content Knowledge (TPACK).[*] Tabach (2011) first used these two complementary frameworks together in research. They have since been combined productively to study the impact that co-designing a virtual manipulative has on the professional development of teachers (Hansen et al., 2016) and to examine the teaching practices and knowledge of mid-adopting mathematics teachers using a digital resource (Drijvers et al., 2013). Together instrumental orchestration and TPACK framed a study on the appropriation of digital resources by mathematics teachers, allowing for the analysis of an instructor's adaptation of the resource to fit his needs and the context of his classroom as well as the evolution of his professional knowledge and practice (Trgalová & Rousson, 2017). While these studies all involve future or current K-12 instructors, our study will further test the appropriateness of bringing together these two frameworks to study the implementation of

---

[*] There is some disagreement in the literature concerning notation. Sometimes this model is abbreviated TPCK and other times TPACK. TPCK is sometimes also used to refer to the intersection of the technological, pedagogical, and content knowledge areas in the Venn diagram, while TPACK refers to the entire model (Ruthven, 2014). In this paper we will follow the convention of using TPACK to refer to the model.

a digital resource from the points of view of the developer and an early-adopting instructor at the undergraduate level.

14.3.1 Instrumental Orchestration

To better understand human–tool interactions, it is helpful to distinguish artifacts and instruments. An *artifact* is any created object or tool; note that an artifact is man-made but may be digital rather than physical. In our study the artifact is a cross product exploration that draws upon the CalcPlot3D multivariable calculus applet. An *instrument* is an artifact for which an individual has identified some use; therefore, there is only an instrument when an individual has appropriated an artifact for a specific purpose. *Instrumental genesis* (Trouche, 2004) refers to the process of an unused artifact becoming a useful, used instrument. Instrumental genesis is dependent on the tool, the individual using it, and the purpose for which the tool is being used as an instrument (Trouche, 2004).

Instrumental orchestration refers to the external steering on the part of an instructor (Trouche, 2004) to help students with the instrumental genesis process. Although, the instructional intentions, organization, and facilitation to support instrumental genesis are so important (Drijvers, Doorman, Boon, Reed, & Gravemeijer 2010), these aspects are often overlooked and not considered in the textbooks and experiments involving digital learning environments (Trouche, 2004).

The instrumental orchestration framework for technology-rich mathematics classrooms classifies classroom practices as either teacher-centered or student-centered. Originally six categories were identified from videotapes of three teachers (Drijvers et al., 2010), but as this framework was tested on different populations new categories have emerged (Drijvers, 2011; Drijvers et al., 2013; Tabach, 2011, 2013). Although the global taxonomy of orchestrations is still under development (Drijvers et al., 2013), a sufficiently detailed taxonomy exists for us to use to identify instructor preferences in employing the CalcPlot3D cross product exploration. We will examine the following categories of instrumental orchestration:

- *Technical-demo*: demonstration of tool techniques (Drijvers et al., 2010).

- *Link-screen-board*: instructor explains the relationship with what happens in the technological environment with representations on the board, paper, or in the book (Drijvers et al., 2010).

- *Explain-the-screen*: instructor explains to the whole class what happens mathematically on the screen in an example (Drijvers et al., 2010).

- *Discuss-the-screen*: whole-class discussion of what happens mathematically on the screen in an example (Drijvers et al., 2010).

- *Spot-and-show*: instructor brings the work of a student or group of students to the attention of the class to facilitate discussion (Drijvers et al., 2010).

- *Sherpa-at-work*: a student "Sherpa" uses the technology to present his/her work to others in the class or to carry out actions dictated by the teacher (Trouche, 2004).

- *Work-and-walk-by* (a.k.a. *monitor and guide*): students work in pairs or individually as the teacher circulates the classroom, engaging at times with individuals as the need arises (Drijvers, 2011; Tabach, 2013).

- *Technical support*: teacher supports the students with technical problems that extend beyond the tool itself (e.g., login problems, hardware issues, etc.) (Drijvers et al., 2013).

- *Discuss tech without it*: instructor discusses the digital resource but without accessing the technology, such as in a classroom with no computers (Tabach, 2013).

14.3.2 TPACK

Tabach (2011) suggests complementing instrumental orchestration with TPACK as a means to better understand teacher implementation of technological tools. While "(t)he orchestration model provides means to describe what the teachers actually do in their technology-rich lessons, the TPACK model helps to identify the skills and knowledge needed to be able to exploit these orchestrations" (Drijvers et al., 2013, p. 996). TPACK is a technology integration framework that identifies three types of knowledge instructors must combine to successfully implement a digital resource in their teaching. This framework, introduced by Mishra and Koehler (2006), is an extension of Shulman's (1986) work. One goal of Shulman's (1986) work is to characterize the complex ways that teachers think about teaching by confronting both the content and pedagogy. It describes how a teacher makes mathematical content accessible to students by identifying different ways to represent the material (Mishra & Koehler, 2006). Shulman suggests three categories of teacher knowledge: content knowledge, pedagogical content knowledge, and curricular knowledge.

Since then, the abundance and diversity of digital resources entering the classroom have presented new challenges that warrant a reconsideration of Shulman's three categories. Mishra and Koehler (2006) offer a model that extends Shulman's framework. This model, displayed in Figure 14.2, involves three categories (and their intersections): knowledge of content, pedagogy, and technology.

Operationalizing TPACK as a framework. The TPACK framework has been criticized (e.g., Cox & Graham, 2009) for lacking clear and operational definitions of the knowledge constructs (i.e., technological knowledge, pedagogical knowledge, and content knowledge) and their intersections. Ruthven (2014) argues that the full system of seven TPACK categories is appropriate for designing professional development courses for teachers and as a lens to raise research questions about the interaction between technology, pedagogy and content; however, in order to be used as a research tool, TPACK must be supplemented by other frameworks. Additionally, for economy of scale, research on TPACK is often organized around prototypical teaching situations (Ruthven, 2014) or learning activities (e.g., Harris et al., 2010b). Therefore, the focus of this study will not be on measuring the sources of TPACK as shown in Figure 14.2, but will be on assessing TPACK for meaningful learning with information and communication technologies (ICT) as applied to specific learning activities.

TPACK for meaningful learning with ICT. TPACK for meaningful learning with ICT considers how technological, pedagogical, and content knowledge can be leveraged

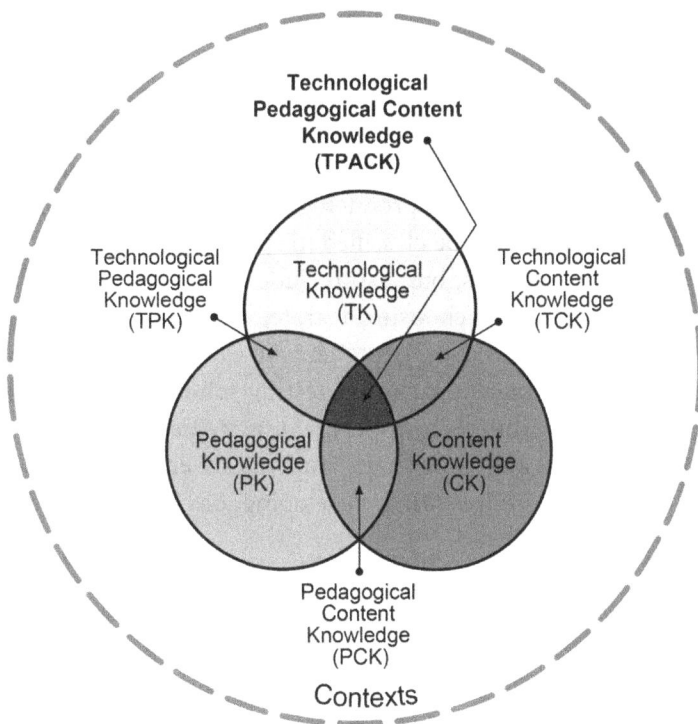

FIGURE 14.2 Technological Pedagogical Content Knowledge. (Reproduced by permission of the publisher, © 2012 by tpack.org.)

to create meaningful learning experiences for students through the selection of appropriate learning activities (Koh, 2013). "What drives [meaningful] learning, more than anything else, is understanding and persisting on some task or activity. The nature of the tasks best determines that nature of the students' learning" (Howland, Jonassen, & Marra, 2012, p. 2). An instructor's TPACK manifests itself with his or her selection, sequencing, and redesign of learning activities, and the choice of learning activities is highly content-dependent (Harris et al., 2010b). Research on technology integration should be similarly focused on learning activities and how these learning activities promote meaningful learning (Howland et al., 2012). In this study, we consider not only the type of the learning activity, but also the pedagogical dimension of each activity.

Taxonomy of learning activities. While there is no known taxonomy of learning activities for undergraduate mathematics instruction, a taxonomy for K-12 mathematics instruction, based on NCTM process standards, has been developed and organized into seven synergetic categories: *consider, practice, interpret, produce, apply, evaluate,* and *create* (Grandgenett et al., 2011; Howland et al., 2012). Although created from K-12 data, we feel that these activity categories are expansive enough to capture the potential learning activities in undergraduate instruction of vector cross products, the focus of our study.

Activities categorized as *consider* involve lower-level student engagement with foundational knowledge. They may include reading a text, recognizing a pattern, or attending a demonstration. *Practice* activities drill students on computational or algorithmic skills. Activities classified as *interpret* help students internalize the meaning of abstract concepts and relationships. They include posing a conjecture, developing an argument, and interpreting a representation. *Produce* activities are characterized by students producing mathematical works involving the communication of mathematical content. For example, describing a concept mathematically, producing a representation, giving a demonstration, and generating text would all be classified in this category. *Apply* activities not only encompass real-world applications, but also include applications of knowledge to solve problems. These activities include choosing a strategy, taking a test, and applying a representation. *Evaluate* activities encourage students to evaluate the mathematical work of others or themselves by making comparisons, testing solutions or conjectures, and integrating feedback from others into their work. *Create* activities require some of the highest levels of student engagement and knowledge. Students are engaged in imaginative thinking processes in which they are inventing, developing, or creating content (Grandgenett et al., 2011).

Pedagogical dimensions of learning activities. Learning activities can be analyzed for meaningful learning across five pedagogical dimensions: *active, constructive, intentional, authentic,* and *cooperative* (Howland et al., 2012). Building on Howland and colleague's work, Koh (2013) created a rubric for TPACK for meaningful learning for each of these (pedagogical) dimensions to emphasize the knowledge of the subject matter (content) needed to be supported by the ICT (technology) in learning activities. These pedagogical dimensions are summarized in Table 14.1.

14.3.3 Research Questions

Now that we have established the theoretical framework for this study, we can articulate the two research questions guiding our work:

1. What instrumental orchestrations do different instructors employ in conjunction with the cross product exploration; in other words how do individual instructors use the cross product exploration in the classroom?

TABLE 14.1   Pedagogical Dimensions of Learning Activities for TPACK for Meaningful Learning

Pedagogical Dimension	Description
Active	Is measured by the percentage of the activity duration that a student used and manipulated the technology to learn the material.
Constructive	Captures the extent to which students used the technology to engage in divergent thought (i.e., organization of, integration of, and reflection upon the content).
Authentic	Takes into account the degree to which the technology helps students to make connections between the content and real-world phenomena and/or their personal experiences.
Intentional	Captures the extent to which the technology provided opportunities for the students to engage in self-diagnosis and remediation of learning gaps during the activity.
Cooperative	Measures the extent to which the technology allowed for group work during the lesson activity that involved divergent discussion either around or through the computer.

Note: Information on table from (Koh, 2013).

2. How have the type of learning activities and pedagogical dimensions of the cross product exploration evolved as a result of the changing TPACK of the developer and early-adopting instructor?

14.4 RESEARCH CONTEXT

The research questions in this study are addressed in the context of a larger project on a digital resource for visually exploring multivariable calculus, the CalcPlot3D applet. CalcPlot3D is a freely available, interactive, online applet, which has been available and in continual development for over ten years (Seeburger, 2020). The CalcPlot3D applet helps students and instructors to visualize multivariable calculus content in three dimensions. The applet* receives over 150,000 page views per year. The CalcPlot3D applet provides visual representations of many multivariable calculus concepts including vectors, parametric curves, multivariable functions, parametric functions, and vector fields. Accompanying the applet are several discovery-based activities that facilitate the exploration of geometric interpretations of the content. The focus of this paper will be on the design, development, and implementation of just one of these explorations, the CalcPlot3D cross product exploration, and not on the applet in general.

The cross product exploration was initially designed in 2008 by Paul Seeburger and was first tested by five professors at four universities. Since then, professors and high school teachers at several other institutions have added the cross product exploration to their instructional materials. The CalcPlot3D cross product exploration consists of a series of questions about the geometric relationship between the cross product and the two vectors that form it, and it makes use of some features of CalcPlot3D. This artifact is what is being investigated; it has undergone two major redesign phases: the first using a Java applet with exploration questions administered in SurveyMonkey and the second using a JavaScript app with exploration questions administered in either WeBWorK[†] or Blackboard.[‡] Throughout each of these redesign phases, the questions on the pre-test, exploration, and post-test have also evolved.

Screenshots of the cross product exploration in the Java and JavaScript versions of the applet appear in Figures 14.3 and 14.4, respectively. In both of these artifacts, users can change the direction and length of the red and blue vectors in the two-dimensional graph on the left, and the cross product along with the red and blue vectors is automatically graphed in the three-dimensional plot to the right. The angle between the red and blue vectors and the length of the cross product is also given at the top of the screen. In the Java version, the red and blue vectors could not be moved off of the xy-plane, but in the JavaScript version users have the option of changing the z-components of the red and blue vectors. In both versions of the artifact, all three vectors are always graphed from the origin.

The CalcPlot3D cross product exploration activity has developed over time. The activity is grouped into pre-test questions, exploration questions, and post-test questions, which

---

* c3d.libretexts.org/CalcPlot3D/index.html
† WeBWorK is a comprehensive, open-source online homework system.
‡ Blackboard is a learning management system which has limited homework capabilities.

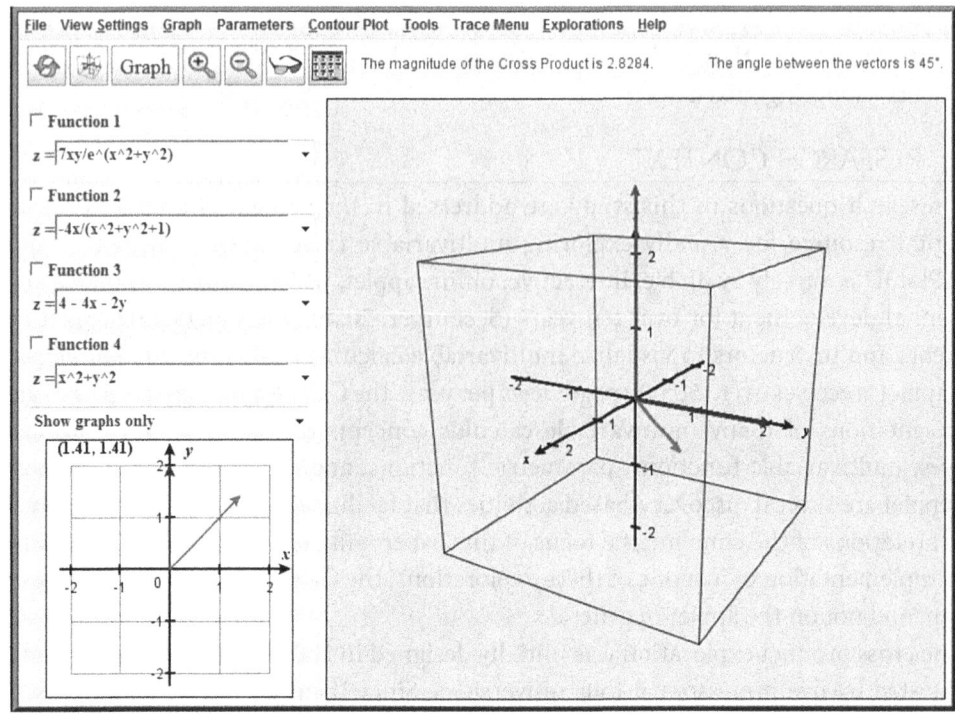

FIGURE 14.3   Screenshot of the Java version of the cross product exploration in CalcPlot3D.

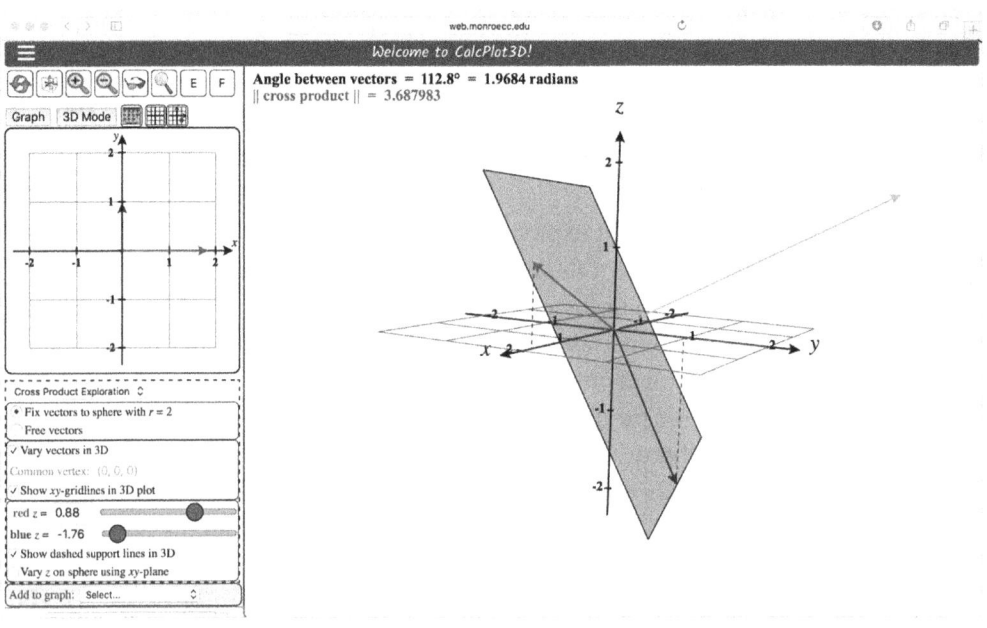

FIGURE 14.4   Screenshot of the JavaScript version of the cross product exploration in CalcPlot3D.

are identical to the pre-test questions. The intent is that students answer the pre-test questions online before engaging with the applet; students use the applet to answer the exploration questions; and students answer the post-test questions without the applet after they have completed the activity questions. Student answers are sent to the instructor electronically. In the SurveyMonkey version, students did not receive feedback on their responses until after the instructor downloaded the answers and provided written comments. In the WeBWorK and Blackboard versions, some questions were automatically graded in real-time. Tables 14.2 and 14.3 list the questions that were administered in SurveyMonkey during the initial phase of the cross product exploration implementation. Over time, the questions have been modified and imported into WeBWorK and Blackboard.

14.5 METHOD

14.5.1 Self-Study

This research is framed as a self-study. A self-study allows the developer and an early-adopting instructor to act as participant-observers and examine notes, plans, and reflections about the process of planning and teaching. As such the teaching styles and other factors of the developer and instructor become part of the study instead of variables to be controlled. The five methodological characteristics of a self-study identified by LaBoskey (2004) apply to this study: self-initiated and focused; improvement-aimed; interactive; multiple, primarily qualitative methods; and exemplar-based validation.

Self-initiated and focused. We frame our research as a critical self-study because it allows us to closely examine how the knowledge and experiences of the developer and an instructor-adopter affect the advancement of one of the CalcPlot3D applet exploration activities. Beyond studying the developer and the instructor-adopter, we also examine others involved in the practice, including other instructors and students.

Improvement-aimed. In a self-study,

> it is important to be cognizant of the continual interplay between research and practice within the practice setting (i.e., as the research unfolds so the learning through the research influences practice and, because the practitioner is the researcher, practice inevitably changes through feedback, thus influencing what is being researched) (Loughran, 2007, p. 15).

As our knowledge of mathematical, pedagogical, and technological knowledge has changed, so have plans for improvements of the exploration and ways in which it was introduced in our classroom settings. Reflection on instrumental orchestration and TPACK is a crucial step in the design cycle of the CalcPlot3D software and exploration activities.

Interactive. LaBoskey (2004) suggests not only collaborating with others, but also using research literature as a means for researchers to interact with others. To this end, we have consulted the literature in framing our interpretations and challenging our assumptions. Also, the second author of the paper is neither the developer nor a multivariable calculus instructor, but an educational researcher. Her collaboration helps the instructor and

TABLE 14.2   Fall 2009 Cross Product Exploration Activity Pre-Test and Post-Test Questions Classified by Learning Activity

Question	Learning Activity Type(s)
1. Which pair of [unit] vectors below will have the cross product with largest magnitude?	*Consider*
2. Which pair of [unit] vectors below will have the cross product with smallest magnitude?	*Consider*
3. Given two vectors of fixed length, and allowing the direction of one vector to vary, what angle between the two vectors will produce a cross product with maximum magnitude? (Select all that apply.) a. 90° b. 45° c. 0° d. 180° e. I don't know	*Consider*
4. Given two vectors of fixed length, and allowing the direction of one vector to vary, what angle between the two vectors will produce a cross product with minimum magnitude? (Select all that apply.) a. 90° b. 45° c. 0° d. 180° e. I don't know	*Consider*
5. What can cause the magnitude of the cross product to be zero? (Select all that apply.) a. The angle between the vectors is 0°. b. The angle between the vectors is 180°. c. The angle between the vectors is 90°. d. The angle between the vectors is 45°. e. One of the vectors is the zero vector. f. One of the vectors is a unit vector.	*Interpret, Evaluate*
6. What is the geometric relationship between the cross product vector and the two vectors that form it?	*Consider, Produce*

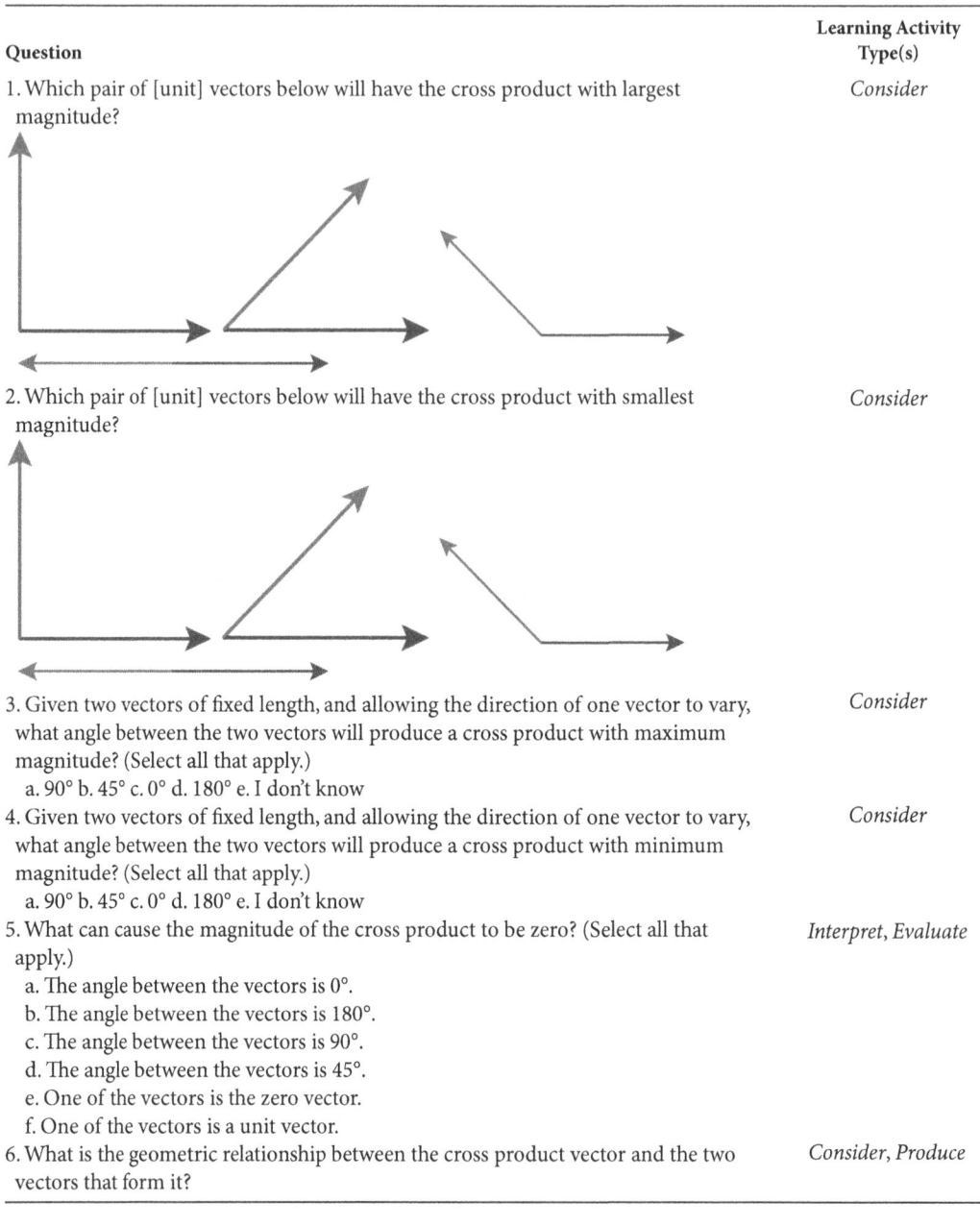

developer to improve and understand their practices and reveal their biases. Additionally, the first and third authors interacted with other power-users of CalcPlot3D in both mathematics and engineering courses to inform the continual development of CalcPlot3D and their own teaching.

Multiple, primarily qualitative methods. To assess an instructor's TPACK, ideally several sources of data including planning documents, instructional actions, interactions with students, and reflections should all be examined (Harris, Grandgenett, & Hofer, 2010a). Therefore, this study draws on several sources of data to triangulate emergent themes.

TABLE 14.3   Fall 2009 Cross Product Exploration Activity Questions Classified by Learning Activity

Exploration Question	Learning Activity Type(s)
1. How is the cross product vector related to the two vectors that form it?	*Consider, Produce*
2. For vectors of fixed length, but varying the direction of one of the vectors, when is the magnitude of the cross product at a *maximum*?	*Consider, Interpret, Produce, Evaluate*
3. For vectors of fixed length, but varying the direction of one of the vectors, when is the magniitude of the cross product at a *minimum*?	*Consider, Interpret, Produce, Evaluate*
4. When does the cross product vector point in the positive *z*-direction?	*Consider, Interpret, Produce, Evaluate*
5. When does the cross product vector point in the negative *z*-direction?	*Consider, Interpret, Produce, Evaluate*
6. What two things can cause the magnitude of the cross product to be 0?	*Interpret, Produce, Evaluate*
7. Which order is the cross product in this exploration? a. Red × Blue b. Blue × Red	*Practice*
8. Which of the geometric properties from questions 1–6 are made clear by the following formula? Please specify these by question number. Explain. $\|\mathbf{a} \times \mathbf{b}\| = \|\mathbf{a}\| \, \|\mathbf{b}\| \sin t$, where t is the angle between vectors $\mathbf{a}$ and $\mathbf{b}$.	*Interpret, Produce, Apply*

These include (a) curricular content (e.g., iterations of the cross product exploration questions that were developed and changed over time); (b) student responses to early formats of exploration questions; (c) reflective accounts and notes of the developer and the instructor-adopter; (d) outcomes assessment reports of the instructor-adopter; and (e) email communication between the developer and early-adopting instructors. Further detail is provided below in the Data Collection and Data Analysis sections.

Exemplar-based validation. LaBoskey (2004, p. 821) defines this as "[advancing] the field through the construction, testing, sharing, and retesting of exemplars of teaching practice." By creating new curricular materials, testing them, making our work public and open to scrutiny, and retesting the material, we aim for the outcomes of the study to have an impact beyond the individual self (Loughran, 2007).

14.5.2 Data Collection and Analysis

This self-study primarily focuses on the developer (the third author) and an early-adopter (the first author) over eight years of interaction. During the eight years, we shared assignments and student work; communicated about experiences using the cross product exploration via email, at conferences, and over Skype; collaborated on joint research proposals; and we revised and reanalyzed the ways we used the exploration. The developer is situated at a public community college where he regularly teaches both in-person (once a year) and online (twice a year) multivariable calculus sections of 20–30 students. The early-adopter is also an instructor teaching 15–30 students per semester in an in-person course with access to a computer lab at a medium-sized, private four-year college.

Collectively, we analyze the data through two theoretical lenses: instrumental orchestration theory and TPACK for meaningful learning. We use an iterative process, identifying new themes as they emerged, challenging and analyzing our findings throughout.

Instrumental orchestration analysis. Our analysis focuses on the teacher's intended learning objectives and their selected instrumental orchestrations in relation to a specific digital artifact, the cross product exploration. In short, we study how each used this artifact as an instrument; this was recorded as part of the reflective lesson plans and outcomes assessment reports.

TPACK for meaningful learning. Versions of pre-test, exploration, and post-test questions from the artifact from 2008 to present are categorized along two dimensions: learning activity type and pedagogical dimension. Outcomes assessment reports, student responses to exploration questions, emails between the developer and instructors, and notes from meetings between the three authors provide rationale for the changes made in the content, ordering, and wording of the questions and for choices in instrumental orchestrations of the exploration over time.

14.6 RESULTS

14.6.1 Instrumental Orchestrations of the Cross Product Exploration

This self-study reinforces a prevailing observation in the literature that even though digital tools often pre-structure how individuals interact with and consider mathematical ideas (Rabardel, 2001), how a tool is ultimately used impacts how effective it is for the individual who is using it and the purpose for which it is being used (Owston, 1997). In fact, a tool may not be used in a way that the designer originally intended (Rabardel, 1995). This is evidenced in the present study in not only the differences in intended learning objectives of the first and third author, but also in their exploitations of instrumental orchestrations of the cross product exploration.

Exploitations of instrumental orchestrations of the cross product exploration. The most noticeable difference between the lesson plans of the developer and the early-adopting instructor is the positioning of the cross product exploration in their curriculum.

When designing this cross product activity, the developer envisioned it as an out-of-class exploration. In both his on-ground and online courses, after providing written instructions and/or a demonstration, he assigns the cross product exploration to be completed individually outside of class. The developer teaches the cross product content in class before assigning the cross product exploration. He presents the properties of the cross product and a proof of the fact that the magnitude of the cross product of two vectors is equal to the area of the parallelogram determined by the two vectors forming it. Additionally, after demonstrating the computation of the cross product, he asks students to observe the fact that the cross product is orthogonal to the vectors that form it by graphing the cross product vector along with the two vectors. Following the completion of the assignment, he discusses the assignment in class with the students without the technology. The developer's instrumental orchestrations can be classified as *technical demo* (Drijvers et al., 2010) and *discuss tech without it* (Tabach, 2013).

On the other hand, the early-adopting instructor uses *technical demo, work-and-walk-by* (Drijvers, 2011), *technical support* (Drijvers et al., 2013), and *explain-the-screen* (Drijvers et al., 2010) instrumental orchestrations. The early-adopting instructor only provides a geometric definition of the cross product to her students before taking them to a computer lab during class time to work through the exploration. After one semester in which she gave this exploration as an out-of-class assignment and fielded student complaints of having difficulty with using the applet and answering the questions, she ceased assigning this as out-of-class work. Instead, she now assigns this activity in class so that she can float around the room to assist students with both the content and the technology, occasionally stopping the class and discussing a problem on the computer projected on the screen. In her class, students are allowed to work together in groups and are encouraged to discuss their answers with one another.

One other distinction between the instrumental orchestrations of the first and third authors is that the third author assigns a dot product exploration and the cross product exploration to the students at the same time, while the first author has her students complete the dot product exploration in class prior to introducing the cross product content and exploration. The sequencing of these activities may have an impact on student understanding. Students tend to confuse the dot product and the cross product and confound their properties, such as interpreting the dot product as a vector (Van Deventer, 2006) or the cross product as a scalar (VanDieren et al., 2017; Zavala & Barniol, 2010). Whether or not this is mediated better by separating the dot product and cross product assignments over time or connecting the assignments together is unknown.

Learning objectives of the cross product exploration. Linking educational technologies more directly with learning objectives and pedagogical practice can support an artifact's successful implementation (Harris et al., 2010b). Because the developer and the early-adopting instructor have differing expectations of pre-requisite content knowledge of their students when engaging with the cross product exploration, they also aim for different learning objectives. Since the first author provides minimal introduction to the cross product before her students engage with the exploration, her learning objectives leverage pre-existing knowledge about scalar multiplication and include the objective of relating the symbolic equation $(c\mathbf{u}) \times \mathbf{v} = \mathbf{u} \times (c\mathbf{v}) = c(\mathbf{u} \times \mathbf{v})$ with a visual or geometric representation. She intends for students to begin to visualize the orthogonal relationship of the cross product with the two vectors that form it, while the third author expects that students recognize the order of the cross product using an understanding of the right-hand rule.

14.6.2 TPACK for Meaningful Learning with ICT of the Developer and Instructors

Throughout the timeframe of this study, the TPACK of both the first and third authors has developed. In addition to teaching multivariable calculus, differential equations, and calculus, the third author has over a decade of extensive experience developing visualization tools for calculus, multivariable calculus, and differential equations during which time his perspectives and knowledge of technology, pedagogy, and mathematical content have evolved. Notable among the changes in TPACK of the developer was the necessity of learning JavaScript when browsers halted support of Java.

After attending a mini-course on CalcPlot3D taught by the third author at the Mathematical Association of America's MathFest over eight years ago, the first author began using the cross product exploration activity in her multivariable calculus course. She began developing other exploration activities using CalcPlot3D to help students visualize other multivariable calculus concepts. This led her to transition her model theoretic research program into one that included research on undergraduate education. Both the knowledge she acquired while learning how to conduct research on undergraduate education and the research that she conducted on student understanding affected change in her perspective of the pedagogical, technological, and mathematical issues related to successful implementation of the cross product exploration in her classroom.

Before describing the transformation of the cross product exploration over time as the developer and the early-adopting instructor's TPACK changed, we first summarize the pedagogical dimensions of the activities communicated in their lesson plans and reflections. Some of the differences in the pedagogical dimensions have led to changes in the cross product exploration.

Pedagogical dimensions of the activities. Since both the developer and early-adopting instructor assign the same pre-test questions, exploration questions, and post-test questions in their implementation of the cross product exploration, they both use the same learning activities inherent in the questions: *consider, interpret, produce, apply*, and *evaluate* (see Tables 14.2 and 14.3). However, when further analyzing the lesson plans of the developer and early-adopting instructor, differences in two of the five pedagogical dimensions of these activities become apparent. There is no evidence for differences between the first and third author with regard to the *active* and *constructive* pedagogical dimensions, but distinctions in the *cooperative, intentional*, and *authentic* dimensions surface.

The *cooperative* pedagogical dimension is more sustained in the first author's instrumental orchestration than the third author's. Her lesson plans indicate that students engage in the cross product exploration in groups in class while she walks around the room to facilitate discussion throughout the completion of the assignment. The third author also engages in this dimension by discussing the results in class following the assignment completion to be sure everyone is clear on the relationships they were supposed to have observed.

Differences in the *intentional* pedagogical dimension between the first and third authors become apparent in early discussions about the delivery and grading of the exploration questions, and later materialize when the first and third authors take different approaches to implementing the questions on WeBWorK and Blackboard, respectively. In particular, the first author prefers to use the technology to help students self-diagnose their learning gaps during the activity by providing immediate and automatic feedback on student responses, while the third author encourages students to discover and explore without intervention, delaying feedback and grading until after the assignment is completed.

Finally, the *authentic* dimension, which measures the degree to which technology helps students make connections between the content and their personal experiences (Koh, 2013), varies between the first and third authors. How a visual representation can be used to deepen student understanding depends on the learners' knowledge and experience with the material (Ainsworth, 2008). Because the third author situates the cross product exploration assignment after he provides the students with a substantial introduction to the topic while the first author only provides students with a brief introduction to the topic, the *authentic* dimension of the learning activities differ. Additionally, because the third author assigns the cross product and dot product explorations to be completed concurrently, his students may have more opportunity to explore the dissimilitude of the cross product and dot product.

Evolution of the cross product exploration resulting from changing TPACK. While every digital tool is anticipated to transform over time as new hardware and software make older versions obsolete, the cross product exploration has undergone not just technological upgrades to adapt to the times, but also the mathematical content and pedagogical dimensions of the exploration have shifted.

Technological updates to the cross product exploration. Technological problems have directed many changes to the cross product exploration. Some modifications relate to the delivery of the exploration questions, and others involve the translation from Java to JavaScript.

During her first semester using the cross product exploration in class, the first author observed that the students in her section required more than the allotted 50 minutes to complete the assignment. Since there was no mechanism for students to save their work, the delivery of the exploration questions in SurveyMonkey required that students complete their work in one sitting. This presented a technological problem that inhibited her ability to implement the exploration to fit her classroom needs and restrictions. When contacted about this problem, the third author was able to provide a work-around that would allow students in her class to save their answers and return to them later to finish the assignment. Unfortunately, it was too clumsy to implement on a large scale to all potential users of the exploration.

The lack of a save feature was one of many limitations of SurveyMonkey that prompted the developer to consider alternative delivery mechanisms. Another problem is that in order for instructors to receive their student work through SurveyMonkey, they are required to email the developer and ask for him to run a report to generate their students' responses in one large PDF file. This file format is awkward for grading and re-distributing back to students. This extra step discouraged more instructors from using the cross product exploration in their classes. In fact the majority of users of the CalcPlot3D applet do not use the explorations in their classes. To avoid these issues, the first author and the third author have imported the cross product exploration questions into WeBWorK and Blackboard, respectively. This not only avoids the problems presented by SurveyMonkey, but it has advantages because these platforms are designed for student–teacher communication. We discuss these advantages in this chapter in a subsequent subsection.

The large-scale revision of the cross product exploration stems from the shift from Java to JavaScript among web browsers. When it was announced that the mainstream web browsers would no longer support Java, the developer hurried to rewrite the CalcPlot3D applet in JavaScript. This granted the developer not only the opportunity to reflect on the cross product exploration and make additional pedagogical and mathematical improvements described in the next sections, but, also, it afforded students and teachers the option of accessing the applet not just on desktop computers, but also on tablets and cell phones, thereby increasing the assortment of potential instrumental orchestrations of the cross product exploration.

Revision of the mathematical content of the cross product exploration. Since the developer found himself rewriting the cross product exploration tool, he was able to incorporate his own ideas and the suggestions both from the first two authors and from other instructors. Some changes have been made to the wording, content, and type of questions in the exploration, while other significant modifications have been made to the visualization applet. The revisions are informed both by instructor feedback and by research on student understanding of vectors.

Revision to the applet. Research indicates that students have trouble with the right-hand rule. Furthermore the positioning of the vectors (e.g., the ease of applying the right-hand rule without needing to twist one's arm; the plane on which the vectors are situated; and the alignment to a local reference frame) have measurable effects on students' ability to apply the right-hand rule (Kustusch, 2016). One of the limitations of the original cross product exploration is that vectors are always situated on the *xy*-plane. This may have led to students incorrectly over-generalizing responses to question 1 in the exploration (see Table 14.3). For instance, one common response to this question was "[The cross product] point[s] in the z axis direction" (VanDieren et al., 2017). Since Marton and Booth's Variation Theory suggests that activities be structured to ensure students experience a diversity of examples (Lo, 2012), students should experience vectors in a variety of orientations, not just those on the *xy*-plane. Therefore, the developer has added a feature (and questions) to the cross product exploration designed to engage students with examples of vectors off of the *xy*-plane (see Figure 14.3).

Another limitation to the original applet was that it always graphed the cross product of the red vector with the blue vector (Red × Blue). For students who are not too familiar with the right-hand rule, it is not apparent that there is a difference between the cross product Red × Blue and the cross product Blue × Red. To address this, the developer created another feature that allows students to toggle the order of the cross product that is graphed (between Red × Blue and Blue × Red).

Another change, indicative of many others, resulted from email correspondence between the developer and another instructor. This instructor required his students to use radians when referring to angles between vectors, and he, therefore, requested that radians be supplied in the applet. The developer incorporated this request and included both radians and degrees in the new instance of the applet.

Revision to the questions. Throughout the years several questions on the original activity in Tables 14.2 and 14.3 have been reworded, and others added. Early on, the first author suggested adding two questions about the effect of scalar multiplication on the cross product to better align with her learning objectives:

- Keeping the red vector fixed, vary the length of the blue vector but do not change its direction. What happens to the magnitude of the cross product of the red and blue vectors when the blue vector's magnitude is doubled? Tripled? Cut in half?

- How do the geometric interpretation of the magnitude of the cross product and your exploration in the previous problem help explain the formula:

$$c(\mathbf{u} \times \mathbf{v}) = (c\mathbf{u}) \times \mathbf{v} = \mathbf{u} \times (c\mathbf{v})?$$

Another change has been made to the questions because students seem to have the most difficulty answering questions 4 and 5 in Table 14.3. Often the student responses involve a negative angle (VanDieren et al., 2017); so, in addition to rewording these questions, another question about the possible angles between two vectors has been added. Even with these modifications, students still struggle with questions 4 and 5 (VanDieren et al., 2017), which is consistent with the research literature (e.g., Kustush, 2016). To address this, the first author has added another question to the exploration to potentially help students to focus on the orientation, as opposed to the location or angle, between two vectors: can you find two sets of vectors (two red and two blue) in the xy-plane for which the cross product of one pair is in the positive z-direction and the cross product of the other is in the negative z-direction, but the angle between each pair of vectors is the same? (Give an example or explain why this is not possible.)

Although the most demanding aspect of the cross product for students seems to be the right-hand rule, the original pre- and post-tests do not directly question students about this topic. Furthermore, question 7 (see Table 14.3) may not validly measure student understanding of the right-hand rule since it is a multiple-choice question involving only two options which encourages guessing and does not capture the common misunderstanding of students that the cross product is a commutative operation. Therefore, in later versions of the cross product exploration, question 7 in the exploration has been rewritten to include more incorrect options, and new right-hand rule questions have been added to both the pre- and post-test to better align with this learning objective.

Finally, research indicates that the positioning of the vectors (relative to one another and to an axis) can affect how well a student can execute the right-hand rule (Kustusch, 2016). By reflecting on the wording of the questions in the pre- and post-test it has become apparent that the multiple-choice options only include vectors in the first two quadrants. To better test student understanding and to encourage students to think about other configurations of vectors outside of the first two quadrants, a greater variety of multiple-choice incorrect options have been added to questions 1 and 2 of the pre- and post-test which depict vector pairs outside of the first two quadrants.

Variety of pedagogical options for implementing the cross product exploration. With the changes to the delivery method, the activity questions, and the applet, there is an increased variety of pedagogical options for implementing the cross product exploration.

By moving the questions from SurveyMonkey to Blackboard and WeBWorK, more agency is given to instructors to select, redesign, and create new questions and learning activities. Additionally, in both WeBWorK and Blackboard more intentional learning activities may be assigned as these platforms allow for instant grading, automated feedback, and instructor comments—even if there is concern the students do not access instructor comments after the assignment is turned in.

Giving instructors agency to create new problems leads to a greater diversity of types and levels of pedagogical dimensions of learning activities. For instance, the new question—asking students to find two pairs of vectors with the same interior angle, but cross products in opposite directions—may be considered a *produce* and/or *create* learning activity with a significant degree of *constructive* and *active* student engagement.

Finally, several changes have been made to the applet itself that may lead to new pedagogical implementations. Up until the translation from Java to JavaScript, instructors using the cross product exploration were limited to in-class demonstrations on a projected lectern computer, in-class assignments in which students had access to laptops or desktop computers (e.g., a computer lab or 1:1 laptop schools), or out-of-class assignments. Now that the applet runs on cell phones and tablets, teachers have more options for classroom implementation. Another feature of the new version of the applet is the "encode view in URL" option. This feature allows students and instructors to save their work in CalcPlot3D as a URL. This gives instructors more options for designing new classroom activities and provides students with a way to submit images from the applet as part of their solutions to learning activities. The first author has implemented this feature in several learning activities for other multivariable calculus content. For example, students are asked to find a Bezier curve which describes a path of an object that begins at the point (2,0,0), ends at the point (–2,1,1), and does not intersect the sphere of radius 1 centered at the origin. Since there are infinitely many solutions to this, an easy assessment mechanism for the instructor is to ask students to graph their solution in CalcPlot3D to visually verify that it is correct and then copy their encoded URL into an essay prompt box. The instructor can follow the link to see and assess the student's solution.

14.7 DISCUSSION AND CONCLUSION

This study aims to determine the instrumental orchestrations different instructors employ in conjunction with the cross product exploration and how the cross product exploration has evolved as a result of the changing mathematical, pedagogical, and technological understanding of the creator and early-adopting instructor. Although the study examines documentation of use from only two individuals, it provides insights into the cyclic process of planning, designing, and implementing an online digital resource, in general, and on the future development of the CalcPlot3D applet, in particular. Furthermore, the study has tested the synergy of the two theoretical frameworks (TPACK for meaningful learning with ICT and instrumental orchestration) in an undergraduate-level mathematics context; it has also highlighted limitations of the instrumental orchestration framework in an online context.

While a self-study has inherent limitations in scope and generalizability, there are some benefits to this approach over the existing literature. Our analysis of eight years' worth of correspondence and documentation covers a broader timeframe than most other studies on larger samples, allowing us to examine not just one instance or version of an online resource but how one resource has changed over time. Additionally, although much research on the appropriation of technology into the classroom focuses on the instructors and developers, viewing students as passive recipients of change (Lin, 2001), our study incorporates student feedback (documented informally through email correspondence or more formally in outcomes assessment reports and research studies) in the form of responses to exploration questions and in the form of comments on the usability of the exploration. The feedback from students has not only contributed to the assessment of the effectiveness of the cross product exploration, but has been an impetus for changes in the activity and its instrumental orchestrations over the years. Furthermore, undertaking this study has served as a catalyst to further the discussion and research among the authors on not only how to improve the appropriation of the cross product exploration activity and CalcPlot3D in general, but also on how to improve student understanding of multivariable calculus concepts.

14.7.1 Further Development of Online Resources

This study points to the variation in instrumental orchestrations of instructors implementing an online resource in their classroom. In particular, the prerequisite knowledge of the students and the positioning of the artifact in instruction cannot be assumed to be the same across all appropriations of a given resource. This has, in part, inspired the developer to create an online help manual* for the CalcPlot3D applet, which may provide instructors with a variety of options and information (e.g., useful prerequisite knowledge) for implementing the exploration in their classes.

Another difference in instrumental orchestrations observed in this study involves the *intentional* pedagogical dimension. This may be useful to consider as we develop other exploration activities for complex concepts using multiple representations that may require several layers of understanding (e.g., Ainsworth's (2006; 2008) six types of understanding). The *intentional* dimension can be realized by providing immediate feedback as to whether a student has responded correctly to a problem. This feedback can prompt students to search for alternative solutions and refine their thinking iteratively as they work, rather than at the end of the process, thereby minimizing the formation of misconceptions (Olive et al., 2009). As we build new explorations in Blackboard and WeBWorK we will critically examine the option to provide immediate feedback to students as they work through problems.

This study also demonstrates that providing instructors agency in adapting the resource to fit their needs and circumstances can improve the resource overall. As the CalcPlot3D applet continues to evolve, the developer and early-adopting instructor will continue to communicate with each other and with other instructors so that they may incorporate instructors' experiences with the applet into future revisions and provide instructors more opportunities for adapting the resource. This study may also encourage others as they develop and revise digital resources to communicate with early-adopting instructors.

14.7.2 Synergy and Limitations of the Theoretical Frameworks

Because the TPACK framework draws criticism for not having effective operational definitions, we used the TPACK for meaningful learning with ICT framework to analyze our data instead of identifying specifically the pedagogical, technological, and content knowledge of the developer and the early-adopting instructor. With this approach, the TPACK is implicit in the choice of the learning activities and pedagogical dimensions. Overall the

---

* c3d.libretexts.org/CalcPlot3D/CalcPlot3D-Help/index.html.

two frameworks employed in this study worked well together and allowed us to reflect on the planning, design, and revision of the cross product activity. Although the taxonomy for learning activities was developed for K-12 mathematics content, it was well-suited for our undergraduate-level context.

One drawback of the frameworks that we observed was that the categories for instrumental orchestrations may be insufficient to capture the types of orchestrations that instructors use in hybrid or online courses or in resources accessed on cell phones or tablets. The categories were originally developed through observations of on-ground teaching, but there appear to be differences in orchestrations between on-ground and online courses. For example, the developer has used this cross product exploration activity in his online courses for several years. His instrumental orchestration of providing students a video lesson or PDF description of how to use the applet may not squarely fit into the *technical demo* category, and there may be significant differences between this orchestration and another instructor showing the students in person how to use the instrument. Another example concerns the *work-and-walk-by* orchestration. Would being available for students via email to answer a question be considered the same as an instructor walking around the room guiding students on work that she observes and ready to answer a student who raises his/her hand? These examples indicate the need for further research on instrumental orchestrations in online learning environments.

Since the ability to use the applet on cellphones and tablets is relatively new, the early-adopting instructor and developer have not yet had the chance to explore how this may change their instrumental orchestrations and their students' instrumental genesis of the artifact. Orchestrations using the cell phone or tablets may not be represented in the current taxonomy. Future research on a larger sample of instructors may be needed to identify new categories or clarify existing categories of instrumental orchestrations when using a cell phone or tablet instead of a computer.

REFERENCES

Ainsworth, S. (2006). DeFT: A conceptual framework for considering learning with multiple representations. *Learning and Instruction, 16*, 183–198.

Ainsworth, S. (2008). The educational value of multiple-representations when learning complex scientific concepts. In J. K. Gilbert et al., (Eds.), *Visualization: Theory and Practice in Science Education* (pp. 191–208).

Borys, Z. & Choppin, J. (2017). Tensions between resource perspectives and trends in design and dissemination of digital resources. Paper Presented at the CERME 10 - Tenth Congress of European Research in Mathematics Education. https://keynote.conference-services.net/resources/444/5118/pdf/CERME10_0258.pdf.

Cox, S. & Graham, C. R. (2009). Diagramming TPACK in practice: Using an elaborated model of the TPACK framework to analyze and depict teacher knowledge. *TechTrends, 53*(5), 60–69.

Drijvers, P. (2011). Teachers transforming resources into orchestrations. Paper Presented at the CERME 7 - Seventh Conference of European Research in Mathematics Education. http://www.cerme7.univ.rzeszow.pl/WG/15a/CERME7-WG15A-Paper06_Drijvers.pdf.

Drijvers, P., Doorman, M., Boon, P., Reed, H. & Gravemeijer, K. (2010). The teacher and the tool: Instrumental orchestrations in the technology-rich mathematics classroom. *Educational Studies in Mathematics, 75*, 213–234.

Drijvers, P., Tacoma, S., Besamusca, A., Doorman, M. & Boon, P. (2013). Digital resources inviting changes in mid-adopting teachers' practices and orchestrations. *ZDM Mathematics Education*, 45, 987–1001.

Govender, N. & Gashe, B. (2016). Framework categorization of pre-service physics teachers' conceptions of vector-kinematics. *Journal of Baltic Science Education*, 15(3), 325–339.

Grandgenett, N., Harris, J. & Hofer, M. (2011). *Mathematics Learning Activity Types*. Retrieved from College of William and Mary, School of Education, Learning Activity Types Wiki: http://activitytypes.wm.edu/MathLearningATs-Feb2011.pdf.

Greenslade, T. B. (1980). Ancestors of the right-hand rule. *The Physics Teacher*, 18(9), 669.

Gueudet, G. & Trouche, L. (2009). Towards new documentation systems for mathematics teachers? *Educational Studies in Mathematics*, 71, 199–218.

Hansen, A., Mavrikis, M. & Geraniou, E. (2016). Supporting teachers' technological pedagogical content knowledge of fractions through co-designing a virtual manipulative. *Journal of Mathematics Teacher Education*, 19, 205–226.

Harris, J., Grandgenett, N. & Hofer, M. J. (2010a). Testing a TPACK-based technology integration assessment rubric. In D. Gibson & B. Dodge (Eds.), *Proceedings of Society for Information Technology and Teacher Education International Conference 2010* (pp. 3833–3840). Chesapeake, VA: AACE. http://scholarworks.wm.edu/bookchapters/6.

Harris, J. B., Hofer, M. J., Schmidt, D. A., Blanchard, M. R., Young, C. Y., Grandgenett, N. F. & Van Olphen, M. (2010b). "Grounded" technology integration: Instructional planning using curriculum-based activity type taxonomies. *Journal of Technology and Teacher Education*, 18(4), 573–605.

Harris, J., Mishra, P. & Koehler, M. (2009). Teachers' technological pedagogical content knowledge and learning activity types: Curriculum-based technology integration reframed. *Journal of Research on Technology in Education*, 41(4), 393–416.

Heid, K. (2005). Technology in mathematics education: Tapping into visions of the future. In *Technology-supported Mathematics Learning Environments: NCTM 67th yearbook*. Reston, VA: NCTM.

Howland, J. L., Jonassen, D. & Marra, R. M. (2012). *Meaningful Learning with Technology* (4th ed.). Boston MA: Allyn & Bacon.

Koh, J. H. L. (2013). A rubric for assessing teachers' lesson activities with respect to TPACK for meaningful learning with ICT. *Australasian Journal of Educational Technology*, 29(6), 887–900.

Kosheleva, O. & Giron, H. (2006). Technology in K-14: What is the best way to teach digital natives? *Proceedings of the 2006 International Sun Conference on Teaching and Learning*, El Paso, TX, March 3–4, 2006, http://www.cs.utep.edu/vladik/2006/olg06-11.pdf. Accessed August, 2018.

Kozma, R. (2003). The material features of multiple representations and their cognitive and social affordances for science understanding. *Learning and Instruction*, 13, 205–226.

Kustusch, M. B. (2016). Assessing the impact of representational and contextual problem features on student use of right-hand rules. *Physical Review Physics Education Research*, 12(1), 010102. doi:10.1103/PhysRevPhysEducRes.12.010102.

LaBoskey, V. K. (2004). The methodology of self-study and its theoretical underpinnings. In J. J. Loughran, M. L. Hamilton, V. K. LaBoskey & T. Russel (Eds.), *International Handbook of Self-Study of Teaching and Teacher Education Practices* (pp. 817–869). Dordrecht: Kluwer.

Lawless, K. A. & Pellagrino, J. W. (2007). Professional development in integrating technology into teaching and learning: Knowns, unknowns, and ways to pursue better questions and answers. *Review of Educational Research*, 77(4), 575–614.

Lin, X. (2001). Reflective adaptation of a technology artifact: A case study of classroom change. *Cognition and Instruction*, 19(4), 395–440.

Lo, M. L. (2012). *Variation Theory and the Improvement of Teaching and Learning: In Gothenburg Studies in Educational Sciences: Vol. 323*. Göteborg: Acta Universitatis Gothoburgensis.

Loughran, J. (2007). Researching teacher education practices: Responding to the challenges, demands, and expectations of self-study. *Journal of Teacher Education, 58*(1), 12–20.

Miller-Young, J. E. (2013). Calculations and Expectations: How engineering students describe three-dimensional forces. *Canadian Journal for the Scholarship of Teaching & Learning, 49*(1), 1–11.

Mishra, P. & Koehler, M. J. (2006). Technological pedagogical content knowledge: A framework for teacher knowledge. *Teachers College Record, 108*(6), 1017–1054.

Nguyen, N. L. & Meltzer, D. E. (2003). Initial understanding of vector concepts among students in introductory physics courses. *American Journal of Physics, 71*(6), 630–638.

Nieminen, P., Savinainen, A. & Viiri, J. (2013). Gender differences in learning of the concept of force, representational consistency, and scientific reasoning. *International Journal of Science and Mathematics Education, 11*, 1137–1156.

Olive, J., Makar, K., Hoyos, V., Kor, L. K., Kosheleva, O. & Sträßer, R. (2009). Mathematical knowledge and practices resulting from access to digital technologies. In C. Hoyles & J. B. Lagrange (Eds.), *Mathematics Education and Technology - Rethinking the Terrain*. New ICMI Study Series (Vol. 13). Boston, MA: Springer.

Owston, R. (1997). The world wide web: A technology to enhance teaching and learning? *Educational Researcher, 26*(2), 27–33.

Rabardel, P. (1995). *Les hommes et les technologies. Approche psychologique des instruments contemporains*. Paris: Armand Colin. [English translation: Rabardel, P. (2002). *People and Technology: A Cognitive Approach to Contemporary Instruments*.]

Rabardel, P. (2001). Instrument mediated activity in situations. In A. Blandford, J. Vanderdonckt & P. Gray (Eds.), *People and Computers XV - Interactions Without Frontiers* (pp. 17–30). Berlin: Springer-Verlag.

Runesson, U. (2006). What is it possible to learn? On variation as a necessary condition for learning. *Scandinavian Journal of Educational Research, 50*(4), 397–410.

Ruthven, K. (2014). Frameworks for analysing the expertise that underpins successful integration of digital technologies into everyday teaching practices. In A. Clark-Wilson, O. Robutti & N. Sinclair (Eds.), *The Mathematics Teacher in the Digital Era: An International Perspective on Technology Focused Professional Development: Mathematics Education in the Digital Era* (Vol 2, pp. 373–393). Dordrecht: Springer.

Scaife, T. M. & Heckler, A. F. (2010). Student understanding of the direction of the magnetic force on a charged particle. *American Journal of Physics, 78*(8), 869–876.

Seeburger, P. (2020). CalcPlot3D [Computer software]. Available from https://c3d.libretexts.org/CalcPlot3D/index.html

Shulman, L. S. (1986). Those who understand: Knowledge growth in teaching. *Educational Researcher, 15*(2), 4–14.

Tabach, M. (2011). A mathematics teacher's practice in a technological environment: A case study analysis using two complementary theories. *Technology, Knowledge, and Learning, 16*, 247–265.

Tabach, M. (2013). Developing a general framework for instrumental orchestration. Paper Presented at the CERME 8 - Eighth Conference of European Research in Mathematics Education. http://www.cerme8.metu.edu.tr/wgpapers/WG15/WG15_Tabach.pdf.

Trgalová, J. & Rousson, L. (2017). Model of appropriation of a curricular resource: A case of a digital game for the teaching of enumeration skills in kindergarten. *ZDM Mathematics Education, 49*, 769–784.

Trouche, L. (2004). Managing the complexity of human/machine interactions in computerized learning environments: Guiding students' command process through instrumental orchestrations. *International Journal of Computers for Mathematics Learning, 9*, 281–307.

Van Deventer, J. (2006). *Comparing Student Performance on Isomorphic Math and Physics Vector Representations* (Master's thesis). https://library.umaine.edu/theses/pdf/restricted/VanDeventerJ2008.pdf.

Van Domelen, D. J. (1999). Artificial right-hand rule device. *The Physics Teacher, 37*(8), 500.

VanDieren, M., Moore-Russo, D., Wilsey, J. & Seeburger, P. (2017). Students' understanding of vectors and cross products: Results from a series of visualization tasks. In A. Weinberg, C. Rasmussen, J. Rabin, M. Wawro & S. Brown (Eds.), *Proceedings of the 20th Annual Conference on Research on Undergraduate Mathematics Education* (pp. 991–1000). San Diego CA.

Zavala, G. & Barniol, P. (2010). Students' understanding of the concepts of vector components and vector products. In C. Singh, M. Sabella & S. Rebello (Eds.), *AIP Conference Proceedings* (Vol. 1289, pp. 341–344).

CHAPTER 15

Enhancement of Mathematics Learning through Novel Online Tools

Zohreh Shahbazi

CONTENTS

15.1 INTRODUCTION

The Math and Statistics Learning Centre (MSLC) at the University of Toronto Scarborough (UTSC) has been established as a collaboration between the departments of Computer and Mathematical Sciences and the Centre for Teaching and Learning both to support students taking undergraduate mathematics courses and to enrich their learning experience in mathematics in general. Various small group sessions, review modules, and seminars have been developed and offered. Special attention has been given to understanding the sources of students' challenges in learning mathematical concepts and exploring new methods to engage students. It is our goal to create a vibrant, welcoming environment for students to come to appreciate the beauty and utility of mathematics.

Of the various learning activities developed at the MSLC, the series of review modules has proven to be particularly successful. The material covered in these review modules include algebraic manipulations, inequalities, and functions (including trigonometry and inverse functions). These review modules have been developed in different formats over the past several years. They are offered at the beginning of each term on a voluntary basis for students who wish to improve their skills in mathematics fundamentals. Additionally, a short non-credit summer course has been created which covers much of the content of the review modules. The impact of this short course on students' learning has been studied in collaboration with the UTSC psychology department. It was discovered that in addition to improving students' mathematical skills, the students also found the course enjoyable and rated it positively. Students also indicated that the course helped them adjust to the transition from high school to university [1]. In the spring of 2015, the MSLC received funding from the Council of Ontario Universities (COU) to develop online video versions of the review modules. We were able to develop 12 online modules by the end of the summer of 2015. Modules 1–8 cover foundational concepts, and modules 9–12 cover advanced concepts. Currently, the online modules are used as self-directed learning support in various UTSC calculus courses. In addition, UTSC offers an Online Mathematics Preparedness Course during summer, which makes use of the online modules. The design and implementation of the various components of the learning modules in multiple support contexts are discussed in Sections 15.2–15.5 of this chapter. The resources may be accessed at: www.utsc.utoronto.ca/math-instruction/

In order to develop students' writing and communication skills in mathematics, the MSLC launched its annual magazine and Math In Action Journal in 2014 and 2017 respectively. MSLC Magazine is a resource for what is happening at UTSC's Math & Statistics Learning Centre. Math In Action (MIA) is an undergraduate research journal developed using a Teaching Enhancement Grant. The journal aims to provide a platform for mathematics undergraduates to share their work with peers and academics alike. Math In Action may also be used for research assignments in senior undergraduate and graduate mathematics courses. It is our hope that this journal will help to foster greater student engagement in mathematics programs. It will also give researchers the opportunity to interact with students and inspire greater interest, with the hope of creating stronger generations of future researchers. Submissions to MIA are accepted in the form of video presentations, along with a two-page extended abstract. Students are encouraged to use creative approaches in presenting their work. The design and implementation of MSLC Magazine and Math In Action Journal will be discussed in Sections 15.6–15.7. The journal may be found at: www.mathinactionjournal.com/.

15.2 PEDAGOGICAL APPROACHES

In the design stage of the online modules and MIA, we have considered well-established pedagogical approaches, theories, and best practices such as constructivism, backward design, educated assessments, promotion of active learning, and creativity. Below I will explain the implementation of such approaches in modules and MIA design in more detail.

15.2.1 Integrating Technology in Teaching Mathematics

The Principles and Standards for School Mathematics state that: "Technology is essential in teaching and learning mathematics; it influences the mathematics that is taught and enhances students' learning" [2, p. 11]. There are many opportunities to integrate technology into teaching methods. Tablets, smartphones, and internet access can all be effectively incorporated into teaching. Selwyn [3] indicates that the use of technology will individualize and personalize learning. Teachers can create videos of their lessons and stream them online to allow multiple viewings for students [4]. Videos should cover the main topics for 10 to 15 minutes, and can be posted on the teacher's course website [5]. Teachers can verify students' viewing of videos by embedding questions, tasks, or quizzes [6]. Moreover, students' development of questions while watching a video lesson will increase their conceptual understanding of the subject [5, 6]. Effective integration of technology allows students to use their current understanding of topics to acquire more advanced knowledge. This notion is consistent with the constructivist theory of learning in mathematics education. Constructivism advocates an active learning environment in which students construct their own knowledge through senses and social negotiations, with the aid and guidance of their mentors [7]. In this model, instructors are considered facilitators rather than dispensers of knowledge, and students construct their own knowledge through realistic exercises and social interactions. In the 1980s, the National of Council of Teachers of Mathematics [2] published documents which described the importance of encouraging students to construct their own mathematical knowledge in order to become proficient and effective learners of mathematics. This new approach to teaching and learning mathematics is referred to as reform in mathematics education. In the design of the modules, we attempted to model an active learning environment to better guide students in the process of building their new knowledge. Consequently, we created diagnostic tests that students could use to examine their knowledge. Once a weakness is identified, students can watch the relevant video lessons to enhance their skills. The video lessons typically start with a question. While the instructor tries to solve the problem, she covers the background concepts and formulae. The message here is that questions are the most valuable starting points for building new knowledge.

15.2.2 Backward Design

Research has shown that relevant learning objectives promote deep learning [8, 9, 10, 11]. Learning objectives are statements that describe knowledge or skills that students should have acquired by the end of a particular assignment, class, course, or program. Properly selected learning objectives can help to challenge students to learn concepts beyond the level of basic factual recall. In the design stage of the modules, we used a backward design process by identifying content objectives with reference to the cognitive domain of Bloom's Taxonomy [12]. Bloom's Taxonomy describes a way to organize skills into three domains: cognitive, psychomotor, and affective. Each domain consists of six or seven levels, from the most basic to the most complex (see Figure 15.1 for more details). The backward design process emphasizes the instructor's role as the designer of student learning processes, linking learning goals to corresponding assessments of student understanding, all of which are supported by effective, scaffolded learning activities.

For example, the content goals for Module 1 are listed below:

Module 1: Algebraic Manipulations
At the end of this unit students will be able to:

- Recognize different types of numbers (natural numbers, integers, rational and irrational numbers) based on a criterium or a simple justification

- Use proper interval and set notations and operations for expressing a given set of numbers

- Apply proper algebraic manipulations and operations for simplifying an expression

Dee Fink presented an alternative taxonomy of learning objectives in 2003, which shifts the emphasis from content to the skills that students will retain from the course [13]. It argues that learning objectives should include certain humanistic dimensions, such as caring about a subject, personal and social implications of knowing a subject, etc. Fink's Significant Learning Taxonomy integrates cognitive and affective learning domains in a non-hierarchical form, and can be used to complement Bloom's Taxonomy.

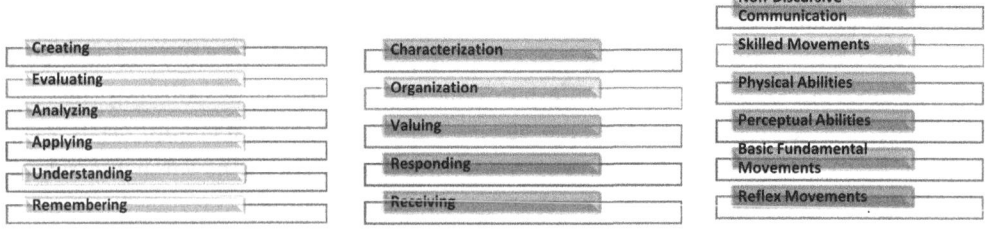

FIGURE 15.1   Cognitive, affective, and psychomotor hierarchies in Blooms Taxonomy.

With Fink's Significant Learning Taxonomy in mind, a list of human dimensional goals were identified in the module design stage to support mathematical thinking and understanding:

- Provide students with a set of resources that would help with increasing students' interest in the subject.

- Provide a set of resources for students to enable students to easily monitor their own progress and identify their own weaknesses.

- Help students to realize and apply better learning strategies.

- Develop mathematical thinking and understanding in students by guiding them towards deep thinking rather than "memorizing all of the rules" by providing challenge questions in the assessment tests.

While the Bloom's Taxonomy was useful for choosing the modules content, Fink's Taxonomy was more useful for identifying successful methods in mathematical learning and modeling such methods. For example, the idea of creating an animated overview of each module helped to raise students' interest in the subject

To have a product with a strong design that helps students to learn topics deeply, we attempted to come up with teaching and learning activities that were well-aligned with our learning goals and assessment tools. Then we could hope that even years later, students would be able to recall what they had learned from the modules. The diagram in Figure 15.2 emphasizes this idea based on the Integrated Design model [13] (Figure 15.2).

We attempted to design the learning support modules in such a way that instructors would have the flexibility to use them anytime, anywhere, at whatever pace suits their specific teaching goals. In order to achieve this objective, the online materials are provided as a set of sequenced resources, illustrative of a sequence of scaffolded lesson units. In other

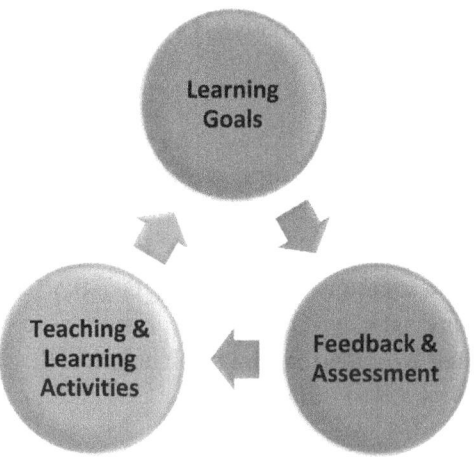

FIGURE 15.2   Integrated design model.

words, the lesson modules are in order from most basic to most advanced. We have also considered accessibility throughout the entire design process in order to maximize the usefulness of these modules for all learners.

15.2.3 Educated Assessments

The module structure reflects a student-centric design. One key component is the orientation stage, which links to additional guides and supporting resources. Materials and assignments are organized into "chunked" units to support ease of access and students' progression through sequenced and scaffolded learning activities. Screencast segments present content, but also model approaches to problem solving. To maximize student engagement, the activities emphasize individual learning, but may also be adapted by instructors for group activities using synchronous learning environments or discussion boards, or for use within academic success/skills centers. In addition to self-directed activities, the modules include optional activities for individual or group assignments in online courses or in the classroom component of hybrid courses. An instructor's guide accompanies each of these activities. Both the content and activities are supplemented with a flexible assessment framework, allowing course instructors to adopt assessments as iterative formative assessment, or as part of their overall summative grading scheme [14].

15.2.4 Promotion of Active Learning

Research shows that students who have active, inquiring minds are more likely to be successful in learning mathematics [11, 15]. Student retention rate will be significantly higher if a course makes use of appropriate activities [16]. In each module, we model the inquiry process and guide students in asking questions, as well as answering them. The modules are designed to strengthen the foundational skills needed for success in all standard first-year calculus courses, as well as to provide thorough explanations of the relevant calculus topics. Students may use these modules independently or follow the guidance of their instructor. A scaffolded progression through the various activities will support the development of the active level of mathematical thinking appropriate for university undergraduate courses. The aim of these activities is to improve students' skills, enthusiasm, motivation, and confidence.

15.2.5 Promotion of Creativity

In many mathematics programs, the course design is usually focused on lower levels of thinking such as memorization, understanding, and application of concepts and methods. However, we need to consciously develop students' higher-order thinking processes such as analysis, evaluation, and creation. Consider the fact that creativity sits at the top of Bloom's Pyramid. What *is* creativity? Can we train ourselves to be more creative? Can we teach creativity to students? And how can we teach *creatively*?

According to Tony Buzan,

> creativity, by its very nature, implies getting away from the norm. Normal is that to which your brain has become accustomed; that which gives you no surprises; that

which remains the same; that which no longer shocks, startles, surprises or provokes you; that which does not stretch your imagination. To create means virtually the opposite: to bring into existence something new; to give rise to; to establish an association that has never been established before

[17].

The primary way that our brains learn is by association and connection. The way a central idea connects to related concepts shapes how we think about a subject. In order to generate new ideas, we need to make more and more connections between different ideas, just like a mind map. Creativity is a conscious and associative process. Our minds' potential for generating new ideas is all but limitless, considering the infinite possibilities of connecting details to one another. A practical deliberate effort toward thinking creatively could help us to design more innovative lectures and activities in order to stimulate our students' creative thinking in turn.

Another way to look at creativity is illustrated by Edward Bono's Six Hats Thinking Model. According to Edward Bono, one can break down the process of thinking into six stages and represent each stage with a colored hat. We can imagine ourselves constantly changing these hats until we find an appropriate solution to a given problem. In this model, the blue hat is responsible for defining a problem clearly, managing our use of time and flow of ideas, and assisting in communication between wearers of different hats. The hat is blue like the sky, which demonstrates its leading role. Some of the types of questions that we might ask while wearing this hat could be: "what is the problem that we are facing? What are our goals and desired outcomes?" The white hat looks like a blank piece of paper, ready to record whatever data our inquiry uncovers. The types of questions that we might ask while wearing this hat might be along the lines of: "what do I know and what don't I know about this problem?" The yellow hat represents the initial optimistic stage of thinking about a problem. It provides us with a roadmap to our ultimate goal, and gives us the motivation to work towards it. It deals with questions like: "what are the benefits and potential positive outcomes of our proposed solutions?" The black hat represents the pessimistic stage of thinking, which is every bit as necessary as the optimistic stage, because it can help us to avoid making mistakes as we carry out our solutions. The types of questions we might expect from the black hat could include: "what are some possible flaws of this way of thinking? What are the drawbacks of our method?" The red hat is all about emotions and intuition. While wearing this hat, we are concerned with internal conflicts, which is why the hat is the color of the blood that courses through our veins. The types of questions that we might ask at this stage could include "is this the right approach? Intuitively, what seems to be the best way to solve this problem?" Finally, the green hat is concerned with creativity itself, which is why it bears the color we most associate with the natural world. While wearing this hat, we strive to bring new methods and ideas into the foreground, in a way that surprises the wearers of other hats. While wearing this hat, we might ask questions like: "could we solve this problem more easily using another method? What other approaches might we consider?"

As discussed in [18, 19], we allow our students to wear the white hat more often than the others, and in my opinion, we ought to strive for a more complete wardrobe of hats. This is why we created MIA—to encourage students to wear their green hats more often, both as undergraduates and beyond.

15.3 ONLINE CALCULUS MODULES: DESIGN OBJECTIVES AND ORGANIZATION

It is common for first-year students to have difficulties passing their first-year calculus courses. One of the main reasons for this is that they have generally not been taught the foundational skills required to succeed in these courses very well in high school. In order to help students to understand required concepts, we have developed review modules for students who wish to brush up on foundational concepts, which are offered at the beginning of each term. Over the past several years, first-year calculus instructors have included these modules in their course syllabi in various ways. For example, students have been offered bonus marks for attending extra-curricular sessions involving the review modules. Through using the modules for numerous semesters, we have been able to identify common issues that students have had in learning foundational concepts. This opportunity has helped us to refine our methods for teaching the key topics with a particular emphasis on active learning and carefully chosen examples. Both students and course instructors express that the modules have been very useful in improving students' understanding of basic calculus topics. In addition, we have developed a short non-credit summer course using the content found in the review modules. The impact of this course on students' learning has been studied in collaboration with the UTSC psychology department. We found that in addition to improving students' mathematical skills, students found the course enjoyable and rated it positively. Students also expressed that the course helped them to adjust to the transition from high school to university [1]. After the study was completed, we started to think about creating an online version of the modules with a similar structure that students could use on their own. In the spring of 2015, the MSLC received funding from the Council of Ontario Universities (COU) to develop these online video modules. We were able to develop 12 online modules by end of the summer of 2015. Modules 1–8 cover foundational concepts, and Modules 9–12 cover advanced concepts. The design and implementation of the various components of the learning modules in multiple support contexts were discussed briefly in [20]. In this section, we review the process in more depth.

15.3.1 Module Components

We have developed 12 online modules: Modules 1–8 cover foundational concepts, and Modules 9–12 cover advanced concepts. Each module contains the following:

Introductory Animation: Each module begins with a three- to four-minute animation, totaling 39 minutes across all modules. The main purpose of the animations is to introduce the module topics in an engaging way. The animations are also useful for the purposes of visualizing more the abstract topics. Our goal was to create a learning tool that would help improve students' (often negative) attitudes toward learning mathematics. We formed a team consisting of an educational developer, a lead instructor, and a creative advisory team which included two students. The team worked together to write the mathematical content as well as the animation scripts. The videos were created in collaboration with the animation company Flikli Video Content Studios. Our team contributed to several rounds of revisions of the storyboards, provided feedback to the animation company, and sent the animations to collaborators and reviewers from other universities for additional feedback. For accessibility purposes, the animations are provided with audio transcripts and descriptive video scripts.

The final product impressed the reviewers; they remarked that the cartoons were not only educational, but also fun to watch. In my own calculus courses, I often use the animations to introduce topics or reinforce key concepts. I usually start calculus courses for mathematics and computer science majors by explaining proof techniques, making use of a number of examples. I show Module 12 as a part of my introductory lecture. Students tend to assume that I found the animation somewhere online, until they see the little cartoon figure of me appear on screen, which never fails to get a laugh (Figure 15.3).

The details of the scenes and visual tools that we used to create the animation for Module 12 are provided in the appendix.

Lessons: each module includes two to five videos, which model approaches to problem solving as relates to the module's content. The combined length of the videos is around ten hours. All that was used to create these lessons was a USB camera, a microphone, printed card stock, pencil, and lined paper. Each instructional video begins with a few questions. For example, in the module which covers solving inequalities, we used five examples of different types of inequalities. We listed important rules and laws relating to a specific concept on a couple of pieces of card stock. The instructor used the card stock to write out solutions to the practice questions for the unit. The learning atmosphere of the videos is meant to be simple, just like traditional teaching on the blackboard. The purpose of the lessons is to engage students as they learn mathematics in a simple, natural, and constructive manner. The lessons have been organized into "chunked" units for ease of access, and the lessons and activities are sequenced in such a way that the students' skillsets improve gradually and fluidly.

These videos were also sent to University of Toronto Mississauga (UTM), Ontario Institute for Studies in Education (OISE), and McMaster collaborators and reviewers for additional feedback.

FIGURE 15.3   Animations characters Susie and Tommy.

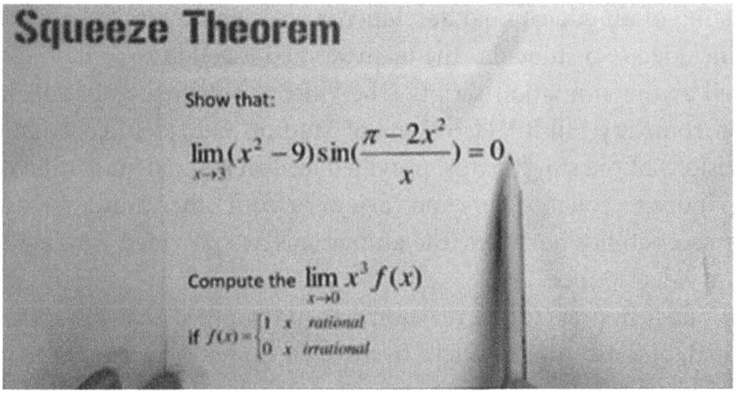

FIGURE 15.4   A lesson snapshot.

For accessibility purposes, the lessons are provided with audio transcripts (Figure 15.4).

15.3.1.1 Modules 1–8

Diagnostic Test + Solutions: There is an initial diagnostic test in order to determine the students' understanding of basic mathematical skills. The test is in multiple-choice format and includes questions related to Modules 1–7. The test includes 56 questions (eight questions from each of Modules 1–7). Further diagnostic assessments are provided at the end of Module 7, which is a transition point between foundational and advanced concepts [21, 22].

Practice Questions + Solutions: Additional questions (and their solutions) are provided on a variety of topics for each module. These questions were designed using Bloom's Taxonomy. Some more challenging problems have been added to the problem sets to stimulate students' creativity and innovative thinking in mathematics. Our hope is that students will learn how to be critical thinkers through attempting to solve these problems. From there, we hope that students will apply these skills to other courses, and experience more "Eureka!" moments as they continue to learn. Students are referred to the instructional videos for additional support if they have difficulties with the practice questions.

Assessment Test + Solutions: Once students feel they have achieved a solid understanding of the module's content, and are ready to test their skills, they can try the assessment test. Ideally they won't look at the answers until *after* they have taken the test. If students do well on the test, they are likely prepared to excel in this topic when it arises in their coursework. If they find the test difficult, it likely means that they should spend some more time with the videos and other materials provided.

15.3.1.2 Modules 9–12

Calculus Notes: This is a 40-page handbook explaining selected topics through carefully chosen examples, including an additional set of practice questions (and their solutions). This is expected to help students to gain a better understanding of the concepts in general and improve their ability to solve related problems.

LaTeX files are available for accessibility accommodation purposes.

15.3.2 Implementation

Since the fall of 2015, the modules have been used extensively in MATA31 (Calculus for Computer and Mathematical Sciences) at UTSC as an additional resource. In addition, Modules 1–7 were used as primary resources for a new online mathematics preparedness course offered at UTSC, a course designed for students lacking a solid background in high school mathematics. In this section, we discuss the structure of this course, which was offered in the summer of 2016.

15.3.3 Online Mathematics Preparedness Course—2016

The course covered seven modules in three weeks during the summer. The weekly structure was as follows:

Tuesday and Wednesday—students watched the animations and lessons within the assigned modules.

Thursday—students did the practice questions within each module as well as the diagnostic tests.

Friday—students completed the assessment test for each module by 5 pm.

Sunday—grades were posted on Blackboard.

The course assessment was based on seven assignments worth 5% each and a final exam worth the remaining 65%. The assignments and exam were all available online, and students were required to submit their solutions online by a certain time. The final exam was a three-hour test with ten multiple-choice and nine problem-solving questions. Students were required to submit their solutions on the course Blackboard page. Since the course was not for credit, we were not particularly concerned with any possible cheating. However, students who received a grade of more than 65% in the course were permitted to use it as a pre-requisite instead of the equivalent high school course. Eighteen students completed the course assignments and exam, 12 of whom passed the course with a grade of more than 65%.

The facilitator of the course was assigned a total of 25 hours for the following course activities:

- Preparation of course management system
- Monitoring students' progress
- Answering students' questions online
- Marking three sets of assignments and the final assessments
- Moderating students' posts on the discussion board

In my undergraduate mathematics courses, I employ active learning techniques such as Think-Pair-Share, Say Something, and Ticket Out The Door. My observations over years of using these techniques affirm the idea that active learning increases interest among students and promotes critical thinking. Additionally, active learning leads to a deeper understanding of the content, higher grades, improved self-esteem, and a stronger tendency to stay on task. Effective usage of active learning techniques can help to better engage students. In fact, as shown in Figure 15.9, active learning and motivation are two main components of student engagement. In the Venn diagram, student engagement is described as a product of motivation and active learning. In the Double Helix model, however, active learning and motivation interact and intensify each other's effect [16]. However, designing these types of activities is both difficult and time-consuming for the instructor. It's just like Sandra Griffiths says: "Successful small group teaching does not happen by chance. Planning for effective small group teaching is as important as planning any other teaching activity" [23] (Figure 15.5).

Our main concern going forward is which types of activities we can design for the Online Mathematics Preparedness Course to engage students even further. One solution we have tried in the summer of 2017 is using an online discussion board. During each step of the process, the instructor encouraged students to post their questions on the Blackboard discussion board. She closely monitored the discussion and responded to students' questions. Students were reminded of the value of peer-to-peer learning. In addition, a participation grade was assigned for these Blackboard posts which were worth a total of 5% of the course grade. The final exam's weight in the syllabus was reduced to 60% to accommodate this. Figures 15.6 and 15.7 are screenshots of the discussion board and a sample student answer to one of the posted questions. For this summer, we extended the time line of the course from three weeks to one month to allow students more time to absorb the concepts (Figure 15.8).

The total page views or on-click actions for videos from May 1, 2017 to April 22, 2018 were 20,801, while 13,848 unique page views were recorded during the same time slot.

15.3.4 Challenges

Working with a team of people with diverse skills, experience, and attitudes from various institutions helped us to achieve our goal. Gathering information on the students'

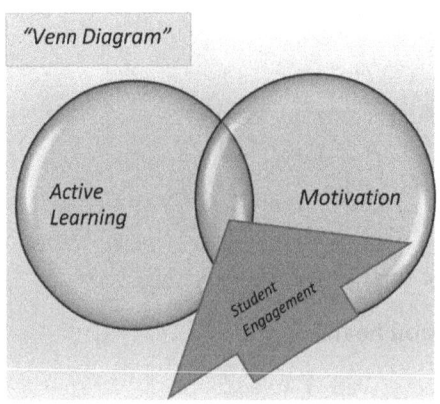
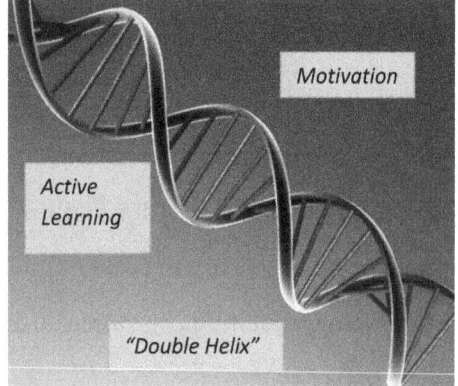

FIGURE 15.5   Student engagement models [1].

Discussion Board
Create Forum
Welcome and Introductions
Questions/Concers
Question #1
Vectors - Introduction
Slope - Whats the point of it all?
Properties of Vectors

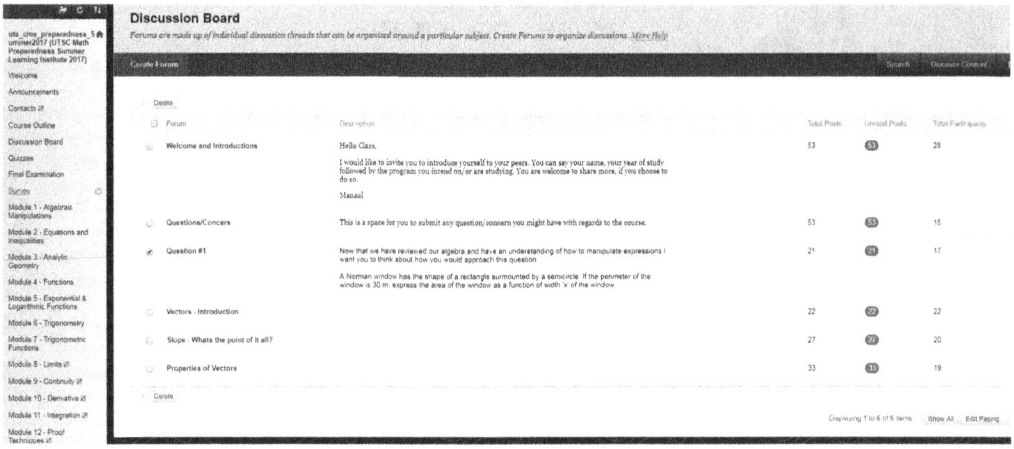

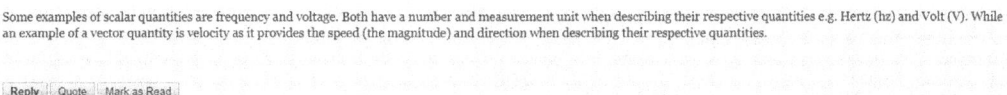

FIGURE 15.6 Discussion board snapshot.

Some examples of scalar quantities are frequency and voltage. Both have a number and measurement unit when describing their respective quantities e.g. Hertz (hz) and Volt (V). While an example of a vector quantity is velocity as it provides the speed (the magnitude) and direction when describing their respective quantities.
Reply
Quote
Mark as Read

FIGURE 15.7 A sample student post.

Pageviews
Unique Pageviews
5,000
2,500
4,000
2,000
July 2017
October 2017
January 2018
April 2018

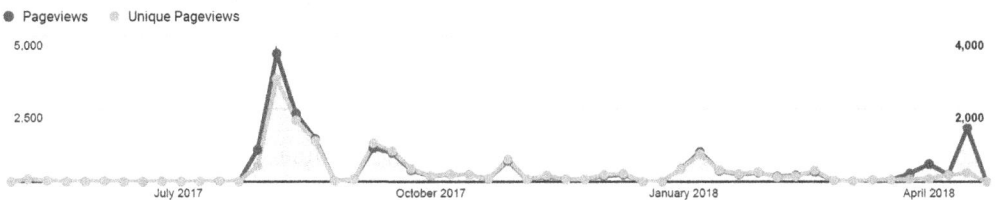

FIGURE 15.8 Stats on video hits or views.

perspectives enriched the project, and we believe it has helped to foster student engagement and will perhaps even help students to learn other concepts more easily in the future. At the beginning of the project, it was difficult to accurately estimate the effort required to meet our quality standards. Constant effort was required to rein in the pacing and scope of the project. We could have made good use of an additional two to three months to complete the project at a more comfortable pace.

The idea of creating video animations for calculus courses was quite ambitious, since none of us had any prior experience with video production. Because of this, we had to spend quite a lot of time making sure that the videos were mathematically accurate while also being interesting and engaging to watch. Despite the challenges, it was a lot of fun to see how the animation company used our scripts and turned them into animations.

Another challenge arose from the fact that the course instructor did not have any experience teaching online. It was difficult to find tools to help her feel comfortable teaching in this way. Because of this, we decided to use simple tools such as a USB camera, microphone, card stock, pencil, and paper to simulate a teaching environment more familiar

FIGURE 15.9   The homepage of MIA.

to the instructor. It also took us quite a long time to tape the first few video lessons, as we were often interrupted by unwanted background sounds—birds chirping, loud colleagues passing by, and so on.

15.4 MATH IN ACTION JOURNAL

We have used what we learned developing these online modules to develop a new online journal called Math In Action (MIA). The journal is meant to provide students studying mathematics and statistics with a platform to share their research with colleagues and faculty alike. MIA aims to serve as a stepping stone for aspiring mathematical researchers on their path to academia. It may also be used for research assignments in senior undergraduate and graduate mathematics courses. This journal will help to foster greater student engagement in mathematics programs. It will also give researchers the opportunity to interact with students and inspire greater interest, with the hope of creating stronger generations of future researchers. The journal may be found at: www.mathinactionjournal.com/.

MIA is first and foremost a tool meant to foster students' creative thinking.

I came up with the idea for this journal while I was teaching a fourth-year undergraduate course called "Classical Geometries and their Transformations." One of the assignments for the course is a research project on modern topics in geometry. My goal in including this assignment on the syllabus is to give students the opportunity to apply their highest levels of thinking to mathematics, such that they can produce creative work in a mathematical context—something that does not happen often at the undergraduate level. For the assignment, students are divided into groups of three or four and assigned a research topic in modern geometry. The purpose of the assignment is to write a ten-page research report in the format of a standard academic mathematics journal. In addition to the reports themselves, students must give ten-minute presentations of their work four times during the semester. On a single presentation day, there are three distinct topics per group and the group members will be randomized each presentation day to allow students to listen to a variety of topics. At the end of these short presentations, peers ask questions and provide feedback to each other for five minutes. This active exchange of ideas helps students to understand topics on a deeper level, keeps them motivated, and ultimately helps them to improve their own final reports. My role as course instructor during these presentation days is to monitor the students' progress and facilitate conversation. The atmosphere of the room is rich and exciting on these days. Some students bring physical models, some use digital images, and some use pencil and paper to illustrate concepts (Figure 15.10). The combined hustle and bustle of the class's 40 students can get quite noisy, but this particular type of noise makes me proud, as I scan the room and see every student contributing. This learning environment reminds me of professional mathematics conferences, and I am glad that I can give my fourth-year students a taste of this sort of environment while they are still undergraduates. Students are instructed to listen carefully to each other's presentations and provide written feedback to each other on the evaluation papers I provide them. Each presentation session is graded by a former student of the course using these evaluation papers, which are worth 2.5%.

I make use of other tools to improve students' confidence in writing about mathematics as well, such as the assignment rubric or the cue cards that I hand out to students to jot down notes during the lectures. For example, one student visited me after class because he was confused about models in hyperbolic geometry. As a result of his concern, I decided to create a page of images of the four models with their interpretation of points, lines, and parallel lines (see Figure 15.11). Talking with students like this has helped me to understand their difficulties better, and consequently to develop better solutions.

FIGURE 15.10   A student working with a model.

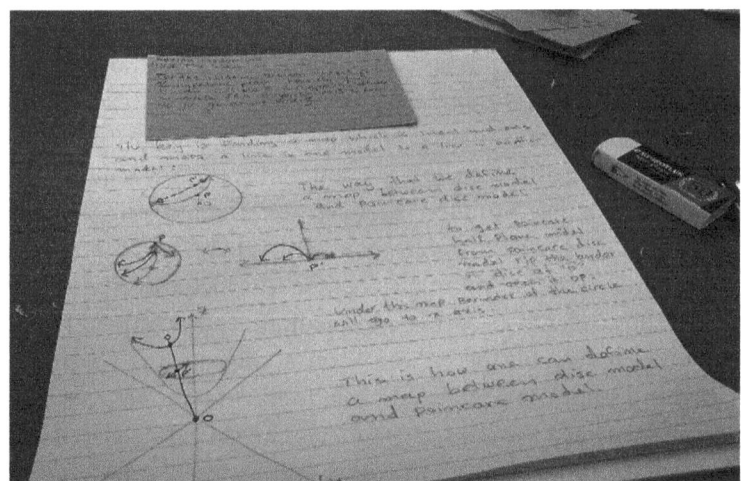

FIGURE 15.11   A cue card dialogue.

In addition, I designed a hands-on workshop in collaboration with The Writing Centre on academic writing skills in mathematics. We used student work from previous years as examples, as well as a sample geometry topic from a book to discuss the qualities of a strong academic report. At the end of workshop, students were better able to understand the role of various components of an academic report in communicating ideas, how to better organize their thoughts, the value of explaining concepts clearly, how to communicate goals and objectives, and how to reference and label figures and tables, as well as learning about various writing styles. We also asked students to identify the strengths of the sample student paper and to suggest ideas to improve its organization or flow. Lastly, we gave them a number of tips for producing better work. To understand students' learning needs further, we designed a pre- and post-workshop survey. I will discuss the details of this study in Section 15.7 of this report.

15.4.1 Survey Results

Written communication plays an important role in the development of the field of mathematics. Students can practice learning challenging topics by writing about them. Writing exercises enable them to think on a deeper level as well as practice using the correct notation and conventions. I also believe that writing academic reports can help students to appreciate and enjoy mathematics more, and yet so few instructors employ this tool in their course designs. I believe that student anxiety toward writing is the reason that so many instructors choose not to explore this option. The main question that we need to answer is why do mathematics students experience writing anxiety so much more commonly than students in other fields, and what methods can we employ to help students to feel more comfortable writing about mathematics?

Informal interviews with students, TAs, and colleagues have led me to believe that lack of practice and support is the main reason for math students' shaky confidence in writing. To help students to improve their writing skills, I provide many opportunities for students to write in my courses, including short written exercises on cue cards and research reports in my upper-level courses. In addition, in collaboration with The Writing Centre, we have provided tailored support for individual students' specific needs.

Based on our observations and experiences, we have designed pre- and post-workshop surveys to better understand students' learning needs related to writing skills. We used a seven-point Likert scale for items 1, 3, and 4 to assess the level at which they are comfortable writing academic papers, as well as formulating questions and comments about mathematics both verbally and in writing. For items 2, 5, and 6 we asked how many mathematics courses they had taken in which they were required to write a research paper, as well as their perceptions of the value and challenges of such writing assignments.

Thirteen students completed both the pre- and post-workshop surveys, which assessed their writing experiences in upper-year undergraduate mathematics courses. The mean grade of these students was 79.38 with a standard deviation of 8.52. Their exam grades had a mean of 74.46 with a standard deviation of 14.35. Their research assignments had a mean of 75.08 with a standard deviation of 6.28. The responses to items 1, 3, and 4 on the survey showed that on average, students were somewhat more comfortable writing academic papers (M = 3.23, SD = 1.48) after the workshop, and they were somewhat more comfortable formulating questions and comments about mathematics verbally (M = 3.54, SD = 0.97), and in writing (M = 3.62, SD = 0.87) after completing their geometry courses. About 47% of the students reported that they did not have a research paper assignment in any of their previous undergraduate mathematics courses.

Correlation analyses revealed that students' grades on research assignments had increased marginally and had a statistically significant and positive correlation with their exam mark (r = 0.55, p < 0.10). This result suggested that those students who obtained high marks on their research assignments also received high marks on their exams. There were statistically significant high and positive correlations among the three survey items (items 1, 3, and 4). The correlation between students' level of comfort in writing academic papers and their level of comfort formulating questions and comments about mathematics verbally was r = 0.72 (p < 0.05); the correlation between students' level of comfort in writing academic papers and their level of comfort formulating questions and comments about mathematics in writing was r = 0.72 (p < 0.05); and the correlation between students' level of comfort formulating questions and comments about mathematics verbally and in writing was r = 0.66 (p < 0.05). All together, these results indicated that students who felt comfortable in writing academic papers also felt comfortable formulating questions and comments about mathematics verbally and in writing. Moreover, students who felt comfortable formulating questions and comments about mathematics verbally also felt comfortable formulating questions and comments about mathematics in writing.

None of the three survey items had statistically significant correlations with course grades, exam grades, or research assignments grades. These results indicated that neither students' perceptions of their level of comfort in academic writing, nor their level of comfort formulating questions and comments about mathematics verbally or in writing contributed to their research assignment grades, exam grades, or overall course grades.

In terms of students' responses to how they perceive the value of research assignments in their mathematics courses, an analysis of qualitative data revealed five major themes, which are listed below:

1. Feeling comfortable and confident exploring mathematical ideas independently

2. Being able to communicate mathematical ideas verbally (e.g. group discussions, working cooperatively)

3. Acquiring mathematical knowledge through gaining a deeper understanding of their respective research topics

4. Realizing the importance of mathematics beyond coursework (e.g. future endeavors)

5. Appreciating the beauty of mathematics

In terms of students' perceptions of the challenges and benefits of completing a research assignment in their mathematics courses, an analysis of qualitative data revealed the following three major themes:

1. Realizing that writing a research paper in mathematics is a difficult but worthwhile task

2. Being able to communicate complex mathematical ideas in simple ways

3. Developing (new) skills in academic writing, particularly in the field of mathematics

Many students noted in the survey that just one research assignment in one course can only go so far in improving their writing skills in mathematics. This inspired me to design and develop new tools to provide opportunities for students to develop their skills further. One such tool is MSLC Magazine, which was launched in 2014. The magazine includes articles submitted by UTSC faculty members and students related to mathematics or statistics, their personal experiences teaching or learning mathematics, probability challenge problems, interviews, and more (Figure 15.12). The magazine may be found at: www.utsc.utoronto.ca/mslc/mslcmagazine.

Another such tool is Math In Action Journal (MIA), which was developed using a Teaching Enhancement Grant. Submissions to MIA take the form of short video presentations, along with a two-page extended abstract. Students are encouraged to be creative in how they present their work. The journal allows mathematics undergraduates to share their work with peers and academics alike. For example, students in my geometry course are encouraged to submit video versions of their mini-presentations to MIA. The experience of creating the online calculus modules inspired me to create a digital magazine consisting of short videos, rather than a more traditional (but less engaging) print journal. In addition, guidelines for how to write mathematics papers and sample video submissions are available on the MIA website for those who are interested in submitting.

Math In Action Journal has enriched my geometry course considerably. Those students who have done well on their research papers now have a legitimate platform to share their

FIGURE 15.12   The cover page of the first issue.

work, which they can use on their CVs as a stepping stone to academia. Take for example one of my students from the fall 2017 semester, who chose to study Pick's Theorem. In her work, she first discusses the history of the problem, and then defines key concepts and components clearly using the appropriate mathematical notation and relevant images. Next, she states and proves the theorem for basic geometrical shapes and extends the proof for more complex objects. The progression of the work from simple shapes such as rectangles to more complicated lattice shapes is what makes her work both elegant and exciting. It is interesting that she was able to discover the link between Pick's Theorem and Euler's Characteristic Invariant in Algebraic Geometry for the case of simple lattice polygons. All of her proofs are accurate, and she references credible resources such as The American Mathematical Monthly. She uses IEEE reference style and properly links the content to her references. She received the UTSC Research Library Award for her work, which will appear in MIA.

REFERENCES

1. Shahbazi, Z., & Ken, T., "A Short Preparatory Calculus Course: Is it Effective?" *International Journal for Cross-Disciplinary Subjects in Education (IJCDSE)* Special Issue, 3.4 (2013). ISSN 2042 6364 (Online).
2. National Council of Teachers of Mathematics. *Curriculum and Evaluation Standards for School Mathematics.* Reston, VA: National Council of Teachers of Mathematics, 1989.

3. Selwyn, *Ch 1: Revisiting the Promise of Digital Technology in Schools. In Schools and Schooling in the Digital Age: A Critical Analysis*. Abingdon, UK: Routledge, 2011.

4. Hamdan, N., McKnight, P., McKnight, K., & Arfstorm, K. M., "The Flipped Learning Model: A White Paper Based on The Literature Review Titled: A Review of Flipped Learning", 2013. Retrieved from http://researchnetwork.pearson.com/wp-content/uploads/WhitePaper_FlippedLearning.pdf.

5. Bergmann, J., & Sam, A., *Flip your Classroom: Reach Every Student in Every Class Every Day*. Eugene, OR: International Society for Technology in Education, 2012.

6. Roehl, A., Shannon, G. J., & Reddy, S. L., "The Flipped Classroom: An Opportunity To Engage Millennial Students Through Active Learning", *Journal of Family and Consumer Sciences* 105.2 (2013): 44.

7. Corte, Erik De, "Mainstreams and Perspectives in Research on Learning (Mathematics) from Instruction", *Applied Psychology* 53.2 (2004): 279–310. Web. 17 Jan, 2019.

8. Big Heart Media, in Becoming Outstanding – Secondary, *Learning Objectives: Sharing Learning Objectives*. Video Clip: 2 mins, Catalog Number W/5073/001, 2010.

9. Big Heart Media, in Becoming Outstanding – Secondary, *Learning Objectives: Smarter Learning Objectives*. Video Clip: 4 mins, Catalog Number W/5071/002, 2010.

10. Big Heart Media, in Becoming Outstanding – Secondary, *Learning Objectives: Self-Assessment*. Video Clip: 3 min., Catalog Number W/5074/001, 2010.

11. Brown, S., *Conceptions of Inquiry*. Routledge, 1981.

12. Bloom, Benjamin S., *Taxonomy of Educational Objectives, Handbook 1: Cognitive Domain*. First edition, New York, NY: Longmans, 1956.

13. Fink, Dee L, *Creating Significant Learning Experiences: An Integrated Approach to Designing College Courses*. San Francisco, CA: Jossey Bass, 2003.

14. Picone-Zocchia, J., & Martin-Kniep, G. O., *Supporting Mathematical Learning: Effective Instruction Assessment, and Student Activities*. San Francisco, CA: Jossey-Bass, 2008.

15. Hunsburger, W. F., *Inquiry Learning: A Narrative Inquiry Into the Experiences of three Teachers*. University of Toronto. ProQuest Dissertations Publishing, NR39806, 2008.

16. Barkley, Elizabeth F., *Student Engagement Techniques: A Handbook for College Faculty*. San Francisco, CA: Jossey-Bass, 2010, Print.

17. Buzan, T., *Make the Most of Your Mind*. Rev.ed, London: Pan Books, 1988.

18. Jackson, N. *Tackling the Wicked Problem of Creativity in Higher Education*. 2008. http://imaginativecurriculumnetwork.pbworks.com/f/WICKED + PROBLEM + OF + CREATIVITY+ IN + HIGHER + EDUCATION.pdf.

19. Petocz, P., Reid, A., & Taylor, P., "Thinking Outside the Square: Business Students' Conceptions of Creativity", *Creativity Research Journal* 21.4 (2009): 409–416. doi:10.1080/10400410903297998.

20. Shahbazi, Z., & Irani, A. "Online Calculus and Pre-Calculus Modules", In *Proceedings of EdMedia: World Conference on Educational Media and Technology 2016* (pp. 42–51). Association for the Advancement of Computing in Education (AACE), 2016.

21. Hay, I., *Inspiring Academics: Learning with the World's Great University Teachers*. Maidenhead, UK: McGraw-Hill, 2011.

22. National Council of Teachers of Mathematics. *Principles and Standards for School Mathematics*. Reston, VA: National Council of Teachers of Mathematics, 2000.

23. Griffith, S., *Teaching and Learning in Higher Education*. 2009. P75.

APPENDIX

Details of Scenes and Visual Tools

General production comments

Happy, light music plays in the background.

Scene 1

Voice over   n/a

Visuals
Module 12: Proof Techniques

At the end of this module you will be able to:

- Understand the justification of calculus facts and relations by applying methods of direct proof, proof by contradiction, proof by contrapositive, and proof by induction

- Have a deeper understanding of math as a discipline of study

Scene 2

Voice over   The history of mathematics goes a long way back into the past. Since antiquity math has taken a central role in advancing the development of the world, as we know it!

Ancient Greeks were using pebbles for counting and doing simple algebra. They called a "pebble" a "calculus."

Playing with pebbles led their curious minds to think about very small and very big quantities.

Visuals   Something with pebbles. Perhaps a bunch drop on the screen to create a pile, and then a few roll to one side to be a smaller batch in line with VO "very small." Then have a whole lot fall onto other pile in line with VO "very big."

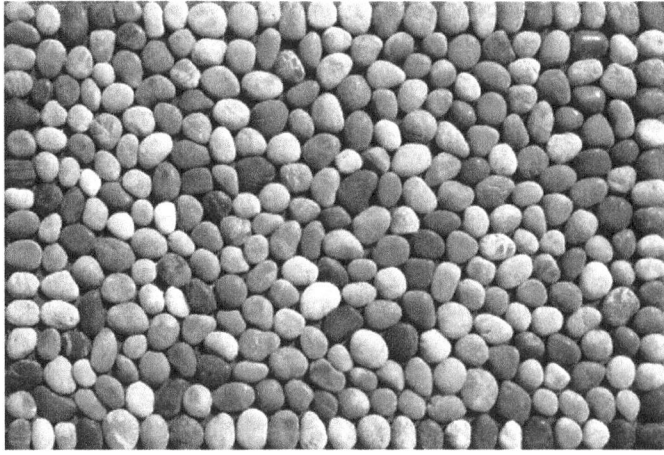

FIGURE A.1

Scene 3

Voice over   Although the modern calculus that we study today is able to answer mathematical questions with much more precision, it is amazing to know that these accomplishments are the result of hard work of individuals working in different parts of the world, over the centuries.

Visuals   Susie jetpacks past to create timeline in line with VO, "over the centuries."

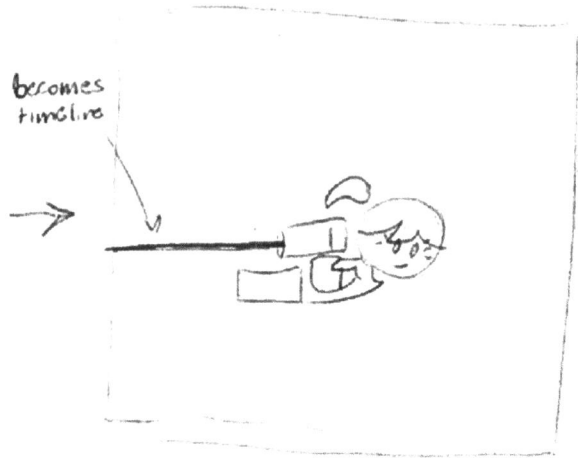

FIGURE A.2

Scene 4

Voice over   Our number system with digits from 0 to 9 originates from ancient Egypt.

The Pythagorean Theorem and its understanding of geometry comes from ancient Greece.

In third-century China, we find the early forms of integral and differential calculus and a calculated value of pi correct to five decimal places!

By the sixth century in India, we find definitions of trigonometric functions and an accurate approximation of pi that recognized it as an irrational number.

By the end of the first millennium, a Persian mathematician was the first to demonstrate proof by mathematical induction, proving the binomial theorem.

Throughout medieval Europe and during the scientific revolutions, the study of math exploded with numerous discoveries!

Visuals   All based in timeline, with boxed information showing on screen including dates and details as provided.

Visuals on this is tricky—we don't want to do any stereotypical (cultured) depiction. So I'm thinking text only. But I also don't want to have VO read everything below—so I've provided a summarized retelling in the voice over section above.

Big question is on pacing: we can use the stacked boxes approach like you've done on other modules (like #10). But how long can they stay on screen? We want this module to be short (as other modules are longer than expected) but it's important that we don't compromise the readability of module. Possibly this is a good candidate for the interactive component.

*(Note: we'll need to source this information. But likely to do so outside of the video itself— unless we're doing the interactive vid in which case we can definitely embed this. Source: www.storyofmathematics.com/mathematicians.html) **I wrote the contact from this site to clear copyright.*

Egyptian

2700 BC (Egyptian): earliest fully developed base 10 number system in use

Greek

570–495 BC (Pythagoras): expansion of geometry, rigorous approach building from first principles, square and triangular numbers, Pythagoras' theorem

Chinese

220–280 AD (Liu Hui): solved linear equations using matrices, leaving roots unevaluated, calculated value of π correct to five decimal places, early forms of integral and differential calculus

Indian

476–550 AD (Aryabhata): definitions of trigonometric functions, complete and accurate sine and versine tables, solutions to simultaneous quadratic equations, accurate approximation for π (and recognition that π is an irrational number)

Islamic/Persian

953-1029 AD (Muhammad Al-Karaji): first use of proof by mathematical induction, including to prove the binomial theorem

Medieval European

1170–1250 (Italian, Leonardo of Pisa – a.k.a. Fibonacci): Fibonacci Sequence of numbers, advocacy of the use of the Hindu-Arabic numeral system in Europe, Fibonacci's identity (product of two sums of two squares is itself a sum of two squares)

Renaissance mathematics

1522–1565 (Italian, Lodovico Ferrari): devised formula for solution of quartic equations

Scientific revolutions in 17th and 18th century

1596–1650 (French, René Descartes): development of Cartesian coordinates and analytic geometry (synthesis of geometry and algebra), also credited with the first use of superscripts for powers or exponents

1643–1727 (British, Isaac Newton): development of infinitesimal calculus (differentiation and integration), laid ground work for almost all of classical mechanics, generalized binomial theorem, infinite power series

Modern math

1792–1856 (Russian, Nikolai Lobachevsky): developed theory of hyperbolic geometry and curved spaces independently of Bolyai

1928– (American, John Nash): work in game theory, differential geometry, and partial differential equations, provided insight into complex systems in daily life such as economics, computing, and military

Scene 5

Voice over   Even to the present day math is continuing to develop.

Here's Professor Zohreh Shahbazi [zoh (like sew) reh … shah bah zee]. You may recognize her from the instructional videos in these modules!

Because it is important to learn the techniques that are used to develop our knowledge in calculus, this module happens to be Zohreh's favorite topic!

Proofing methods are logical tools that are essential for expanding mathematics.

In this module you will learn the proof techniques such as direct proof, proof by contradiction, proof by contrapositive, and proof by induction.

Visuals   Cartoon-ified version of Zohreh (Prof. Shahbazi) waving, or holding some math items (like in end of module 10) Also, in line with VO: "In this module you will learn the proof techniques such as direct proof, proof by contradiction, proof by contrapositive, and proof by induction." → Show thought bubble or something like listing the four different proofing techniques covered, shown in text.

Scene 7

Voice over   Whether you are hoping to become the next big mathematician, or just looking to survive your first-year calculus courses, remember that *you* are the future of math!

Visuals   Zohreh character slides/shifts to left (still in frame), with timeline showing revealing future dates, and revealing Susie and her jetpack. Susie does a bit of a loop to remain on screen and line traced from jetpack fizzles out showing just Susie. Somehow Tommy should be there too, maybe she's holding his hand carrying him along. End frame has the three characters on screen.

With VO: "*you* are future of math," all three characters wave. Perhaps have applause and fireworks or something to adequately end the series!

CHAPTER 16

Making Online Mathematics Method Courses Interactive and Effective with OER

Bhesh Raj Mainali

CONTENTS

16.1 INTRODUCTION

This is the age of Information and Communication Technology (ICT). In the last few decades, the roles of technology in the educational sector have been increasing tremendously. Because of the availability and accessibility of various technological tools and Open Educational Resources (OERs), the online education system is rapidly expanding around the world. Teaching and learning via online delivery formats have emerged as a popular alternative to face-to-face classroom instruction (Crawford-Ferre & Wiest, 2012) because of their various benefits such as saving time, providing greater flexibility, and eliminating travel distance by offering equal opportunities and resources to every learner (Dede, Ketelhut, Whitehouse, Breit, & McCloskey, 2009; Isman & Altinay, 2006). Therefore, online course offerings are increasing at a faster rate than on-campus course offerings (Beck, 2010), and almost 100% of public institutions report online instruction as a critical part of their long-term plans (Major, 2010). Furthermore, demand to develop online mathematics method courses in teacher education programs has also increased extensively (Ko & Rosen, 2010). However, teaching mathematics method courses in online environments can be difficult and challenging. In this chapter, an effort has been made to explore some of the main OER, which can be utilized to make online courses as effective and interactive as on-campus courses.

Despite some variations, there is a consensus among scholars in defining the meaning of interactive in the educational arena. Kahveci and Imamoglu (2007) define interaction as a kind of action that occurs as two or more objects have an effect upon one another. Sessoms (2008) relates interactive with student-centered instruction, where tools and technologies help learners to engage actively in their learning. He further states that "the transformative nature of technology integration changes the process of teaching and learning to an interactive learning environment" (p. 88). Wagner (1994) associate interaction with learner control, self-regulation, and motivation. Thus, interactive learning environment refers to a student-centered instructional approach, where learners actively engage in their learning process. Furthermore, in an interactive learning environment, learners get the opportunity to explore knowledge while interacting with teachers and or with technology. For the purpose of this chapter, interactive refers to learners' active engagement in their learning process by utilizing Open Educational Resources, in particular Web-based tools and technology, including educational audio-video resources.

16.2 CHALLENGES AND ISSUES

Despite the various benefits, there are challenges and issues in teaching courses in an online delivery format. Kebritchi, Lipschuetz, and Santiague (2017) state that it is challenging and difficult to effectively transfer what is taught in the face-to-face classroom to an online delivery format. Furthermore, three main issues and challenges they suggest are online learners, instructors, and content development. Issues related to content consist of the integration of multimedia and the role of instructional strategies in content development. Whereas Fein and Logan (2003) suggested that instructors face challenges at three phases with online courses: the design, the delivery, and the follow up. Similarly, Anderson, Imdieke, and Standerford (2011) suggested that one of the main challenges in online education is how the course content must be delivered effectively in online environments. Furthermore, the challenges and issues can vary from course to course since the nature of courses and subjects of study are different. For example, teaching a mathematics method course in an online format certainly would be different from teaching a history course in an online environment.

Generally, mathematics method courses aim to focus on pedagogical content knowledge. How to teach mathematics in effective ways is one of the main goals of mathematics method courses. Similarly, focus on conceptual understanding and meaningful learning is the other important aspect of method courses. Therefore, mathematics instructors tend to utilize different types of materials, hands-on activities, tools, and technologies in order to create learning environments where learners get a chance to explore and discover mathematical ideas in meaningful ways. Additionally, mathematics instructors demonstrate effective teaching by utilizing various hands-on activities, manipulatives, tools, and technologies. Thus, instructors need to create and design various material as well as integrate different tools and technologies in method courses regardless of the format of the course delivery: online or on-campus. Apparently, preparing various hands-on activities and other relevant materials for mathematics method courses certainly needs a lot of time. But more importantly, creating various materials and adjusting them from face-to-face courses to online courses can be even more time-consuming and challenging (Li & Irby, 2008).

Another important aspect of the usage of hands-on activities in mathematics method courses is to promote learning environments where students get opportunities to explore, experiment, and discover mathematical ideas. Indeed, such learning environments would enhance the theory of constructivism since learning would be an active meaning-seeking process. For example, both physical and virtual manipulatives have various benefits such as (1) providing immediate feedback to learners so that students avoid misconception, (2) providing connection and visualization between numeric and visual representation, (3) helping mathematizing, and (4) offering opportunities to teach in and represent mathematical ideas in nontraditional ways (Suh, Johnston, & Douds, 2008). Similarly, technological tools such as Dynamic Geometry Software (DGS) (GeoGebra) can be used for demonstration and visualizations of objects, as well as for discovering various mathematical concepts (Hohenwarter et al., 2004), which offers students opportunities to formulate theories and to draw their own conclusions (Hannafin et al., 2001). Thus, in order to enhance mathematics learning, different hands-on learning activities including, but not limited to, Geoboard, Tangram, Base Ten Blocks, Algebra Tiles, AngLegs, mathematical software, and mathematics apps are commonly utilized in face-to-face mathematics method courses. However, the questions come in online courses: how can instructors utilize similar materials in the online learning environment? Thus, it is challenging to use similar activities in the online courses since, regardless of the format of delivery, online or on-campus, the learning outcomes need to be same for a specific mathematics method course.

Teaching online mathematics method courses at a college and university level is even more challenging due to the nature of the content of method courses. Furthermore, engaging distance learners in methods courses is not easy due to the limitation of the online learning platform. Mathematics instructors who teach both on-campus and online mathematics method courses may find it difficult to apply the same teaching strategies and learning activities that can be used in face-to-face classroom to online. The fact is that different instructional strategies that instructors tend to integrate in online courses from their on-campus sections might not work (Coppola, Hiltz, & Rotter, 2001). Thus, math instructors who teach both on-campus and online courses may find it more difficult to engage distance learners in their learning as compared to their on-campus sections. Additionally, there can be various other challenges including, but not limited to, how to translate instructional strategies and activities designed for traditional face-to-face courses to the online courses (Baran, Correia, & Thompson, 2011).

Educators and other related experts have been trying to implement various strategies as well as resources in an online learning environment in order to resolve various challenges and issues that are encountered in online learning environments, and one of the strategies is the utilization of OER. There are different types of OERs that can be integrated in mathematics method courses in order to make online courses more effective and interactive. Moreover, OERs will help to save time for instructors since they are easy to integrate in the online courses. Some of the OERs that can be utilized particularly in online mathematics method courses will be discussed.

16.3 OPEN EDUCATION RESOURCE (OER)

According to the United Nations Educational, Scientific and Cultural Organization (UNESCO) (2017), an OER is any type of educational material that is in the public domain or introduced with an open license. The nature of these open materials is that anyone can legally and freely copy, use, adapt, and re-share them. OERs range from textbooks to curricula, syllabi, lecture notes, assignments, tests, projects, audio, video, animation, technological tools, and software packages. According to the William and Flora Hewlett Foundation, an OER is:

> Open Educational Resources are teaching, learning, and research resources that reside in the public domain or have been released under an intellectual property license that permits their free use and re-purposing by others. OER include full courses, course materials, modules, textbooks, streaming videos, tests, software, and any other tools, materials, or techniques used to support access to knowledge.
>
> (Hewlett Foundation, [N.D.])

Whereas the Organization for Economic Co-operation and Development (OECD, 2007) defines as: "Open Educational resource are digitized materials offered freely and openly for educators, students and self-learners to use and reuse for teaching, learning and research." Such resources are accumulated assets that can be enjoyed without restricting the possibilities of others to enjoy them. This means that they should be non-rival (public goods), or that the value of the resource should be enlarged when used (open fountain of goods).

It is clear from the above two definitions that OERs include various ranges of resources as long as they are freely available. Furthermore, OER also includes various types of virtual manipulatives and mathematics applets, which can be utilized in mathematics lesson activities. Similarly, OER also includes different types of mathematics lesson-teaching videos, which demonstrate problem-solving strategies that students utilize while completing mathematical tasks.

The purpose of the OERs movement is to make education more affordable and help the learners to better achieve the proposed learning outcomes. Hilton, Gaudet, Clark, Robinson, and Wiley (2013) suggested that the utilization and integration of OERs into academic programs coordinated at the educational institution level likely save money for students and make education more affordable. Sapire and Reed (2011) further suggested that careful selection, adoption, modification, and implementation of the OERs by content experts have the potential to help students achieve quality-learning experiences.

The wide range of available OER for instructors to use in their courses is provided by various organizations/institutions or even by individuals. For example, MIT

OpenCourseWare (MIT OCW, https://ocw.mit.edu/index.htm) comprises a wide range of open resources for different types of courses. The resources include, but are not limited to, notes and slides from class lectures, recordings, reading lists, and so forth. Similarly, the University of Nottingham (https://rdmc.nottingham.ac.uk/handle/internal/79), the University of Oxford (http://openspires.oucs.ox.ac.uk/), Humbox (http://humbox.ac.uk/), and the Annenberg Foundation (https://annenberg.org/) also have a repository of wide ranges of OERs. Moreover, Massive Open Online Courses (MOOCs) can be utilized for online learning environments; however, MOOCs resources are only accessible for the duration of the courses. As a part of MOOCs movement, many institutions have been creating platforms, for example YouTube, where resources are available for the use of public. There are also other open resources, which can be found in certain network or repositories in the World Wide Web. MRLOT (www.merlot.org/merlot/index.htm), and OER Commons (www.oercommons.org/) are two examples of OER repositories.

The complex nature of teaching and learning processes, particularly for mathematics education, can be made simple with the appropriate integration of OERs. The OERs, in particular such as technology, are becoming important tools for imparting knowledge and skills to learners. The integration of OERs for teaching and learning purposes, particularly in mathematics education, is a high priority for many nations and different educational systems. Mathematics teachers, educators, and researchers in mathematics education need to revise mathematics curricula and instructional strategies to take advantage of various OER including electronic information technology (Fey, 1989). The National Council of Teachers of Mathematics (2000) also emphasized the importance of the use of technology (OER) in mathematics education by stating that technology is essential in teaching and learning mathematics; it influences the mathematics that is taught and enhances students' learning. Thus, various OERs do not only contribute to students' learning but also facilitate the instructional strategies for teachers in an online as well as on-campus learning environment.

OER can play an important role particularly for online courses for various reasons. Some of the benefits include, but not are limited to, time-saving, easy integration, being interactive in nature, and freely available resources. Moreover, OER also helps to save cost for students since they can get free books for example (Ko & Rossen, 2017), as well as helping to save instructors' time since they can directly incorporate ready-made resources and activities in their lessons. More importantly, the OER replaces similar learning materials, allowing for the same functionalities (Orr, Rimini, & Damme, 2015) as compared to face-to-face learning.

16.4 OER's UTILIZATION

Due to various challenges and issues as aforementioned, instructors may hesitate to teach online mathematics method courses. Despite that unwillingness, however, instructors might have to teach online courses due to the increasing demand for online learning environments. Moreover, university administrators may request instructors to design and teach online courses due to increasing demand. Thus, instructors need to find ways to integrate various activities in online courses that are often utilized in on-campus courses. For example, Base Ten Blocks are commonly used materials while teaching place value concepts, as well as basic operation of whole numbers. Similarly, Geoboard is also widely used while teaching geometry concepts in mathematics method courses. Base Ten Blocks and Geoboard are two examples that instructors tend to use in lesson activities to enhance the conceptual understanding in face-to-face courses. But it might be challenging for instructors to utilize similar lesson activities in online courses because of the nature of the online learning environment. Thus, instructors need to think about how they can utilize lesson activities in online courses that are similar to what is used in face-to-face classes.

OERs can be utilized in online courses to substitute similar functionalities of the face-to-face courses in order to make online courses as effective and interactive as on-campus courses. In order to make teaching online courses effective, however, instructors must be willing to integrate and adopt new technologies and relevant OERs that best address learner needs (Ludlow, 2001). If appropriately selected and utilized, OERs certainly help to equate online courses with different facets of on-campus sections of similar courses.

There are various types of OERs available in the digital age that can be utilized in online mathematics method courses. However, instructors need to be careful in selecting and utilizing them in the mathematics lessons. One of the important aspects of utilizing OER in online courses is that to be aware of the authenticity, quality, and appropriateness, and need to select them carefully based on the learning outcomes and objectives of the online courses. Miller (2016) suggested various criteria while selecting OER. Sometimes the appropriateness of OER also depends on the mathematics lesson being taught. Instructors are the ones who can understand and select the OER that best fits in the mathematics lessons. Thus, an effort has been made to provide insight and ideas as to how we can utilize OERs in online mathematics method courses to make them more effective and interactive, similar to on-campus courses. In that regard, some of the OERs would be discussed in conjunction with teaching online mathematics method courses. However, this chapter is not aimed to describe every OER available in the digital age that can be utilized in online mathematics method courses.

16.4.1 Virtual Manipulative

The utilization of virtual manipulatives has increased in the last couple of decades in the teaching and learning of mathematics. More importantly, in the digital age, its role is becoming even more important for online mathematics method courses. Generally, two types of manipulatives are available for educational purposes: virtual and physical. Virtual manipulatives are computer-based, whereas physical manipulatives are an actual hands-on activity. Moyer and Bolyard (2016) suggest that virtual manipulatives are "an interactive, technology-enabled visual representation of dynamic mathematical object, including all of the programmable features that allow it to be manipulated, that present opportunities for constructing mathematical knowledge" (p. 13). Different types of physical manipulatives include, but are not limited to, Geoboard, Base Ten Blocks, Tangram, Counter Chips, and Algebra Tiles. The physical manipulatives also include different hands-on activities, which can be designed using different materials such as construction paper. However, these different types of physical manipulatives can also be found in the digital format as virtual manipulatives. Normally, the virtual manipulatives can be found on the World Wide Web network, and they are compatible with certain computer system requirements. Spicer (2000) states that the two types of representations on the World Wide Web are called virtual manipulatives—these are static and dynamic visual representations of concrete manipulatives. Static virtual manipulatives are the static representation of concrete objects (physical manipulatives), whereas dynamic visual representation can be manipulated in the same ways as concrete manipulatives. In fact, virtual manipulative is an interactive, Web-based visual representation of a dynamic object that presents opportunities for constructing mathematical knowledge (Moyer, Bolyard, & Spikell, 2002).

In mathematics method courses, instructors generally utilize different types of physical manipulatives in order to teach various mathematical concepts. For example, Base Ten Blocks can be used to explain the concept of the place value system, as well as for the grouping-regrouping concept of addition and subtraction of the whole number system. Similarly, Geoboard can be utilized to explore various aspects of triangles. The utilization of physical manipulatives is almost impossible in online courses, but virtual manipulatives can be employed in online courses to substitute the physical manipulatives used in on-campus courses. Thus, instructors tend to use different types of virtual manipulatives in their online courses. In doing so, distance learners are afforded opportunities and learning environments similar to on-campus learners. Examples of two virtual manipulatives are discussed below in order to demonstrate how mathematics instructors can utilize virtual manipulatives in methods courses.

16.4.2 The National Library of Virtual Manipulatives (NLVM)

The National Library of Virtual Manipulatives (NLVM) began in 1999 to develop a library of uniquely interactive Web-based virtual manipulatives or concept tutorials, mostly in the form of Java applets, for mathematics instruction (K-12 emphasis) (http://nlvm.usu.edu/en/nav/vlibrary.html). The NLVM contains various types of virtual manipulatives for wide a range of mathematics content for K-12 grades. The virtual manipulatives are classified based on five mathematics content areas: number and operation, algebra, geometry, measurement, and data and probability, as well as four grade level bands: pre-K–2, 3–5, 6–8, and 9–12. The collection also includes various quizzes, games, and drag-and-drop activities, which can be easily incorporated in mathematics lessons.

Base Ten Blocks are one of the widely used physical manipulatives, particularly in elementary mathematics method courses, in order to teach the basic operation of the whole number system, including the concepts of carry over, grouping, and regrouping. Moreover, Base Ten Blocks are designed to provide conceptual understanding of the basic operation of whole numbers. The fact is that learners can manipulate, arrange, and rearrange the Base Ten Blocks to understand how, for example, addition can be performed without using the standard algorithm. For example, consider an addition problem: 128 + 622 completed using Base Ten Blocks without using the standard algorithm. In Base Ten Blocks, the cube, rod, and flat respectively represent 1, 10, and 100. In order to complete the addition task, we need to start with two groups of Base Ten Blocks that represent 128 and 622. We need to have eight cubes, two rods, and one flat for 128 and two cubes, two rods, and six flats for 622. Combining these two groups of Base Ten Blocks is the same as adding 128 and 622. When we combine these two groups, we want to start by combining cube to cube, and then rod to rod, and finally flat to flat. When we combine eight cubes and two cubes, we get ten cubes. We can trade these ten cubes for one rod since ten cubes exactly make a rod. Thus, trading the ten cubes for a rod can be connected to the concept of carry over as well as the place value concept, while adding 8 and 2 using the standard algorithm approach.

Instructors can easily utilize Base Ten Blocks in their on-campus courses; however, it is difficult to integrate those physical manipulatives in the online courses for the distance learners. The Base Ten Blocks can be easily incorporated into online methods courses as the substitution of physical manipulatives. Moreover, the virtual Base Ten Blocks may even be more manageable, clean, flexible, and extensible than their physical counterpart (Sarama & Clements, 2016). Furthermore, the virtual Base Ten Blocks environment offers greater control and flexibility to students than actual physical Base Ten Blocks (Char, 1989). Thus, the virtual manipulatives can be great resources for online methods courses. A screenshot of the Base Ten Blocks is shown in Figure 16.1, which represent an example of the addition problem 128 + 622. The cubes and blocks can be grouped and regrouped as shown in Figure 16.1. For example, eight and two cubes are in the unit place in Figure 16.1. We can select all cubes to combine together to get ten, which will provide one rod (one ten) as shown in Figure 16.1(b). Then we can drag the rod from the unit place to the ten place.

This activity illustrates the utilization of Base Ten Blocks in the online method courses, where physical manipulatives are not feasible to utilize as in on-campus courses. Instructors can embed the NLVM link or other OER in the online course and require distance learners to play with the virtual Base Ten Blocks and complete mathematical tasks. Furthermore, instructors can provide assignments, where distance learners are required to explain the addition concept based on virtual Base Ten Blocks and connect it with the standard algorithm. At the same time, instructor may require the distance learners to attach a screenshot of the Base Ten Blocks to demonstrate their reasonings.

16.4.3 Illumination (NCTM)

The Illuminations (http://illuminations.nctm.org/) is a project designed by The National Council of Teachers of Mathematics (NCTM), which is aimed to serve mathematics

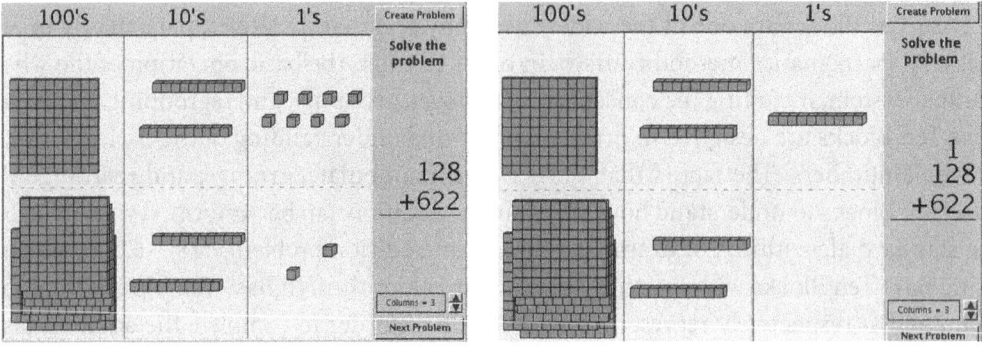

FIGURE 16.1 (a) Base Ten Blocks addition. (b) Base Ten Blocks addition result.

learners, teachers, and educators by increasing access to quality standards-based resources for teaching and learning mathematics, including interactive tools for students and instructional support for teachers. The Illumination virtual manipulatives are chosen based on the grade level, mathematics content area, Common Core State Standards (CCSS), and NCTM mathematics standards. It covers five content areas of mathematics: number and operations, algebra, geometry, measurement, and data analysis and probability. Options are also available to choose a complete lesson or lesson with interactive manipulative. The grade band ranges from pre-K–2 to 9–12.

Algebra Tiles are a commonly used resource at middle and high school mathematics to teach algebraic concepts, including linear and quadratic equations. Instructors can use physical Algebra Tiles in the on-campus courses for preservice teachers to demonstrate how to teach algebraic concepts in effective ways by focusing on conceptual understanding. Algebra tiles are made up of small squares, rectangles, and large squares. The areas of the small square, rectangle, and big square respectively are 1, x, and x^2 units. Each of the tiles are related in terms of area. For example, the length of the small square is the same as the width of the x tile, and the length of the x tile is the same as the dimension of x^2. The Algebra Tiles also consist of two colors: one to indicate positive values and another to indicate negative values. Combining a positive and a negative tile with the same area yields a zero, called zero pairing. For example, to solve a simple linear equation: $2x - 3 = 5$, we need to have two rectangles with positive values and three small squares with negative values on one side, and five small squares with positive values to the other side. In order to solve it, we need to add two small squares tiles with positive values to both sides, which will give us zero pairing between three small squares with positive and negative values. Then we have left over two rectangles on one side and eight small squares on the other side. If we distribute the two rectangles equally between eight small squares, it will yield four small squares. Thus, a rectangle (x) is equal to four units ($x = 4$). However, it is difficult to use physical Algebra Tiles for the online courses in order to teach the same concepts as compared to face-to-face classes. Therefore, instructors can incorporate virtual Algebra Tiles from Illuminations (www.nctm.org/Classroom-Resources/Illuminations/Interactives/Algebra-Tiles/) for the distance learners in order to meet the same learning goals with similar activities. A screenshot of the Algebra Tiles is shown in Figure 16.2. Figure 16.2(a) indicates a linear equation to be solved using interactive Algebra Tiles. Instructors can build an algebraic model using Algebra Tiles to represent the given equation as shown in Figure 16.2(b). We can see there are four modes as shown in the figure: *Solve, Substitute, Expand*, or *Factor*. In Solve and Substitute, vertical lines indicate "is equal to" (=). In Expand and Factor, the large area is the product of the top and left areas, as in a standard multiplication table. In Algebra Tiles, the big square represents x^2 units, the rectangle represents x units, and a small square represents one unit. The different colors represent positive and negative units.

Place tiles equal to the expression to the left and right side of the workspace. For example, the expression is $2x - 3$; place two green x tiles and three red tiles on one left side of the workspace. You will need to flip the tiles to get the red inverse tiles. Place five yellow tiles on the right side of the workspace. Check your model to move on to the

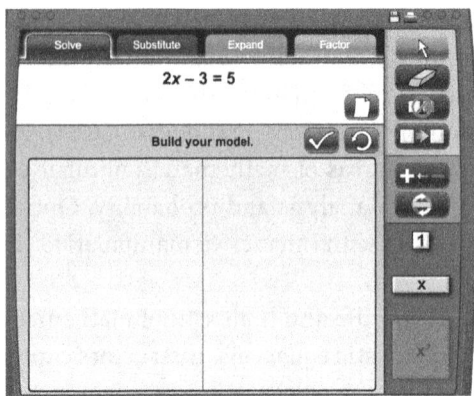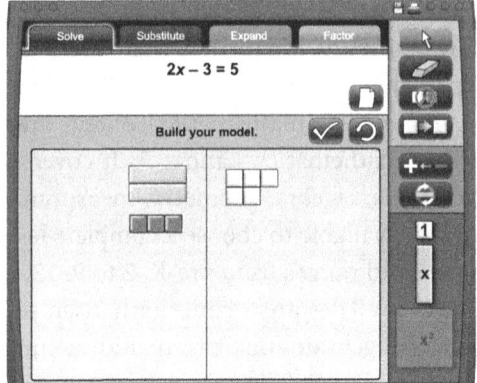

FIGURE 16.2   An example Algebra Tiles. (a) A linear equation. (b) Modeling equation with Algebra Tiles.

next step. To solve the equation, we can rearrange tiles so only a single green tile remains in one half of the workspace. Remember, only equal values can be removed from both sides, such as a yellow tile from the left and a yellow tile from the right. To "move" a tile, first add the inverse value to both sides. If we want to move a red rectangle, add a yellow rectangle to both sides. Then we can remove the zero pair from one side. We need to add three yellow tiles on both sides to eliminate three reds in the left side. Now, we can erase three red and three yellow tiles in the left side. Then, we have two green rectangles in the left side and eight small tiles in the right side. Each green rectangle in the left side represents four tiles in the right side, i.e. $x = 4$. We can solve different problems in a similar way including quadratic equations. For more details, we can go to NCTM-Illumination Web page. Instructors can assign mathematical tasks for online learners, where they are required to solve algebraic problems using virtual Algebra Tiles. Additionally, along with assignments submission, instructors may require the distance learners to explain each step of the given problem utilizing virtual Algebra Tiles including some screenshot of Algebra Tiles.

This is an example to demonstrate how instructors can utilize virtual manipulatives in an online methods course for distance learners. Different types of virtual manipulatives are available in NCTM's Illuminations portal, which can be incorporated in various lesson activities for the distance learners in an online learning environment. In doing so, the online method courses are likely to have the same learning goals with similar activities compared to on-campus courses.

16.4.4 Mathematics Software

Various mathematical software, such as GeoGebra, Desmos, etc., are available to utilize in mathematics method courses. GeoGebra is a free multi-platform dynamic mathematics software for all levels of education that joins geometry, algebra, tables, graphing, statistics, and calculus. It has several features, including an easy-to-use package, useful for learning and teaching mathematics at all levels, and joins interactive geometry, algebra, tables, graphing, calculus, and statistics.

We can utilize GeoGebra to teach various mathematical concepts. For example, we can use it to explore the nature of quadratic equations ($ax^2 + bx + c$) including the function of a, b, and c. The software can be easily used to explore, for example, how the graph of a quadratic equation will change as we change the values of a, b, and c. The software is freely available in a public domain so instructors can easily integrate it in online methods courses. With different available tutorial videos that demonstrate how to design specific mathematics lesson activities using GeoGebra, instructors may require distance learners to design simple math applets for the lesson.

Despite integrating the software itself in the online mathematics method course, different types of GeoGebra applets are also available to use in method courses. There are tremendous numbers of freely available ready-made GeoGebra applets, created by mathematics teachers and educators, which can be easily integrated in mathematics lesson activities for online method courses. A fraction bar is a commonly used physical manipulative to teach the concept of fractions. Instructors can purchase fraction bars, or they can construct their own fraction bars for their lesson activities. In order to substitute the role of a physical fraction bar, instructors easily can employ a virtual fraction-bar applet, which is based on GeoGebra. The various GeoGebra applets can be found at: www.geogebra. org/materials. A screenshot of a fraction bar created (www.geogebra.org/m/U2szXnVe) by using GeoGebra is shown in Figure 16.3. In this interactive applet, the fraction bar can be divided into different parts. The two different fractions bars can be compared and contrasted as shown in Figure 16.3.

Mathematics instructors can create simple GeoGebra applets and require learners to explore the applets to get a conceptual understanding of fractions. Moreover, instructors may require distance learners to watch GeoGebra tutorial videos and can assign them to make GeoGebra applets to explore mathematical concepts, such as demonstration of the slope-intercept form ($y = mx + b$). Instructors may also provide assignments such as a technology-integrated lesson plan, where distance learners are required to find ready-made GeoGebra applets and utilize them in the lesson plan to enhance mathematical understanding. Furthermore, an online discussion board can be formed, where distance learners can discuss and share their ideas in regard to the role of applets to explore given specific mathematics concepts. In doing so, the distance learners are likely to engage actively in

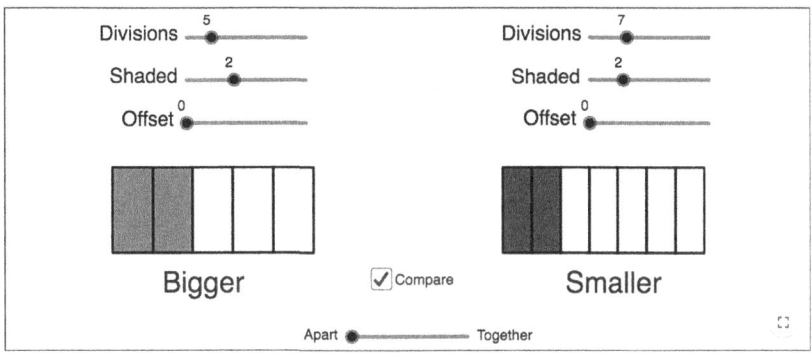

FIGURE 16.3   GeoGebra fraction bar applet.

online courses by sharing ideas and providing feedback to their peers in class. As a result, online learning tends to be more interactive and effective.

Desmos is an advanced graphing calculator implemented as a browser application and a mobile application (www.desmos.com/). It has different features including graph functions, plot data, evaluate equations, etc., that can be used to enhance mathematics teaching and learning. Instructors can easily incorporate Desmos lesson activities into the online courses to explore various mathematical concepts. Furthermore, instructors may require students to solve mathematics problems utilizing Desmos. Since distance learners easily get access to Desmos, instructors can assign mathematics problems to be done by exploring and utilizing Desmos. Moreover, Desmos also has several videos demonstrating various mathematical concepts, which can be easily integrated in the online methods courses.

16.4.5 Mathematical Apps

With the emergent of mobile device technologies, various educational tools, including mathematical apps, have been developed. Namukasa et al. (2016) state, "Mathematical Apps are computer applications in which virtual manipulatives are delivered on touch-screen mobile devices" (p. 276). Because of their mobile-friendly features, apps can be useful resources for online methods courses. Some of the apps are freely available while others need to be purchased. Geoboard is one of the most commonly used physical manipulatives used in mathematics method courses. It is difficult, if not impossible, to use Geoboard in an online mathematics method course. However, freely available mathematics apps can be integrated into online courses in order to achieve the same learning goals compared to the face-to-face method courses. A screenshot of a virtual Geoboard adapted from The Math Learning Center (https://apps.mathlearningcenter.org/geoboard/) is shown in Figure 16.4, where a triangle and a rectangle are constructed.

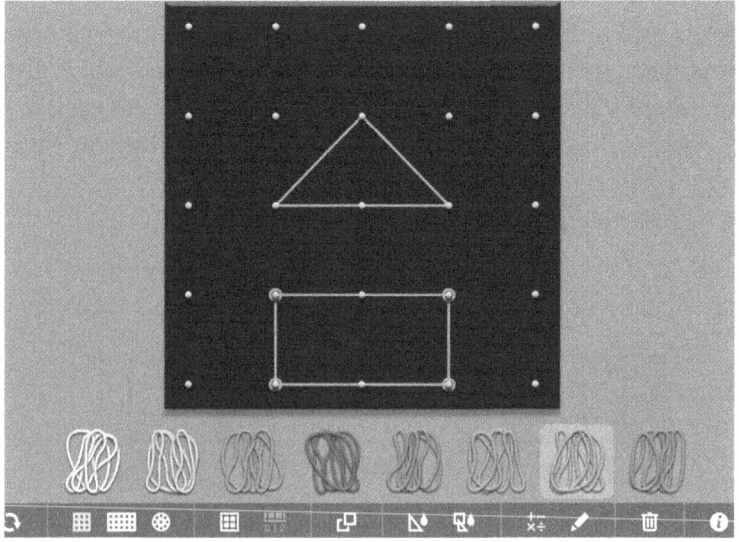

FIGURE 16.4   An example of Geoboard.

Instructors can get different types of mathematics apps that can be easily incorporated both in on-campus and online mathematics method courses. However, we need to be careful in selecting and utilizing the apps for our mathematics lessons. Namukasa et al. (2016) suggested various criteria for app selection, which may help instructors to make a decision for the app choice for their mathematics lessons, whereas Larkin (2013) shares a list of different types of apps for mathematics lessons.

Physical Geoboard is a common and popular manipulative to explore various geometrical concepts, and instructors tend to use it in on-campus courses. The Geoboard apps can be used in online courses to substitute the function of the physical Geoboard that can be used in face-to-face courses. Instructors in online methods courses further assign mathematical tasks, where distance learners are required to use Geoboard to explore and solve the given problems. This is an example to demonstrate how we can incorporate math apps to replace the role of physical Geoboard for the online course.

16.4.6 Mathematics Teaching Video

The teaching and learning strategies vary from course to course since the nature of the content of each course is uniquely different. For example, in mathematics method courses instructors tend to demonstrate some exemplary lesson activities for preservice teachers to provide better insights as to what types of misconceptions and errors they are likely see in their future teaching careers. Moreover, instructors in method courses are likely to present real classroom caveat of school-teaching mathematics for preservice teachers in order to better prepare preservice teachers as how to deal with various types of misconceptions that they will likely encounter in their future mathematics lessons. Additionally, instructors want to explain various mathematics concepts for distance learners. The fact is that instructors can easily explain and demonstrate such math misconceptions in the face-to-face classes; however, it is very difficult for the online courses. Producing several recorded videos of instructors teaching mathematics lessons for their distance learners could be one strategy. However, producing several videos for online classes may be cumbersome and time-consuming as well as technically challenging.

Different types of OERs can be utilized to overcome the difficulty of making video of instructors teaching mathematics lessons because various types of videos teaching mathematics lessons are available in the digital format. Thus, instructors can utilize these OERs in online courses after carefully analyzing them based on the needs of online courses. Two of the resources will be discussed below, which can be integrated into online courses.

Khan Academy: Khan Academy produces short lectures in the form of YouTube videos. Its website also includes supplementary practice exercises and tools for educators. All resources are available to users in the Khan Academy website (www.khanacademy.org/). The short video lectures cover a wide range of mathematics topics for K-12 grades and above. The videos are divided into different sections based on grade level and mathematics content areas. Students can select the exact video they need to watch for certain mathematics content. More importantly, learners can watch them whenever and wherever they want and watch as many times as they want (Parslow, 2012). Depending on the needs of courses, instructors can choose the videos and easily incorporate them into the online methods courses. Rather than producing their own videos, instructors can utilize Khan videos in the online course for distance learners to watch. Furthermore, instructors can assign homework and mathematics problems for distance learners from Khan Academy.

Teaching Channel: Teaching Channel is a multi-platform service delivering professional development videos for teachers over the Internet and on television (www.teachingchannel.org/). The channel has many showcasing, inspiring, and exemplary teaching videos from real classroom settings. Teachers can watch, share, and learn diverse teaching techniques to help every student grow (Teaching Channel). The teaching channel does not only have mathematics lesson teaching videos but also videos for various fields of study.

Understanding different types of instructional strategies that can be utilized to teach mathematics effectively at school level is an important aspect in mathematics method courses. In online mathematics method courses, providing caveat of real classroom scenarios is important for preservice teachers to understand various misconceptions preservice teachers are likely to see in their future math lessons. Thus, instructors may require distance learners to watch exemplary mathematics lessons-teaching videos from the Teaching Channel. More specifically, instructors may require distance learners to watch teaching lesson videos and can pose questions as how they will reteach/fix the mathematics misconception or errors presented in the videos. In doing so, preservice teachers will get a better picture of a real classroom setting with real students making mathematical errors. Moreover, instructors may require distance learners to reflect and write commentary for specific mathematics content after watching Teaching Channel videos. This would likely help distance learners to think about the misconception and errors students made, analyze errors, and come up with ideas/activities to fix the misconception that they notice in Teaching Channel videos.

The Khan Academy and Teaching Channel are just two examples of OER that can be utilized to make online mathematics method courses more effective and interactive. However, there are certainly more resources we can utilize in online courses. One of them, for example, is whyU (http://whyu.org/). WhyU animated videos are designed for K-12 and college-level mathematics. The animated videos contain various mathematical topics including pre-algebra, algebra, topology, and infinite series. We can also utilize different types of YouTube videos, but we need to be careful in selecting them. Not all of them would be reliable, useful, and authentic, and they would not necessarily enhance teaching and learning mathematics in online environments.

16.5 CONCLUSION

The recent development in technology supports online education systems around the world. Thus, academic institutions have begun to offer more online courses, including mathematics method courses. Despite various benefits, there are also issues and challenges in online learning environments and the degree of challenges may vary based on the nature of content of the online courses. Mathematics method courses, for example, can be challenging to teach in online learning environments. However, to embrace the ongoing trend in online education systems, instructors need to find ways to overcome the challenges in online learning environments. In an effort to overcome the challenges, instructors can utilize various OER in online mathematics method courses.

Despite it being a recently emerging field, the growing number of research studies reported different type of benefits of OER. For example, OER has positive impacts on learners' attitudes and perceptions of learning; it enhances the learning interest, and is open access to all (Gourley & Lane, 2009; Weller et al., 2015). It also helps to expand teachers' roles and to make them more active in sharing information and learning from each other (Petrides, Jimes, Middleton-Detzner, & Howell, 2010). Another important facet of distance learning is the independence and increased responsibility for self-learning, which can be enhanced by integrating appropriate usage of OER. Andreatos and Katsoulis (2012) further assert that OER helps distance learners to be more independent of instructors, authors, and textbooks. More importantly, utilizing OER in an online learning environment enables the instructor to bridge the gap between on-campus and online courses in terms of classroom activities and interaction. The role of OER appears to be even more relevant and important to the rapidly growing online educational system. However, more research studies need to be conducted as it relates to the roles and effects of OER for online mathematics method courses in general, since fewer studies have been done in this area. Just simply integrating OER does not necessarily enhance the learning. "There is a need to continue the development of the tools and resources to support the transition to OER" (Miller, 2016, p. 254). The author has not collected data formally yet, but based on course evaluations, students' feedback, interactions with students, and other online activities, the author noticed that OER helps to make online learning more effective, as well as helping to reduce the gap between online and on-campus courses. Thus, more research in this area is recommended in order to investigate the role and effects of OER, particularly in online mathematics method courses.

REFERENCES

Anderson, D., Imdieke, S., & Standerford, N. S. (2011). Feedback please: Studying self in the online classroom. *International Journal of Instruction, 4,* 3–15.

Andreatos, A., & Katsoulis, S. (2012). Using open educational resources in course Syllabi. *The American Journal of Distance Education, 26,* 126–139.

Baran, E., Correia, A., & Thompson, A. (2011). Transforming online teaching practice: Critical analysis of the literature on the roles and competencies of online teachers. *Distance Education, 32*(3), 421–439.

Beck, V. S. (2010). Comparing online and face-to- face teaching and learning. *Journal on Excellence in College Teaching, 21*(3), 95–108.

Char, C., & Education Development Center, Inc., Newton, MA. Center for Learning Technology. (1991). *Computer graphic feltboards new software approaches to children's mathematical exploration.* Report No. 91-1, Washington, D.C.

Coppola, N. W., Hiltz, S. R., & Rotter, N. (2001). Becoming a virtual professor: Pedagogical roles and ALN. System sciences, 2001. *Proceedings of the 34ᵗʰ Annual Hawaii International Conference.*

Crawford-Ferre, H. G., & Wiest. L. R. (2012). Effective online instruction in higher education. *The Quarterly Review of Distance Education, 13*(1), 11–14.

Dede, C., Ketelhut, D. J., Whitehouse, P., Breit, L., & McCloskey, E. M. (2009). A research agenda for online teacher professional development. *Journal of Teacher Education, 60*(1), 8–19.

Fein, A. D., & Logan, M. C. (2003). Preparing instructors for online instruction. *New Directions for Adult and Continuing Education*, *100*, 45–55.

Fey, J. T. (1989). Technology and mathematics education: A survey of recent developments and important problems. *Educational Studies in Mathematics*, *20*, 237–272.

Gourley, B., & Lane, A. (2009). Re-invigorating openness at The Open University: The role of Open Educational Resources. *Open Learning: The Journal of Open, Distance and e-Learning*, *24*(1), 57–65. doi:10.1080/02680510802627845.

Hannafin, R. B., Burrus, J. D., & Little, C. (2001). Learning with dynamic geometry programs: Perspectives of teachers and learners. *The Journal for Educational Research*, *94*(3), 132–144.

Hilton III, J. L., Gaudet, D., Clark, P., Robinson, J., & Wiley, D. (2013). The adoption of Open Educational Resources by one community college math department. *International Review of Research in Open & Distance Learning*, *14*(4), 37–50.

Hohenwarter, M., & Fuchs, K. (2004). Combination of dynamic geometry, algebra and calculus in the software system GeoGebra. Electronically. Retrieved from http://www.geogebra.org/en/wiki/index.php/Publications.

Isman, I., & Altinay, Z. (2006). Self-Perceptions and roles: How eastern Mediterranean university students and teachers view online program and courses. *Turkish Online Journal of Distance Education*, *7*(1), 75–90.

Kahveci, M., & Imamoglu, Y. (2007). *Interactive learning in mathematics education: Review of recent literature*. San Antonio, TX: Society for Information and Technology and Teacher Education.

Kebritchi, M., Lipschuetz, A., & Santiague, L. (2017). Issues and challenges for teaching successful online courses in higher education: A literature review. *Journal of Educational Technology*, *46*(1), 4–29.

Ko, S., & Rosen, T. (2010). *Teaching online: A practical guide*. New York, NY: Rutledge.

Larkin, K. (2013). Mathematics education: Is there an app for that? In V. Setinele, L. Ball, & C. Bardino (Eds). *Mathematics education: Yesterday, today and tomorrow. Proceedings of the 36th annual conference of mathematics education research group of Australasia* (pp. 426–433). Melbourne, VIC: MERGA.

Li, C., & Irby, B. (2008). An overview of online education: Attractiveness, benefits, challenges, concerns, and recommendations. *College Student Journal*, *42*, 449–458.

Ludlow, B. L. (2001). Technology and teacher education in special education: Disaster or deliverance. *Teacher Education and Special Education*, *24*(2), 43–163.

Major, C. H. (2010). Do virtual professors dream of electric students? University faculty experiences with online distance education. *Teacher College Record*, *112*(8), 2154–2208.

Miller, H. (2016). A practitioner's guide to Open Educational Resources: A case study. In P. Blessinger, & T. Bliss (Eds). *Open Education* (pp. 237–255). Open Book Publishers.

Moyer, P., & Bolyard, J. (2016). Revisiting the definition of a virtual manipulatives. In P. Moyer (Ed), *International perspectives on teaching and learning mathematics with Virtual Manipulatives* (pp. 2–24). Switzerland: Springer International Publishing.

Moyer, P., Bolyard, J., & Spikell, M. (2002). What are virtual manipulatives? *Teaching Children Mathematics*, February 2.

Namukasa, I. K., Gadanidis, G., Sarina, V., Scucuglia, S., & Aryee, K. (2016). Selection of apps for teaching difficult mathematics topics: An instrument to evaluate touch-screen tablet and smartphone mathematics apps. In P. Moyer (Ed), *International perspectives on teaching and learning mathematics with Virtual Manipulatives* (pp. 275–300). Switzerland: Springer International Publishing.

National Council of Teacher of Mathematics. (2000). *Principles and standards for school mathematics*. Reston, VA: Author.

Organization for Economic Co-operation and Development (OECD). (2007). *Giving knowl- edge for free: The emergence of open educational resources*. Paris: Author. Retrieved from http://www.oecd.org/education/ceri/38654317.pdf.

Orr, D., Rimini, M., & Van Damme, D. (2015). *Open educational resources: A catalyst for innovation*. Paris: Educational Research and Innovation, OECD Publishing.

Parslow, G. R. (2012). Commentary: The Khan academy and the day-night filled classroom. *Biochemistry and Molecular Biology Education, 40*(5), 337–338.

Petrides, L., Jimes, C., Middleton-Detzner, C., & Howell, H. (2010). OER as a model for enhanced teaching and learning. In *Open ED 2010 proceedings*. Barcelona: UOC, OU, BYU.

Sapire, I., & Reed, Y. (2011). Collaborative design and use of open educational resources: A case study of a mathematics teacher education project in South Africa. *Distance Education, 32*(2), 195–211.

Sarama, J., & Clements, D. H. (2016). Physical and virtual manipulatives: What is "concrete"? In P. Moyer (Ed), *International perspectives on teaching and learning mathematics with Virtual Manipulatives* (pp. 71–93). Switzerland: Springer International Publishing.

Sessoms, D. (2008). Interactive instruction: Creating interactive learning environment through tomorrow's teachers. *International Journal of Technology in Teaching and Learning, 4*(2), 86–96.

Spicer, J. (2000). Virtual manipulatives: A new tool for hands-on math. *ENC Focus, 7*(4), 14–15.

Suh, J. M., Johnston, C. J., & Douds, J. (2008). Enhancing mathematical learning in a technological–rich environment. *Teaching Children Mathematics, 15*(4), 235–241.

United Nations Educational, Scientific, and Cultural Organization (UNESCO). What are Open Educational Resources (OERs). (n.d.). Retrieved from http://www.unesco.org/new/en/communication-and-information/access-to-knowledge/open-educational-resources/what-are-open-educational-resources-oers/.

Wagner, E. D. (1994). In support of a functional definition of interaction. *American Journal of Distance Education, 8*(2), 6–29.

Weller, M., Arcos, B., Farrow, R., Pitt, B., & McAndrew, P. (2015). The impact of OER on teaching and learning practice. *Open Praxis, 7*(4), 351–361.

William and Flora Hewlett Foundation. (n.d.). Retrieved from https://hewlett.org/strategy/open-educational-resources/.

CHAPTER 17

Developing Interactive Demonstrations for the Online Mathematics Classroom

Interactive Diagrams

Mina Sedaghatjou, Harpreet Kaur, and Kelly A. Williams

CONTENTS

17.1 INTRODUCTION

In today's digital era, technology plays an important role in teaching and learning. There is increased demand for online courses in all subjects, including mathematics, due to their flexible scheduling, capability of digital media integration, self-paced learning, lower costs, and ability to provide an inclusive education catering to students with different needs.

Most online learning occurs through an institution's Learning Management System (LMS), which is a software application for delivering, maintaining, sharing, and tracking educational resources. In such an environment (as with in-person mathematics classes), graphs, sketches, and diagrams play a crucial role in teaching and learning mathematics. Stylianou and Silver (2004) write, "indeed, the abstract arguments presented in Euclid's Elements rely heavily on the use of diagrams, and this use of visual representations remained an acceptable practice in mathematics well into the eighteenth century" (p. 354). This emphasizes the vital role of diagrams and visual presentations in geometrical argumentations. In an online mathematics learning environment, argumenting and conjecturing are made possible through using IDs. The term Interactive Diagram (ID) refers to dynamic embedded diagrams that require action and participation from learners. IDs can be preconstructed or constructed; they can be modified, manipulated, or explored in any dynamic geometry environment such as GeoGebra (Hohenwarter, 2002), Cabri (Laborde & Strässer, 1990), Geometer's Sketchpad (Jackiw, 2001), or Cinderella (Richter-Gebert & Kortenkamp, 1999), while also exploring concepts of geometry, calculus, algebra, and other areas of mathematics. In addition, a well-constructed interactive embedded diagram provides an environment enriched by multiple representations as well as direct, continuous, real-time, and interactive manipulation in an online mathematics platform (Abrahamson & Sanchez-Garcia, 2016; Arzarello, Robutti, & Thomas, 2015; Battista, 2008; Kaur, 2015; Nasim, 2008; Sinclair & Moss, 2012; Sinclair & Yurita, 2008).

Studies have shown that a well-integrated Dynamic Geometry Environment (DGE) in a hybrid learning environment such as GeoGebra can be used to explore geometrical content in an interactive and adaptive environment. Additionally, in-session instruction can be intertwined with ID-based activities in a variety of settings, individual or collaborative, synchronous or asynchronous, inside or outside the classroom (Santos & Quaresma, 2012; Santos, Quaresma, Marić, & Campos, 2018). However, it is still unclear how utilizing DGE-based embedded dynamic and interactive diagrams in LMS facilitates learning mathematics in general and geometry in particular.

In this chapter, we use our study implementing interactive diagrams in a hybrid learning environment to present some ideas that facilitate learning geometry in an online environment. This study was conducted in a teacher education introductory level mathematics course for teacher candidates. The overarching question in this study is "how does utilizing DGE-based embedded dynamic and interactive diagrams in LMS facilitate learning geometry in a hybrid mathematics classroom?" In the following pages, we provide examples of embedded dynamic interactive diagrams and analyze their usage in an online LMS. We present tasks that invite both individual and collaborative learning in an online environment. We also discuss how IDs can trigger diagrammatic thinking, reasoning, and argumentation.

17.2 INCORPORATING FEATURES OF AN EFFECTIVE MATHEMATICS CLASSROOM IN A HYBRID SETTING

Researchers have recently highlighted the importance of developing students' mathematical thinking and making it visible (see Suurtamm, Quigley, & Lazarus, 2015) by posing engaging and appropriate questions. They encourage students to work cooperatively and collaboratively with each other and interact in ways that both support and challenge one another's strategic thinking (Artzt, Armour-Thomas, & Curcio, 2008; Davidson, 1990). Activities should be structured in ways that allow students to explore, explain, extend, and evaluate their progress (National Research Council, 1999).

Protheroe (2007) describes an ideal mathematics classroom where students actively engage in doing mathematics, making interdisciplinary connections, sharing mathematical ideas with each other, and using manipulative strategies to solve mathematical problems. The teacher also demonstrates acceptance of students' divergent strategies and ideas, influences learning by posing challenging and interesting questions, and projects a positive attitude towards mathematics. Liljedahl's (2016) research on "thinking classrooms" provides another example of ideal mathematics classrooms where students engage in collaborative problem solving. One can easily imagine this happening in a face-to-face mathematics classroom, but maintaining this approach in an online environment would be challenging.

In face-to-face classrooms, teachers take moment-by-moment actions and make decisions that are informed by evidence of student understanding in order to facilitate student learning (Leahy, Lyon, Thompson, & Wiliam, 2005). Teachers elicit students to think through many tactics, such as observations during problem-solving, informal interviews during class, or using focused questions during mathematical discussions, so that they can adjust their instruction accordingly. However, keeping such strategies up in an online or hybrid classroom becomes challenging for both teachers and learners. It seems difficult to preserve the crucial roles of all forms of mathematizing in students' lived experiences, including reading, writing, speaking, drawing, and doing mathematics, as well as assessment (self or teacher), in an online classroom.

Over the past few years, the concept of hybrid learning has become popular among mathematics teachers and educators. Hybrid learning, or blended learning, is a formal education program that integrates face-to-face learning with technology-based, digital instruction. Learning takes place in online, mobile, or classroom environments, or some combination of the three. The implementation varies from teacher to teacher. Some teachers like to use the flipped classroom model, where students review lecture videos online at home and use the class time for solving problems and doing a portion of homework (Jungić, Kaur, Mulholland, & Xin, 2015; Lo & Hew, 2017). Others give homework assignments using online software like LonCapa, MapleTA, or CourseCompass by Pearson/Addison-Wesley (Burch & Kuo, 2010) that students can attempt and submit individually. Some teachers prefer to blend classroom instruction and online activities using their institute's LMS like Canvas, Blackboard, Brightspace, or Moodle, Sakai, etc. The interactive diagrams can be used through such LMS platforms where students can create, modify, and manipulate mathematical embedded dynamic diagrams to explore and discuss various mathematical concepts. In mathematics education literature, diagramming has been used as a technical term in two different ways: diagrammatic reasoning/thinking and diagrammatic argumentation. Diagrammatic reasoning is seen as informal reasoning. It is basically a heuristic approach to problem solving (Radford, 2008). Diagrams also have prescribed sets of rules that describe how to produce a diagram. These types of diagrams are used as a tool to demonstrate, clarify, warrant, or back up claims in verbal interactions, hence facilitating diagrammatic argumentation (Krummheuer, 2013). Thus, it invites readers to think about what kinds of diagrammatical interactions and communication generally happen in a mathematics classroom, as well as what kinds of interactions and communication are possible in a hybrid dynamic geometry environment in particular.

17.3 INTERACTIONS, COMMUNICATION, AND ASSESSMENTS IN AN ONLINE MATHEMATICS CLASSROOM

There is no doubt that interactions and communication are essential for any kind of learning. In an online classroom (or any classroom), effective teacher–student, student–content, and student–student interactions build community and lead to robust communication. While these interactions seem easy and natural in face-to-face classrooms, in an online environment, purposeful, organized, and systematic discussions and interactions are necessary for successful learning (Vaughan, Cleveland-Innes & Garrison, 2013). The teacher needs to take on a range of roles as students' "guide, facilitator, and teacher" (Ragan, 2009, p. 6) in order to inspire students and build an active, student-centered learning environment in an online setting.

Research indicates that online discussion forums can help in formative assessment and can be an important part of the learning process (Cross & Palese, 2015; DeCosta, Bergquist, & Holbeck, 2015). The idea of formative assessment emerged from Bloom's (1956) work, which noted that formative assessment could become part of the teaching cycle. Formative assessments differ from summative assessments in the sense that they are intended to inform the teacher and the student of their current progress. In addition, "formative feedback actualizes and reinforces self-regulated learning strategies among students" (Clark, 2012, p. 1). It helps teachers to adapt their teaching practices, which in turn may help students move towards understanding. Over the years, different types of techniques for formative assessment have been suggested, and hence different teaching methods for online classrooms emerged. Classroom Assessment Techniques (CATs) are one popular form of formative assessment put forth by Cross and Angelo (1988). CATs have been widely used in face-to-face classrooms and come in many forms, such as assessing prior knowledge to understand mastery of concepts. They are designed to gather information about both the learner and the teacher. Cross and Angelo used different versions of CATs in their research studies, such as Documented Problem Solution CAT, where students document, in written form, how they solve a particular problem as a way to reason through the steps.

Studies about using CATs in online environments are beginning to emerge (Cross & Palese, 2015; Steele & Dyer, 2014). Cross and Palese (2015) conducted a study with five full-time online mathematics instructors teaching intermediate algebra classes to test using discussion forums as a space for formative assessments. They reported that discussion forum posting frequencies and quiz scores were significantly higher for the sections that used CAT as compared to non-CATs sections. Thus, Cross and Palese established the idea that "using classroom assessment techniques … may actually elicit more discussion forum conversation from students and positively impact student quiz scores" (Cross & Palese, 2015, p. 98). In our study, the students were asked to do constructions and explorations with interactive embedded diagrams that resulted in student–content interactions and indirect student–teacher interactions. Students' written responses about a pre-constructed sketch and interactive embedded student constructions served as a tool of formative assessment. Communication is defined as "a process by which information is exchanged between individuals through a common system of symbols, signs, or behavior" (Merriam Webster's Collegiate Dictionary, 2018). Therefore, mathematical communication entails human senses and cognitive skills like comprehension, interpretation, visualization, and expression of mathematical ideas, as well as non-linguistic methods of communication. In other words, mathematics communication and conceptualization have strong embodied components (Arzarello et al., 2014; Arzarello, Robutti, & Thomas, 2015; Campbell, 2010; Charoenying, 2015; de Freitas, 2016; de Freitas & Sinclair, 2012, 2013; Kaur, 2015; Mowat, 2010, Nemirovsky & Ferrara, 2008; Nemirovsky, Kelton, & Rhodehamel, 2012; Nemirovsky, Kelton, & Rhodehamel, 2012, 2013; Radford, 2013, 2014; Sedaghatjou, 2018; Tall, 2006). It means, not only is mathematical communication inherently multimodal and embodied, it is also grounded on gesticulation, pointing, gaze, and body language. In addition, the processes of comprehending, interpreting, and visualizing mathematical and pictorial graphs all play a vital role in learning mathematics in general. One of the central tasks for learning mathematics is to understand, draw, and interpret mathematical graphs and objects (Healy & Fernandes, 2011). The important role of diagramming is explored by Menz (2015), who goes further and suggests diagramming is a place of mathematizing. Understanding given static diagrams or graphing in an online classroom become challenging tasks in the absence of face-to-face interactions with the instructor; however, being in the classroom doesn't necessarily mean understanding what the instructor graphs or draws. In such a situation, DGE, with its dynamic features, seems like a great environment for students and teachers to experience mathematics. IDs help in promoting student–content communication in an online environment through embodied interactions.

In addition, any LMS has some basic components that enable collaborative engagements for teaching and learning mathematics in an online platform. For example, there are forums for sharing ideas and discussions. Each student may also use the LMS Wikis or blogging to write journals or reflections. Current technology makes it possible to use a variety of interactive diagrams/graphs and representations in a desired LMS. GeoGebra, a free ID, can be embedded in all above LMS components as they are Java-enabled. In such an environment, IDs promote mathematical communication through diagrammatic argumentation and/or diagrammatic thinking. Naftaliev and Yerushalmy (2017) set out various functions served by IDs that help in mathematizing, which are described in detail in the next section.

Naftaliev and Yerushalmy (2013, 2017) conducted a semiotic analysis based on Kress and van Leeuwen's (1996) visual social-semiotic theory. They defined three types of ID functions in teaching and learning mathematics as "presentational," "orientational," and "organizational." "Presentational" function is a dynamic illustration of the geometry concept that helps the learner analyze a situation with its given geometrical and mathematical constraints. Presentational function can happen in two different ways. The first way is to provide random examples of the geometrical rules by presenting different presentation information in various times for each learner. For example, the "Mysterious Geometrical Transformation" task that we present later in our study can be sorted in this category. That is because the task keeps changing the examples by generating similar or new examples. The second approach of presentational function is through providing a generic example when the ID serves as a generic view of a concept. In other words:

> "a generic example is an actual example, but one presented in such a way as to bring out its intended role as the carrier of the general; this is done by means of stressing and ignoring various key features, of attempting to structure one's perception of it" (Mason & Pimm, 1984, p. 287).

The "Golden Ratio Activity," which we will discuss later in the chapter, is a generic example. It is important to consider generalities and presentation when designing generic examples (Naftaliev & Yerushalmy, 2017).

"Orientational" function of IDs refers to the subject design decision in terms of "sketchiness" vs. "rigorousness." It means a graph may highlight important elements instead of accurate appearance and complete picture. So, it can provide a variety of final products that share the same geometrical properties. In our study, both "Mysterious Geometrical Transformation" and "Golden Ratio Activity" tasks depict sketchiness and rigorousness features respectively, hence illustrating the orientational function of IDs. Thus, the role of interactive embedded diagrams in the LMS as a form of CATs, and their presentational and orientational functions in online and nonverbal environments, can open another realm to teaching and learning mathematics online.

In the following sections, we share our study of teaching and learning mathematics in a hybrid classroom, where IDs play an important facilitating role. We first present some background information about our study setting, then highlight the main attributes of the interactive and dynamic diagrams as used in the online component of the classroom through an LMS known as Canvas.

17.4 RESEARCH METHOD

The participants of this study were 30 teacher candidates registered in an introductory mathematics class. None of the participants had a science or mathematics background. The teacher candidates' knowledge of geometry and mathematics was minimal. No participants had taken any mathematics courses at a university level prior to this course. The study occurred over 13 weeks in the spring semester of the academic year. The course was designed in the form of a hybrid classroom. This means the teacher candidates had a chance to participate in some three hours of face-to-face class time per week, as well as some online activities and classes. Prior to registration in the course, they were informed that they must adapt to a new software program, GeoGebra, while minimal instruction would be provided (other than some resources). During the study, lessons were conducted either in face-to-face classes or through Canvas.

The classroom was equipped with six large touchscreen-based computers attached to the walls. GeoGebra was installed on all the touchscreen-based computers, and teacher candidates could interact with them using their hands, specialized pens, and/or external keyboards. In addition, all the teacher candidates were asked to install GeoGebra on their own handheld devices. During class time, the teacher candidates were asked to work as groups of five. The main component of online classes was to construct, manipulate, and explore geometrical concepts through interactive and dynamic diagrams, make conjectures, prove or disprove open-ended questions, and/or explain what teacher candidates did observe. In this chapter, as we move forward, we present evidence of how utilizing DGE-based embedded dynamic and interactive diagrams in the LMS facilitates learning geometry. Next, we provide some examples of the use of embedded IDs and analysis of their usage in Canvas based on the semiotic and diagrammatic communication frameworks discussed in previous sections.

17.5 EXPLORATION-BASED ACTIVITIES: MYSTERIOUS GEOMETRICAL TRANSFORMATION

For the online component in the hybrid mathematics classroom, the first author and researcher presented different mathematical tasks to the teacher candidates throughout the course. For example, one such task that took place in an online classroom was the "Mysterious Geometrical Transformation" task. The teacher candidates were asked to explore the interactive worksheets, move the image around point A, remain attentive to the traces that they may create, and identify what kind of geometrical transformation each task was (see Figure 17.1). They were also asked to explore how often the two points (the image and preimage) would ever meet. The teacher candidates were expected to realize that the two points meet on the center/line of transformation.

As shown in Figure 17.1, the ID provides both *sketchy* and *rigorous* diagrams through a "generic" example in a dynamic format. It means that, although coordinate values and scales are intentionally hidden in ID (*sketch*-like drawings on a piece of plain paper), the line and center of symmetries are *rigorously* fixed with the given degree and direction for each transformation. In such a dynamic environment, the ID produces *sketchy* freehand drawings, which are *rigorously* ordered by mathematics rules and laws. So, it produces a variety of the random final products (by starting from scratch), although they all share the same mathematical idea. In addition, these IDs invite and promote diagrammatic thinking and reasoning in an online classroom. They challenge teacher candidates by moving from the prototype and static examples illustrated in the textbook to a freehand drawing in an ID. The discussion forum in this activity was set in a way that the teacher candidates were not given permission to view peers' input before posting their work; thus minimal collaborative discussion on each other's posts was observed. The main aim of this setting was to give a chance for each teacher candidate to explore the task, reflect, and reason individually before jumping to conclusions based on others' responses. In other words, the discussion and interaction here took place between the IDs, the teacher candidates, and mathematics. Table 17.1 illustrates some of the teacher candidates' thematic written explanations on the "Mysterious Geometrical Transformation" task.

The ID in this activity enriched the mathematical environment by providing multiple real-time "representations" that differ from paper-based solutions. We found written

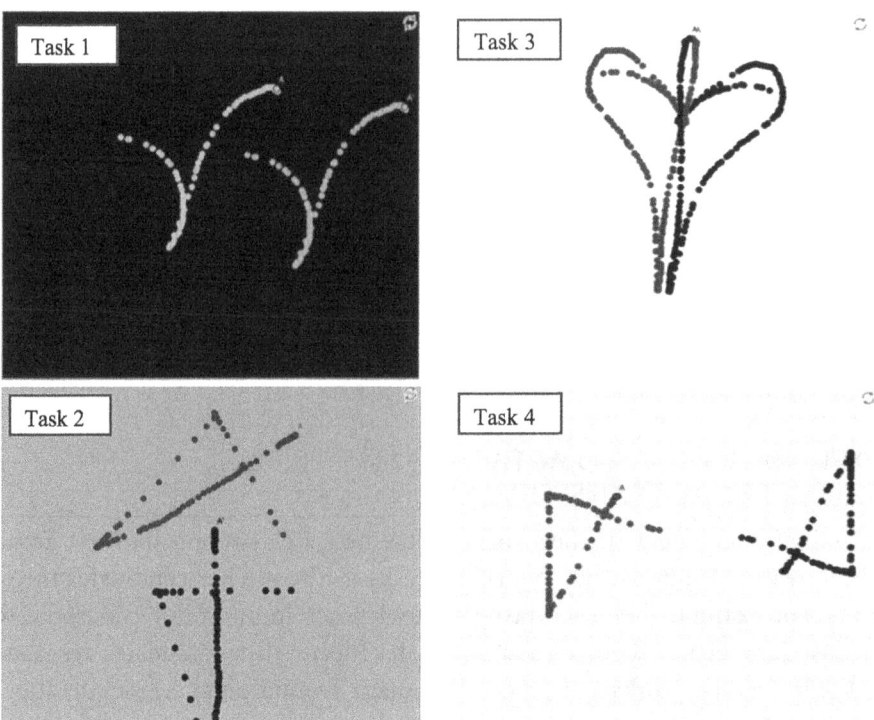

FIGURE 17.1 In this task, only two points were shown on the screen. Students could drag only point A and not the image. In the first task, a translated shape can be constructed with moving point A, and therefore its image, A′, by creating a trace. The points meet on the reflection line. In tasks two and four, the points turn about a fixed center, which is the place that the they also meet. In task three, the two points reflect and meet on the vertical line. (Drawings here are random and made by the authors.)

diagrammatic reasoning in teacher candidates' responses that was rooted in their lived ID exploration. They described geometrical transformation using dynamic verbs like *moves*, *creates*, *mirrors*, and *follows*. For example, teacher candidate A refers to the flexible, free-drawing nature of the exploration while preserving "translation" transformation (where the lines move in unison a constant distance apart from one another). Teacher candidate B acknowledges the role of perseverance in ID exploration. Task 4 is a bit trickier in that it appears to be reflecting; however, after "playing around" with it, she discovered it was, in fact, a rotation. In addition, teacher candidate C refers to the both sketchy and rigorous nature of the ID by mentioning the free but mathematically regulated movement of points using "always" and "exactly" adverbs. Although teacher candidate E did not answer the question as she was asked to, she remained attentive to the invariants and learned that the image of the point moves if, and only if, the original point moves. She also noticed that the movement speed affects the trace formations, which is part of the GeoGebra features.

Thus, in the above activity, the teacher not only invited the teacher candidates to explore the concept of geometrical transformation using the embedded IDs, but also to present and share their explorations in written format in the discussion forum, which was intentionally

TABLE 17.1 Examples of Teacher Candidates' Reflection in an Embedded ID Assignment

	Task 1	Task 2	Task 3	Task 4
A	Translation, **the lines move in unison** a **constant distance** apart from one another	Rotation. The placement again is equal distance apart but rotated.	Reflection. Like a **mirror** :)	Rotation. **The placement again is equal distance apart** but **rotated**.
B	**The relationship** is that they **create** translated figures.	We see a **rotation relationship.**	Task 3 is an example of reflecting.	Task 4 is a bit trickier in that it appears to be reflecting, however **if you play around with it** you discover that it is in fact a rotation
C	Translation. The two points are **always a fixed distance apart.** One point **followed** the other **exactly.** They **always** have the same distance between them laterally and vertically.	Rotation. The lines **can cross** each other at the point they are rotating about. **They retain the same shape** but are rotating about at point at about 45 degrees.	Reflection. **Points mirror each other.** The line over which they are reflected is not at 90 degrees.	Rotation. **These lines are rotating** about a point that is a little way off and not actually touching the origin of either point.
D	This is translation because **it creates** the exact same shape in a different area without resizing anything (simply moves).	Rotation (**creates** same shape at a different angle)	Reflection (**creates** an identical image)	Reflection and rotation (**create** an identical image)
E	A is **the main** because A' cannot be moved. Only whatever A does, A' will do as well right beside it. The **faster** you drag the point A, the more the circles are spread apart. These points will never meet.			
F	Translation: A' **follows** the same path as A. However, it is translated southeast. They are different colors. The two points will never meet each other.	Rotation: A' follows the same path as A. However, it is rotated northeast. These two points will meet one another, but only ever with the red dot above the blue.	Reflection: A' follows the same path as A. However, it is opposite, across a line of reflection. These two points will meet one another, at this line of reflection.	Reflection/Rotation: they follow the same path, however reflected on a line of rotation. They also move in different directions. They meet.

made accessible to other teacher candidates only after they posted their work. This helped the teacher candidates to see each other's mathematical and diagrammatical reasoning based on the same set of dynamic IDs, which in turn facilitated collaborative learning by providing different explanations for the same mathematical task. This strategy is similar to the classroom assessment techniques (CATs) used by Cross and Angelo in their face-to-face classroom in the sense that it helped with formative assessment and informed the teacher about teacher candidates' understanding of the concept.

17.6 DRAGGING AS A CONCEPTUAL TOOL: INTERPLAY BETWEEN DRAGGING AND VARIATION

There are ample research findings in which the affordances of IDs, and dragging in particular, can help to support students' reasoning, as well as their ability to formulate conjectures and proofs. The researchers go so far as to claim that no counterpart in traditional learning environments could raise the same level of conjecturing, thinking, and reasoning as IDs in DGE (Battista, 2008; Hollebrands, 2003; Mariotti, 2000; Yu & Barrett, 2002). For example, as we discussed earlier, the "Mysterious Geometrical Transformation" task merely relies on the dragging feature of the ID, plus the other features of the LMS. This communicative ID offers construction of diverse, dynamic, and complex examples in real time.

> Dragging figures/constructs offers continuous and real-time transformations that maintain the geometrical relationships integrated among the construct's components. In this case, constructs, unlike drawings, move and transform while preserving the invariant geometric properties. For instance, dragging a rhombus may produce any desired orientation, shape, side-lengths rhombus (visually apparent) but the transformed shape is always a parallelogram, even if it transforms to a square.
>
> *(Sedaghatjou, 2017, p. 78)*

This is also in line with the presentational and orientational function of IDs as suggested by a semiotic framework.

The notion of "continuous motion" was introduced by J.V. Poncelet in 1864. The principle states:

> If we suppose a given figure to change its position by having its points undergo a continuous motion without violating the conditions initially assumed to hold between them, the ... properties which hold for the first position of the figure still hold in a generalized form for all the derived figures.
>
> *(as cited in Nasim, 2008, p. 144)*

Battista (2008) postulated that geometric relationships could be perceived "as invariants in the continuous moving of the draggable figures" (p. 350), which supports humans' abilities to notice invariance. Specifically, in absence of an instructor and a face-to-face classroom, an ID facilitates the alteration of a shape's representation through dragging and thus the way it is perceived, while the learner remains attentive to the visual modes of what changes or remains invariant. This is an instance of a generic example as described in the semiotic framework. Furthermore, such a component prompts an environment for reasoning, in addition to explicit descriptions of geometric relationships and shapes. Also, direct manipulation allows students to conceive a construction closer to the theoretical definition of a geometrical figure (Pratt & Ainley, 1997).

As discussed above, prior studies on the use of DGEs have considered dragging to be the main feature of DGE. Also, different types of dragging in a DGE can indicate various levels of cognitive domain of mathematical thinking (Arzarello, Olivero, Paola, & Robutti, 2002). Arzarello et al. (2002) suggest that dragging mediates the relationships between conceptual and perceptual entities: "dragging supports the production of conjectures: exploring drawings by moving them, looking at the ways after which their forms change (or do not change), [and] allows users to discover their invariant properties" (p. 66). The tracing feature of GeoGebra in the "Mysterious Geometrical Transformation" task is a good example of such a mediation. The tracing affordance of GeoGebra also maintains the sketchiness feature of the orientational function of semiotic framework.

Studies have investigated the growth of knowledge of geometrical transformations by identifying changes in mental schemas (Flanagan, 2001; Yanik & Flores, 2009). The fundamental assumption for these studies is that the concepts are mental representations which develop in different stages and are located in individual minds. Therefore, learning means developing a representation (Cobb & Yackel, 1996). However, this approach suppresses the learner's experiences as well as her interactions with the physical and social world by following learning trajectories and looking for mental schemas, mainly done in clinical interviews. We were interested neither in studying what happens in the learner's mind when she solves a task in a DGE, nor in analyzing the learner's utterances, as we did not have access to such data. Instead, we used the dragging and tracing feature of IDs because it provided instant feedback to the learner and helped the learner to assess her constructions, as well as those of her peers. Thus, the assumption is that IDs could work best for an online classroom as they provide instant sketchy or rigorous feedback to learners.

17.7 DIAGRAMMATIC ARGUMENTATION USING IDS

Mathematical proofs, disproofs, and conjectures are at the heart of mathematical understanding. Hoyles (1997) argues that many of the mathematical ideas that a teacher asks students to prove are known to be true by themselves. For example, it seems meaningless for students to prove each angle in a square is 90 degrees. However, in a DGE, it is unavoidable to pay attention to the geometrical properties throughout the construction of a square in a way that the construction passes the drag-test. In such a situation, diagrammatic thinking and reasoning in a tacit way are involved. Furthermore, in an online platform using DGE, a diagrammatic argumentation can occur during object construction or exploring a pre-constructed ID. For example, in a preconstructed equilateral triangle when the learner examines and conjectures that the perpendicular bisector and angle bisector are the same, or when she constructs an equilateral triangle using the same property, she is doing a diagrammatic argumentation.

17.8 CONSTRUCTING A GOLDEN RECTANGLE

As another task, the teacher candidates were asked to construct a golden rectangle, then embed and share the construction in the discussion forum. Both Fibonacci numbers and golden rectangles were taught to the them in the face-to-face classroom. The requested GeoGebra construction had to pass the drag test. It means, if the golden rectangle was constructed correctly, it always remained that same ratio, regardless of how the vertices were dragged. The teacher candidates were also asked to examine their construction geometrically (to find the ratio of the sides).

A quick look at Table 17.2 reveals some ideas about teacher candidates' work. As shown in the table, different solutions are given to the same problem. In fact, different types of communication are involved. For example, the ID itself communicates with the teacher candidates and provides feedback simultaneously during and after the construction process. The drag test feature plays the role of a teacher and enables teacher candidates to reflect on their work. However, when the drag test feature is ignored, the teacher challenges teacher candidates to do a drag test, as with teacher candidate II. In addition, ID construction steps on the left sides of the files provide the teacher with evidence of teacher candidates' understanding of the geometrical concept and the steps that were taken for construction. This feature assists the teacher in evaluating teacher candidates' work, even without any written explanation. The teacher could analyze teacher candidates' submissions based on mathematical objects and relations shown in the diagram even in the absence of written verbal explanations. Below we provide some examples of interpretation of the teacher candidate' golden rectangles constructions.

Teacher candidate I: construction steps are explained clearly, and the shape passed the drag test. It seems that the teacher candidate is proud of herself. However, no algebraic measurement is followed.

Teacher candidate II: it seems this teacher candidate did not understand the meaning of construction. So, the teacher challenged the teacher candidate with a drag test. Also, she did not insert an algebraic formula that shows G_1H_1/H_1B_1 changes while dragging. However, she certainly had examined the rectangle and knows this is a drawing and not a precise construction.

Teacher candidate III: this teacher candidate took a different strategy and used $\dfrac{a+b}{a} = \dfrac{a}{b} = \varphi$ (see Figure 17.2). She also included the construction protocol and algebraic view, so everyone could follow the construction steps (as shown in Table 17.2).

Teacher candidate IV: this teacher candidate did not include a written explanation; the diagram itself demonstrates her understanding in the form of an extended solution. We would call this pure diagrammatic reasoning. This diagram passes the drag test successfully and shows a higher level of thinking by expanding and repeating the construction steps and therefore creating the third golden rectangle. It also demonstrates teacher candidates' tool fluency, which can explain mathematical understanding (Nemirovsky et al., 2013). Sometimes only diagrammatic constructions are enough to illustrate students' mathematical fluency, but it might not be the same for all students. In such a situation, the

TABLE 17.2   Examples of Teacher Candidates' Shared Solutions

I   Here I constructed a golden ratio. I do this by creating a circle where EC is the radius, and C was used as a point of the circle. I then made a line that extends AB to the point that it meets the circle. I then made a line that ran parallel to CB, which fall on the farthest right point of the circle. I then finished of the rectangle by extending DC to the line I had just created. Voila! A rectangle with a golden ratio.

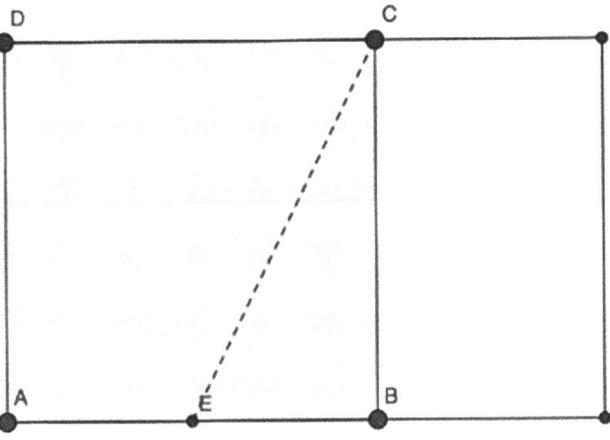

II   The length of the large rectangle (G_1H_1) is 6.6, while the width of the large rectangle (H_1B_1) is 4.08. L/W
The width of the small rectangle (E_1B_1) is 2.52 while the length of the small rectangle (H_1B_1) is 4.08. 2.52/4.08 = 1.6190. This will not happen all the time, but only in a golden rectangle, which this is.

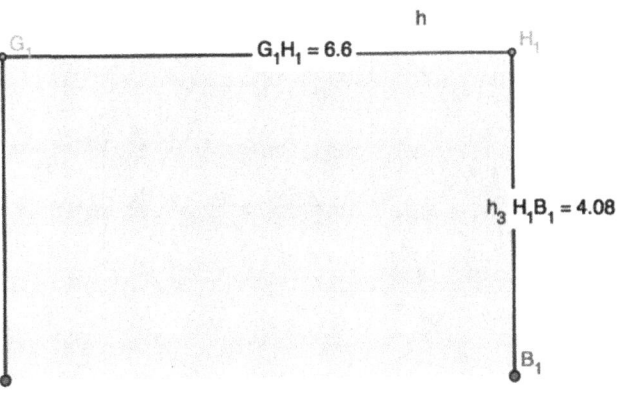

After Drag test

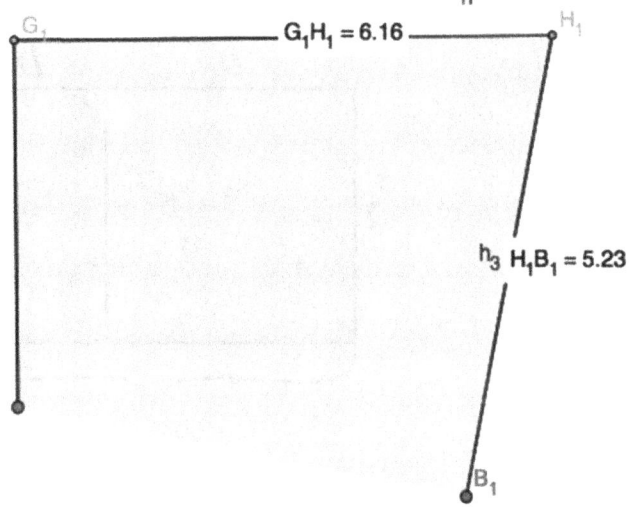

(Continued)

TABLE 17.2 (CONTINUED)   Examples of Teacher Candidates' Shared Solutions

III   To find whether the blue
          rectangle is in golden
          ratio, I used $(a+b)/g =$
          $(6.43+3.97)/6.43 =$
          $1.61741...$
      To find whether the green
          rectangle is in ratio I used
          $g/b = 6.43/3.97 = 1.61...,$
          so both rectangles are in
          ratio
      The ratio of a/b is 1.62,
          which is similar to phi
          (the golden ratio).
      This remains the same
          even if the rectangles
          move or increase in size

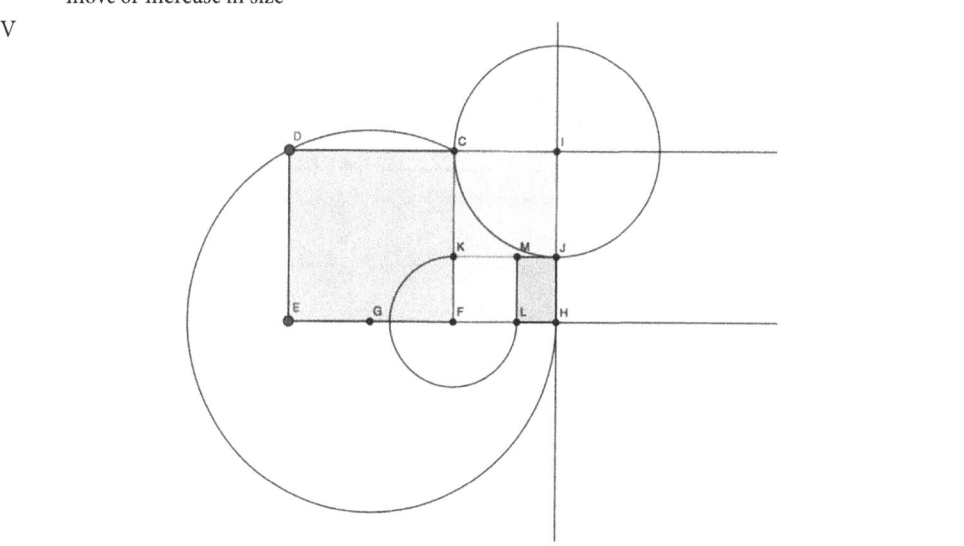

IV

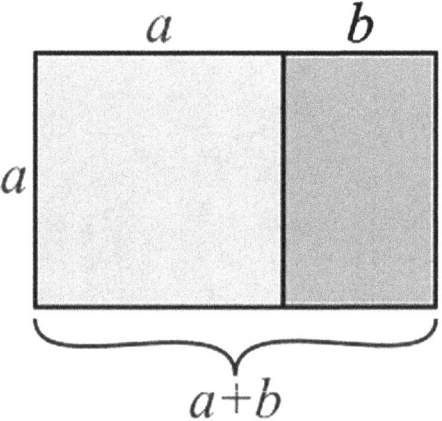

FIGURE 17.2   Golden rectangle.

teacher can explicitly ask students to do a drag test of their construction before submitting their work.

17.9 CONCLUSION

Mathematical diagrams play a crucial role in mathematics and are highly appreciated in geometry. Geometry—as one of the earliest fields of mathematics, and a visual field—keeps succinct, sketchy, rigorous, and expressive diagrams at its heart. Such diagrams involve shapes, solids, and spatial figures, as well as their properties and relationships. Alternatively, interactive and dynamic diagrams play the most crucial role in mathematical diagramming, and hence mathematizing, in an online platform. IDs in the digital format provide a rich environment to identify covariants and/or invariants under variations, as well as forming, examining, and verifying conjectures. This is an epistemological approach to mathematics that emphasizes experiencing mathematics, which is enabled through diagrammatic thinking, arguing, and reasoning.

In this chapter, we discussed why a tool-based environment enriched with IDs is necessary in an online mathematics class. It opens up pedagogical spaces for exploration, examination, and conjecturing, somehow in a total diagrammatic and tacit way. IDs invite learners' bodily engagement in online collaborative mathematizing. In such a situation, task design plays a crucial role. The diagrams then can be turned into a meaningful representation for the uttered or unuttered mathematical processes. IDs also play the role of mediator between mathematics, teaching, and learning. In addition, our study supports Azzouni's (2013) argument and shows how diagrammatic argumentation can be perfectly rigorous in a DGE. IDs provide a rich, dynamic environment that enables the instructor to assess every taken step. IDs also lift the confusion of proof expression procedures when compared with language proofs. The dynamic features of IDs in DGE diminish the impossibility of naturally generalizing a diagrammatic proof procedure, if any chance of that exists.

In this chapter, we argued that not only are IDs in an online mathematics course necessary, they are central. There will be little meaning left for mathematizing if IDs are taken out of the LMS because discussions, object constructions, and geometrical explorations heavily rely on IDs. We strongly recommend integrating IDs in online mathematics classes because of the different roles they can play in thinking, reasoning, arguing, and proving. Further studies incorporating other fields of mathematics and Massive Open Online Courses (MOOCs) using embedded interactive diagrams are suggested.

ACKNOWLEDGMENT

The authors would like to thank Abby Williams for technical editing and proofreading of the manuscript.

REFERENCES

Abrahamson, D., & Sanchez-Garcia, R. (2016). Learning is moving in new ways: The ecological dynamics of mathematics education. *The Journal of the Learning Sciences, 25*(2), 203–239. doi:10.1080/10508406.2016.1143370.

Artzt, A. F., Armour-Thomas, E., & Curcio, F. R. (2008). *Becoming a Reflective Mathematics Teacher.* New York, NY: Lawrence Erlbaum Associates.

Arzarello, F., Bairral, M. A., & Danè, C. (2014). Moving from dragging to touchscreen: Geometrical learning with geometric dynamic software. *Teaching Mathematics and Its Applications, 33*(1), 39–51. doi:10.1093/teamat/hru002.

Arzarello, F., Olivero, F., Paola, D., & Robutti, O. (2002). A cognitive analysis of dragging practises in Cabri environments. *ZDM, 34*(3), 66–72. doi:10.1007/BF02655708.

Arzarello, F., Robutti, O., & Thomas, M. (2015). Growth point and gestures: Looking inside mathematical meanings. *Educational Studies in Mathematics, 90*(1), 19–37. doi:10.1007/s10649-015-9611-5.

Azzouni, J. (2013). That we see that some diagrammatic proofs are perfectly rigorous. *Philosophia Mathematica, 21*(3), 323–338.

Battista, M. T. (2008). Development of the shape makers geometry microworld. *Research on Technology and the Teaching and Learning of Mathematics, 2,* 131–156. *Theses.* University of Guelph (Canada), Canada. Retrieved from http://proxy.lib.sfu.ca/login?url=http://search.proquest.com/docview/304890426?accountid=13800.

Bloom, B. S., Engelhart, M. D., Furst, E. J., Hill, W. H., Krathwohl, D. R. (1956). *Taxonomy of educational objectives: The classification of educational goals.* Handbook I: Cognitive domain. New York: David McKay Company.

Burch, K. J., & Y. Kuo. (2010). Traditional vs. online homework in college algebra. *Mathematics and Computer Education, 44,* 53–63.

Campbell, S. R. (2010). Embodied minds and dancing brains: New opportunities for research in mathematics education, In: B. Sriraman, & L. English (Eds.), *Theories of Mathematics Education: Seeking New Frontiers* (pp. 309–331). Berlin: Springer. doi:10.1007/978-3-642-00742-2.

Charoenying, Timothy. (2015). *Fostering Embodied Coherence: A Study of the Relationship Between Learners' Physical Actions and Mathematical Cognition.* (Doctoral dissertation). UC Berkeley: Education. Retrieved from http://escholarship.org/uc/item/5wz5f6zh.

Clark, I. (2012). Formative assessment: Assessment is for self-regulated learning. *Educational Psychology Review, 24*(2), 205–249.

Cobb, P., & Yackel, E. (1996). Constructivist, emergent, and sociocultural perspectives in the context of developmental research. *Educational Psychologist, 31*(3–4), 175–190.

Communication. (n.d.). *In Merriam Webster's Collegiate Dictionary.* Retrieved 08 08, 2018, from https://www.merriam-webster.com/dictionary/communication.

Cross, K. Patricia (Kathryn Patricia), Angelo, Thomas A., & National Center for Research to Improve Postsecondary Teaching and Learning. (1988). *Classroom Assessment Techniques: A Handbook for Faculty.* Ann Arbor, MI: National Center for Research to Improve Postsecondary Teaching and Learning, University of Michigan.

Cross, T., & Palese, K. (2015). Increasing learning: Classroom assessment techniques in the online classroom. *American Journal of Distance Education, 29*(2), 98–108.

Davidson, N. (Ed.). (1990). *Cooperative Learning in Mathematics: A Handbook for Teachers.* Menlo Park, CA: Addison-Wesley.

DeCosta, M., Bergquist, E., & Holbeck, R. (2015). A desire for growth: Online full-time faculty's perceptions of evaluation processes. *Journal of Educators Online, 12*(2), 73–102.

de Freitas, E. (2016). Material encounters and media events: What kind of mathematics can a body do? *Educational Studies in Mathematics, 91*(2), 185–202. doi:10.1007/s10649-015-9657-4.

de Freitas, E., & Sinclair, N. (2012). Diagram, gesture, agency: Theorizing embodiment in the mathematics classroom. *Educational Studies in Mathematics, 80*(1), 133–152. doi:10.1007/s10649-011-9364-8.

de Freitas, E., & Sinclair, N. (2013). New materialist ontologies in mathematics education: The body in/of mathematics. *Educational Studies in Mathematics, 83*(3), 454–470. doi:10.1007/s10649-012-9465-z.

Flanagan, K. (2001). *High School Students' Understandings of Geometric Transformations in the Context of a Technological Environment.* (Unpublished doctoral dissertation), Pennsylvania State University.

Healy, L., & Fernandes, A. (2011). The role of gestures in the mathematical practices of those who do not see with their eyes. *Educational Studies in Mathematics, 77*(2–3), 157–174. doi:10.1007/s10649-010-9290-1.

Hohenwarter, M. (2002). GeoGebra. Retrieved from http://www. geogebra. org/cms/en.

Hollebrands, K. F. (2003). High school students' understandings of geometric transformations in the context of a technological environment. *The Journal of Mathematical Behavior, 22*(1), 55–72.

Hoyles, C. (1997). The curricular shaping of students' approaches to proof. *For the Learning of Mathematics, 17*(1), 7–16.

Jackiw, N. (2001). *The Geometer's Sketchpad (Version 4.0) [Computer software].* Emeryville, CA: KCP Technologies.

Jungic, V., Kaur, H., Mulholland, J., & Xin, C. (2015). On flipping the classroom in large first year calculus courses. *International Journal of Mathematical Education in Science and Technology, 46*(4), 508–520.

Kaur, H. (2015). Two aspects of young children's thinking about different types of dynamic triangles: Prototypicality and inclusion. *ZDM, 47*(3), 407–420.

Kress, G. R., & Van Leeuwen, T. (1996). *Reading Images: The Grammar of Visual Design.* Psychology Press. London: Routledge.

Krummheuer, G. (2013). The relationship between diagrammatic argumentation and narrative argumentation in the context of the development of mathematical thinking in the early years. *Educational Studies in Mathematics, 84*(2), 249–265.

Laborde, J. M., & Strässer, R. (1990). Cabri-Geometre: A microworld of geometry for guided discovery learning. *Zentralblatt für didaktik der mathematik, 5*, 171–177.

Leahy, S., Lyon, C., Thompson, M., & Wiliam, D. (2005). Classroom assessment minute by minute, day by day. *Educational Leadership, 63*(3), 18–24.

Liljedahl, P. (2016). Building thinking classrooms: Conditions for problem solving. In: P. Felmer, J. Kilpatrick, & E. Pekhonen (Eds.), *Posing and Solving Mathematical Problems: Advances and New Perspectives.* New York, NY: Springer.

Lo, C. K., Hew, K. F. T., & Chen, G. (2017). Toward a set of design principles for mathematics flipped classroom: A synthesis of research in mathematics education. *Educational Research Review, 22*, 50–73.

Lo, C.K. & Hew, K.F. (2017). A critical review of flipped classroom challenges in K-12 education: possible solutions and recommendations for future research. *Research and Practice in Technology Enhanced Learning,* 12 (4). https://doi.org/10.1186/s41039-016-0044-2

Mariotti, M. A. (2000). Introduction to proof: The mediation of a dynamic software environment. *Educational Studies in Mathematics, 44*(1–2), 25–53.

Mason, J., & Pimm, D. (1984). Generic examples: Seeing the general in the particular. *Educational Studies in Mathematics, 15*(3), 277–289.

Menz, P. (2015). *Unfolding of Diagramming and Gesturing Between Mathematics Graduate Student and Supervisor During Research Meetings.* (Published doctoral thesis). Simon Fraser University.

Mowat, E. (2010). *Making Connections: Network Theory, Embodied Mathematics, and Mathematical Understanding.* (Doctoral dissertation). University of Alberta.

Naftaliev, E., & Yerushalmy, M. (2013). Guiding explorations: Design principles and functions of interactive diagrams. *Journal of Computers in the Schools, 30*(1–2), 61–75.

Naftaliev, E., & Yerushalmy, M. (2017). Engagement with interactive diagrams: The role played by resources and constraints. In: A. Leung, A. Baccaglini-Frank (Eds.), *Digital Technologies in Designing Mathematics Education Tasks: Mathematics Education in the Digital Era,* vol. 8. Cham: Springer,

Nasim, O. (2008). *Bertrand Russell and the Edwardian philosophers: Constructing the world.* Basingstoke: Palgrave Macmillan.

National Research Council (NRC). (1999). *How People Learn: Brain, Mind, Experience, and School.* J. D. Bransford, A. L. Brown, & R. R. Cocking (Eds.), Washington, DC: National Academy Press.

Nemirovsky, R., & Ferrara, F. (2008). Mathematical imagination and embodied cognition. *Educational Studies in Mathematics, 70*(2), 159–174. doi:10.1007/s10649-008-9150-4.

Nemirovsky, R., Kelton, M. L., & Rhodehamel, B. (2012). Gesture and imagination on the constitution and uses of phantasms. *Gesture, 12*(2), 130–165.

Nemirovsky, R., Kelton, M. L., & Rhodehamel, B. (2013). Playing mathematical instruments: Emerging perceptuomotor integration with an interactive mathematics exhibit. *Journal for Research in Mathematics Education, 44*(2), 372–415. doi:10.5951/jresematheduc.44.2.0372.

Pratt, D., & Ainley, J. (1997). The construction of meanings for geometric construction: Two contrasting cases. *International Journal of Computers for Mathematics Learning, 1*(3), 293–322.

Protheroe, N. (2007). What does good math instruction look like? *Principal, 7*(1), 51–54.

Radford, L. (2008). Diagrammatic thinking: Notes on Peirce's semiotics and epistemology. *PNA, 3*(1), 1–18.

Radford, L. (2013). Sensuous cognition. In: D. Martinovic, V. Freiman, & Z. Karadag (Eds.), *Visual Mathematics and Cyber Learning* (pp. 141–162). New York, NY: Springer.

Radford, L. (2014). Towards an embodied, cultural, and material conception of mathematics cognition. *Mathematics Education, 46*, 349–361. doi:10.1007/s11858-014-0591-1.

Ragan, L. (2009). Defining competencies for online teaching success. *Distance Education Report, 1*, 3–6. Retrieved from http://www.magnapubs.com/newsletter/distance-education-report/135/defining_competencies_for_online_teaching_success-8536-1.html.

Richter-Gebert, J., & Kortenkamp, U. H. (1999). *CINDERELLA: The Interactive Geometry Software.* Version 1.2. Springer.

Tall, D. (2006). A theory of mathematical growth through embodiment, symbolism and proof. *Annales de didactique et de sciences cognitives, 11*, 195–215. Retrieved from http://homepages.warwick.ac.uk/staff/David.Tall/pdfs/dot2006e-theory-math-growth-annales.pdf.

Santos, V., & Quaresma, P. (2012). Integrating DGSs and GATPs in an adaptative and collaborative blended-learning Web-environment. *arXiv preprint.* arXiv:1202.4833.

Santos, V., Quaresma, P,. Marić, M., & Campos, H. (2018). Web geometry laboratory: Case studies in Portugal and Serbia. *Interactive Learning Environments, 26*(1), 3–21, doi:10.1080/10494820.2016.1258715.

Sedaghatjou, M. (2017). *Mathematical Tool Fluency: Learning Mathematics via Touch-based Technology.* (Published doctoral thesis). SFU. Retrieved from http://summit.sfu.ca/identifier/etd10260.

Sedaghatjou, M. (2018). Advanced mathematics communication beyond modality of sight. *International Journal of Mathematical Education in Science and Technology, 49*(1), 46–65. doi:10.1080/0020739X.2017.1339132.

Sinclair, N., & Moss, J. (2012). The more it changes, the more it becomes the same: The development of the routine of shape identification in dynamic geometry environment. *International Journal of Educational Research, 51*, 28–44.

Sinclair, N., & Yurita, V. (2008). To be or to become: How dynamic geometry changes discourse. *Research in Mathematics Education, 10*(2), 135–150.

Steele, J., & Dyer, T. (2014). Use of KWLs in the online classroom as it correlates to increased participation. *Journal of Instructional Research, 3*, 8–14.

Stylianou, D., & Silver, E. (2004). The role of visual representations in advanced mathematical problem solving: An examination of expert-novice similarities and differences. *Mathematical Thinking and Learning, 6*(4), 353–387.

Suurtamm, C., Quigley, B., & Lazarus, J. (2015). Making space for students to think mathematically. *What Works: Research into Practice, Research Monograph, 59.*

Vaughan, N. D., Cleveland-Innes, M., & Garrison, D. R. (2013). *Teaching in Blended Learning Environments: Creating and Sustaining Communities of Inquiry.* Athabasca University Press.

Yanik, H. B., & Flores, A. (2009). Understanding rigid geometric transformations: Jeff's learning path for translation. *The Journal of Mathematical Behavior, 28*(1), 41–57.

Yu, P., & Barrett, J. E. 2002. Shapes, actions, and relationships: A semiotic investigation of student discourse in a dynamic geometric environment. In: D. S. Mewborn, P. Sztajin, D. White, H. Wiegel, R. Bryant, & K. Nooney (Eds.), *Proceedings of the Twenty-Fourth Conference of PME-NA* (pp. 775–784). Columbus, OH: Clearing House for Science, Mathematics, and Environmental Education.

PART 4

Teacher Education

CHAPTER 18

MOOCs for Mathematics Teacher Education

New Environments for Professional Development

Eugenia Taranto

CONTENTS

18.1 INTRODUCTION

Massive Open Online Courses (MOOCs) are free online courses, open to all and intended for distance learning of a large number of participants coming from very different cultural backgrounds and from different geographical areas. They were first introduced in 2008 and emerged as a popular mode of learning in 2012 (Pappano, 2012). Access to education significantly improves people's lives and can bring changes to communities. With MOOCs, growth opportunities become available to everyone, all over the world: not only does the access to education increase exponentially, but also an unheard-of quality of teaching is offered, thanks to the digital tools available. Students from anywhere with an Internet connection can access free courses facilitated by some of the top educators and experts in every subject area, and they may customize their workload according to their needs (Daza, Makriyannis & Rovira Riera, 2013). "From mathematics to computer science, to philosophy, to business design, MOOCs give learners unprecedented access to some of the most valuable knowledge, from some of the most prestigious universities, for free" (www.teach thought.com/archived/list-75-moocs-teachers-students/). Most of the courses do not offer actual credit towards a degree. However, some MOOCs offer certificates, additional credit options, and other enhanced learning services for nominal fees. Students have also been able to submit course work done through MOOCs to their own universities and be granted credit or research units (*ibid*). The activities provided in these courses range from watching certain videos, posting on forums or blogs, sharing experiences on social media, responding to quizzes, doing learning tasks for individuals or workgroups, and/or conducting peer review activities (Daza et al., 2013; Abeer & Miri, 2014). Students are involved to various degrees: many just want to check out the resources and the new educational model, while others are really motivated and follow every aspect of the course, often interacting with other MOOC participants. Likewise, educators' involvement varies substantially: in some courses, the educators disappear when the course starts; in others, they are intensively involved, injecting dynamism to the proposed activities and providing their students with feedback (Daza et al., 2013).

If the MOOC presents itself as a training opportunity that is easy to access and open to many, the experiences show that it is not easy for everyone to complete the path: in fact, the drop-out rate of these courses is very high (Yang, Sinha, Adamson & Rosé, 2013; Bayne & Ross, 2014; Onah, Sinclair & Boyatt, 2014). As Onah, Sinclair, and Boyatt (2014) observe, "although many thousands of participants enroll on these courses, the completion rate for most courses is below 13%" (p. 5825). Feng, Tang, and Liu (2019) show that the low completion rate can be determined by some factors: "Age is an important factor — young people are more inclined to drop out; Gender is another important factor— roughly, female users are more likely to drop science courses and male users are more likely to give up non-science courses; finally, educational background is also important" (p. 1). The completion rate continues to remain at the thresholds declared in the literature even for MOOCs where the trainees are teachers. For example, the experience of Panero, Aldon, Trgalová, and Trouche (2017) with a MOOC for French-speaking mathematics teachers, with the aim to improve their practices in using technology in their classrooms, shows a completion rate of 12%.

It is worth noting that the emergence and use of MOOCs for professional teacher development are still uncommon, especially in mathematics. In fact, although there is a wide choice of many different topics, when looking specifically for a MOOC aimed at mathematics teacher education the range is limited (Aldon, Arzarello, Panero, Robutti, Taranto & Trgalová, 2017; Avineri, Lee, Tran, Lovett & Gibson, 2017; Borba, Askar, Engelbrecht, Gadanidis, Llinares & Aguilar, 2017). Nevertheless, there is a growing interest in MOOCs involving mathematics teachers as participants, as shown by Topic Study Group 44

(TGS44) work during the 13th ICME.* Therefore, MOOCs for teacher education are on the verge of gaining a foothold. For example, Avineri et al. (2017), in North Carolina, offer some MOOCs for improving mathematics and statistics instruction, more precisely to help K-5 teachers teach fraction concepts and to think about statistics teaching in ways likely different from current practices in middle school through introductory statistics; Borba et al. (2017) describe the use of a MOOC in Costa Rica to meet the challenge of implementing new curricular standards, namely supporting in-service teachers in the gradual implementation of the new curriculum.

Precisely because of the fact that the MOOCs for teacher education are still expanding, not only the literature on this subject is scarce, but the research has not yet developed a theoretical framework sufficient to explain how changes in knowledge (understood as teachers' professional development) can possibly occur as a product of the activities in these new environments totally online. There is, however, a wide literature (Goos, 2005; Campbell, 2009; Joubert, 2013; Clark-Wilson, Aldon, Cusi, Goos, Haspekian & Robutti, 2014; Robutti et al., 2016) that deals with the way in which the teachers can develop their professional learning in traditional, face-to-face courses, particularly when the theme of the update concerns the relationship between education and technology. Based on the fact that there are no theoretical frameworks in the literature aimed at explaining professional development specifically in MOOCs, in my doctoral thesis (Taranto, 2018), I developed a new theoretical framework aimed at understanding the complexities of the learning trajectories of the participants in a MOOC. The protagonists that I considered are the in-service secondary school mathematics teachers enrolled in the MOOC and the mathematics teacher educators involved in the MOOC design and delivery. With this learning trajectory, I mean how these protagonists interact online, both with the platform and with each other, in particular, if and how these interactions change their knowledge and beliefs and generate a perception of change in the practices.

In this chapter, I will give a partial description of the theoretical framework, illustrating the elements that allow the following research question to be answered: What are benefits for professional development that mathematics in-service teachers can derive from attending a MOOC?

In particular, I will show an Italian experience with MOOCs for mathematics teacher education, within the *Math MOOC UniTo* project, in which I am involved as an educator, illustrating some examples taken from the concrete experiences of teachers who have enrolled in these MOOCs. I will also highlight how the completion rates of these MOOCs are different than those reported in the literature, and justify why. Finally, I will mention how a study so set up represents a new research trail in relation to the modalities of interaction that new technologies offer for the teachers' professional development.

18.2 THEORETICAL FRAMEWORK

A MOOC is a very complex environment in which several protagonists alternate their roles. In fact, in the design phase, when the MOOC starts to take shape, it is inhabited

---

* For more information, see www.icme13.org/files/tsg/TSG_44.pdf.

by designers and trainers who propose the materials and/or the resources to be included in it. When the MOOC is ready, it is still at an inert state; then it is opened in order to accommodate the entry of the trainees. They will discover it weekly, moving far and wide. "Far" because they will wait for the next module opening to see new materials and "wide" because every week a new module is opened, but the previous ones are kept open until the end of the course. So, the trainees can discover the novelty of the new module, but at the same time may come back and see in depth the previous material and resources.

In order to describe and analyze the dynamics that take place within a MOOC, I have developed a theoretical framework called MOOC-MDT (Taranto, 2018). It integrates three theoretical frameworks in a new form: the Meta-Didactical Transposition* (MDT: Arzarello et al., 2014), the Instrumental Approach (Vérillon & Rabardel, 1995), and the Connectivism (Siemens, 2005). In what follows, I give a synthetic idea of this framework.

First, a MOOC can be considered as an *artifact* (Vérillon & Rabardel, 1995), that is, a static set of materials. Connectivism allows to picture the *MOOC-artifact* with its own network of knowledge: its nodes are the content, the ideas, the images and videos used; the connections are the links between their node pairs. When a MOOC module is activated, it dynamically generates a complex structure that I call *ecosystem*: "all the relations (exchange of materials, experiences and personal ideas/point of view) put in place by participants of an online community, thanks to the technological tools through which they interact with each other, establishing connections within a given context" (Taranto, Arzarello, Robutti, Alberti, Labasin & Gaido, 2017 b, p. 2481).

In a MOOC there are two communities, a community of designers and researchers (from here and after the trainers) and a community of enrolled and participating teachers (from here and after the trainees). Each trainee has her own didactical praxeologies, namely the concrete practices used by the teachers in their professional activities. The trainers have the objective of transposing a certain piece of knowledge, related to the teaching and learning of mathematics, to favor the professional development of the trainees, according to the reference institutions (national curricula, textbooks, etc.). This trainers' knowledge to be communicated constitutes the trainers' meta-didactical praxeologies (*m-dp*).

Learning within a MOOC happens in a Connectivist modality: each trainee is part of a community, with which she comes into contact and has the opportunity to share her own views, self-organizing information, creating new connections, and questioning existing ones. Moreover, what is shown in the MOOC should encourage trainees to experiment in their classrooms. In fact, during the implementation of the MOOC-artifact network of knowledge, the trainers foster its ecosystem nature, sharing tools and posing appropriate key questions. In addition, since the tasks are designed by the trainers, this more or less explicitly suggests to trainees that they should use the proposed material in their classes. In this way, the MOOC is enriched with reports about trainees' teaching experiences: this

---

* MDT is a model that describes the process of teachers' professional development, involved in a face-to-face educational course, with the aim of grasping its complexity. It is a tool to analyse the dynamic aspects of this process, namely the evolution of teachers and researchers' activity over time. This activity is described through teachers and researchers' meta-didactical praxeologies (Arzarello et al., 2014, pp. 353–355), which consist of the task in which they are engaged in the educational program, with the techniques used to solve it, along with its theoretical justification.

process produces an organic cycle that encourages other trainees to experience the same materials.

The community of trainees is made of subjects who are involved in the MOOC-ecosystem and who transform it into an instrument. As in many social platforms, every participant develops her/his own way through the system, but within the affordances that the system itself supports. The trainee has to solve tasks, through techniques, properly justified. In fact, she must look at the proposed material, share her thoughts through sharing tools, and experience the MOOC activities. These tasks are not predetermined, they depend on the time, approach, and depth with which each trainee addresses them. The techniques are the ways in which the trainee extends and modifies her network of knowledge, drawing on the ecosystem, and influencing it in turn, thus impacting all other trainees. Therefore, the *MOOC-ecosystem* network of knowledge is dynamic: it evolves as the MOOC-artifact network, thanks to each participant's contribution. Also, each MOOC participant's network of knowledge evolves as personal self-organization (Siemens 2005, p. 4) of the ecosystem. The process of transformation from artifact to instrument (Vérillon and Rabardel, 1995) is here reinterpreted by the evolution from artifact to ecosystem/instrument—a process Taranto et al. (2017b, p. 2482) call the *double learning process*. The double learning process has the following components, intertwined and self-feeding each other:

i. *Instrumentation/self-organization* (from the ecosystem to the individual): process by which the MOOC-ecosystem's network expands the individual's network of knowledge. In particular, the instrumentation (Vérillon & Rabardel, 1995) is the phase by which the chaos (in the sense of Siemens, 2005) of the ecosystem network reaches the individual. The many novelties of views and experiences make sure that the individual compares himself with new usage schemes. A phase of self-organization (Siemens, 2005) of the MOOC's information follows: when the individual selects which usage schemes proposed by the MOOC are valuable and which are not.

ii. *Instrumentalization/sharing* (from the individual to the ecosystem): process by which the individual's network of knowledge expands the MOOC-ecosystem's network. The instrumentalization (Vérillon & Rabardel, 1995) is the phase by which the individual, with her/his renewed network of knowledge, independently builds new connections. The individual is stimulated by a task requested by the MOOC and (s)he caters to the ecosystem to turn it according to her own (new) usage schemes. (S)he wants to integrate it with her/his own cognitive structures. Sharing is the phase by which the MOOC welcomes the contribution of the individual and makes it available to all: information goes towards all members.

Since the MOOC participants are massive, the process is iterated: a moment of sharing is followed by a new instrumentation; a self-organization by an instrumentalization. It should be emphasized that the two processes are "intertwined"; there is no moment in which one ends and the next one begins.

Within this complex, iterative learning process lies the inherent difference between the frame of the MDT and the MOOC-MDT. In fact, in the MDT, the trainers shape their proposal according to the practices they consider appropriate (which can also be decided on the moment, depending on how the group to which the training is addressed is responding). The in-presence comparison helps trainers to understand how much trainees learn this proposal. On the contrary, inside the MOOC-MDT the process appears to be more difficult to control. The trainers do not know "what" the user has really looked at among the presented materials, nor they can know how (s)he interpreted them. At the same time, the trainees benefit from material provided not only by trainers but also by other trainees who share some of their ideas using the space for online communication. The process evolves stochastically: a determining role is played by the individual trainees, and by their feeling as a community with whom to collaborate, to inspire, and to share results.

18.3 MATH MOOC UNITO AND ITS METHODOLOGICAL CHOICES

Math MOOC UniTo is an Italian project for the development of MOOCs aimed at the education of Italian secondary school mathematics in-service teachers. It was born in spring 2015 at the Department of Mathematics "G. Peano" in Torino. The MOOCs are designed, implemented, and delivered through the collaboration between some researchers in mathematics education (to which I belong) from the mentioned department and some in-service teachers (from here and after experienced teachers), graduated from the second level Master "Trainers in Mathematics Education" held at the same department from September 2013 to June 2015. In the following, in the light of the theoretical framework, I refer to this entire group of engaged people as trainers.

Four MOOCs were designed, one for each of the main topics in the official Italian programs for secondary school: Geometry, Arithmetic and Algebra, Change and Relations, Uncertainty and Data. So far, the first three have been delivered (Table 18.1). The last one is work in progress, and it will be delivered at the beginning of 2019.

These MOOCs are open, free, and available online on a Moodle platform called DI.FI.MA (i.e., Didactics of Physics and Mathematics: http://difima. i-learn.unito. it/), managed by the mentioned department. They offer access to materials (e.g., mathematical activities, media resources like GeoGebra applet) through technological tools, which foster communication and sharing.

All the trainers are involved in the design, the courses' delivery, and monitoring their evolution in terms of interaction among participants and educational resources made available. In particular, a group of the experienced teachers creates the activities delivered

TABLE 18.1 MOOCs Duration

MOOCs of Math MOOC UniTo	From ... to ...
MOOC Geometria, based on Geometry content	October 2015–January 2016
MOOC Numeri, based on Arithmetic and Algebra content	November 2016–February 2017
MOOC Relazioni e Funzioni, based on Change and Relations content	February 2018–May 2018

in the MOOCs, adapted from the m@t.abel project* and revised by the university research-ers. Moreover, the trainers help MOOC learners (namely, the trainees) to solve technical problems (thanks to some tutorials too), to recall the tasks to be done week by week with weekly emails.

The mathematical resources inside the MOOCs provide trainees with suggestions to support their teaching. The mathematics curricula, to which the activities refer, are in line with the Italian curriculum. The activities do not exhaust all the topics of the curricula but have the goal to provide detailed methodological indications on how to deal with some conceptual nodes of particular importance for the mathematical education of the students.

Each week, trainees worked individually to become familiar with different approaches. In our MOOCs, these activities included watching videos where an expert introduced the mathematical topic of the week or reading about mathematical activities based on a labora-tory methodology, explicitly suggested by the Italian curriculum, and, optionally,† experi-menting with these in their classroom. The trainees were invited to share thoughts and comments about the activities and their contextualization within their personal experi-ence, using specific communication message boards (CMBs, i.e., the forum, as we will see in the following, but not only this). In each activity, MTEs inserted a specific question to be answered or a title that served as a talking point. In this way, the trainees collected their weekly badges, which acknowledged their different kinds of participation. In fact, once all the module requests were accomplished, the platform released a badge. In this way, it is quite easy for the trainers to monitor the progress of the trainees, knowing the number of badges they had collected.

On the one hand, choosing resources that support interactions (e.g., the CMBs) that increase the birth of new connections and/or nodes in the trainees' network of knowl-edge is a methodological choice that fosters the development of the instrumentation/self-organization phase. On the other hand, inserting specific stimulus questions or titles in the CMBs or inviting trainees to experiment with the activities with their own students is a methodological choice that promotes and increases the interactions among trainees, hence the development of the instrumentalization/sharing phase. Moreover, the trainers chose to limit their own interventions in the CMBs to a minimum in order to support the birth of a "trainees-only" online community. In fact, our MOOCs methodology aims to create collaborative contexts for teachers' work, where trainees can learn from these kinds of practices.

At the end of the whole MOOC, each trainee is asked to design a teaching activity (Project Work) and to review another activity prepared by a colleague (Peer Review). For all those who complete the course in all its stages, namely if all the badges are collected, a participation certificate is issued. Only for the people who have registered in the MOOCs, the materials remain available to allow future consultation. The communication spaces are closed. A reflective analysis of all the problems encountered during the evolution of this

---

* A plurennial Italian program that promoted innovation in mathematics teaching, based on concrete activities proposed to teachers and discussed with them in suitable professional learning programs (https://goo.gl/Q30Dn0).
† If the trainees liked them; if the trainees were explaining at that time topics close to those proposed in their own classes.

experience of MOOC can be consulted in Aldon, Arzarello, Panero, Robutti, Taranto, and Trgalová (2018).

In the following, we concentrate the analysis on the first two delivered MOOCs—*Geometria* and *Numeri*—because the data relative to the third one are under investigation.

18.4 MOOCs' MATERIALS AND TRAINEES' INTERACTIONS

In Table 18.2 there are some significant data to describe the trainees of *MOOC Geometria* and *MOOC Numeri*.

Focusing only on the level of design, the MOOC-artifact, which is the place where only the trainers have access, is the container of specific products, i.e., materials rich in

TABLE 18.2   The Trainees of *MOOC Geometria* and *Numeri*

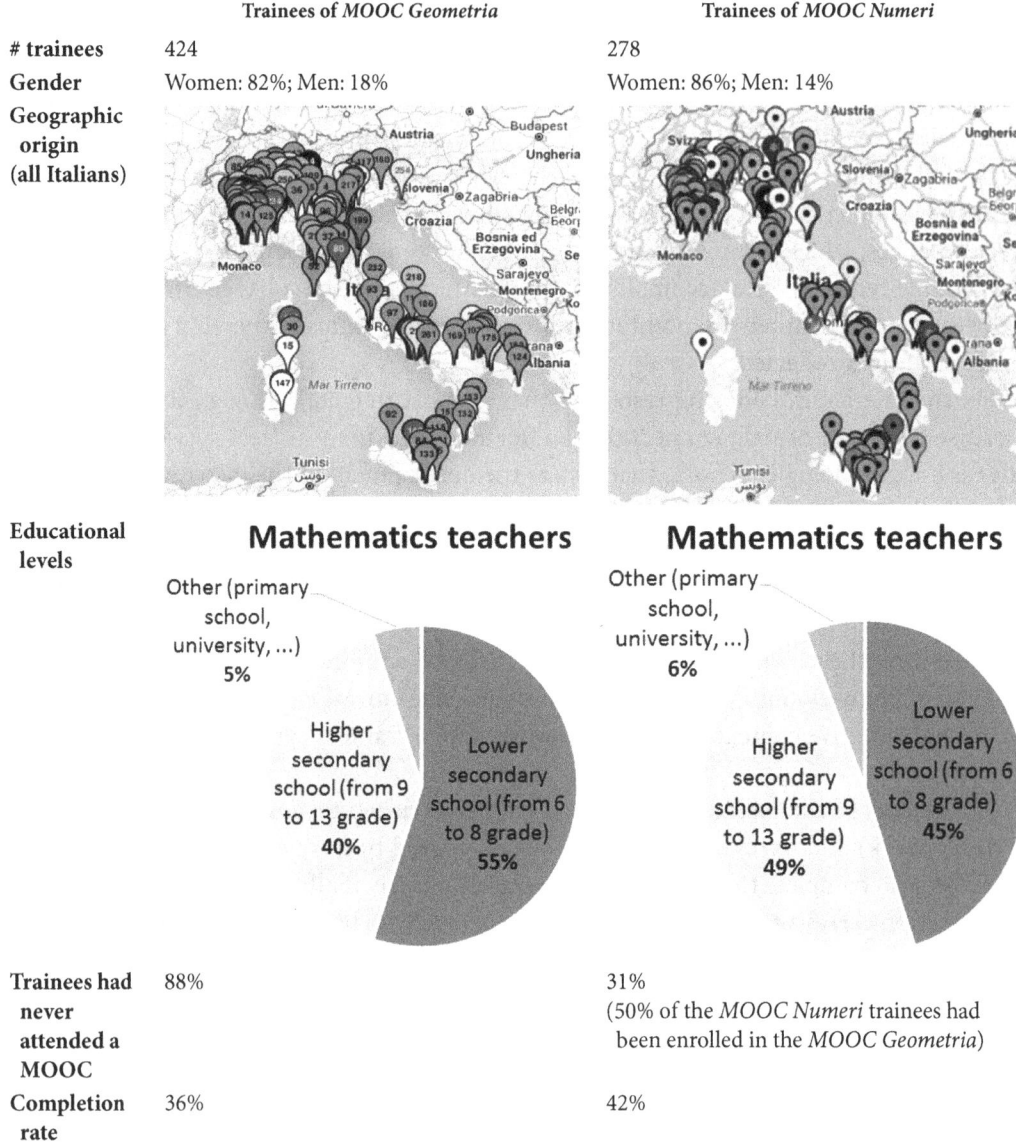

	Trainees of *MOOC Geometria*	Trainees of *MOOC Numeri*
# trainees	424	278
Gender	Women: 82%; Men: 18%	Women: 86%; Men: 14%
Geographic origin (all Italians)		
Educational levels		
Trainees had never attended a MOOC	88%	31% (50% of the *MOOC Numeri* trainees had been enrolled in the *MOOC Geometria*)
Completion rate	36%	42%

TABLE 18.3 *MOOC Geometria* and *Numeri* Modules

	MOOC Geometria		*MOOC Numeri*	
	Title	Content	Title	Content
Module 1	Ramps, sails, park and folding paper	Approaching distance with laboratory activity, using GeoGebra as well	Meteorites, bacteria and rice grains: the numbers and their meaning	Approaching order of size and number sense math laboratory activity, using GeoGebra as well
Module 2	From watches, pinwheels, skaters to the Christmas show	Approaching angle with laboratory activity, using GeoGebra as well	MERLO methodology	Recognizing different representations with the same meaning
Module 3	Heritage, a Polyaag problem and that demonstration?	Arguing, conjecturing, prying, using GeoGebra as well	Assessment & INVALSI	Evaluating different skills
Module 4	Assessment & INVALSI	Evaluating different skills	Climbing stairs	Approaching the concept of inductton'''recurs1on, using GeoGebra as well
Module 5	MERLO methodology	Recognizing different representations with the same meaning	Arithmetic, algebra and mathematical languages	Approaching arithmetic and algebraic language and meaning of the symbols, using GeoGebra as well
Final module	Project work and peer review	Designing a teaching situation with a specific web-based tool and reviewing an activity designed by another colleague	Project work and peer review	Designing a caching situation with a specific web-based tool and reviewing an activity designed by another colleague

innovative teaching methods and specific technological tools. Therefore, we can understand it as a repository from which teachers can draw inspiration (Taranto et al., 2017a).

Both in *MOOC Geometria* and *MOOC Numeri*, five modules, with captivating titles, were created on specific contents, plus a final one (Table 18.3).

In each module specific methodologies of mathematics education are proposed, such as laboratory-based methodology (Anichini et al., 2004) with attention to considering activities that are inspired by real contexts, therefore familiar or easily imaginable for students; group work to incite a collaborative perspective and to encourage mathematical discussion (Bartolini Bussi, Boni & Ferri, 1995); MERLO* methodology (Arzarello, Robutti & Carante, 2015); use of technology (especially GeoGebra with guided examples of construction), but also simple tactile materials (cardboard, string, etc.). The duration of each module varies from one to two weeks, depending on the content proposed and the estimate of time needed for their assimilation according to the trainers' opinion.

In our project, we have collected and analyzed data coming from trainees' posts from the CMBs, questionnaires, interviews with a sample of trainees, and resources that they

---

* MERLO stands for Meaning Equivalence Reusable Learning Objects.

designed and uploaded in the MOOC platform. In the following, we will report some of the mathematical activities proposed by trainers in the MOOCs modules, and we will focus more on some data that come from trainees' intervention on the forum regarding these materials. Then we will see how these elements of the MOOC-artifact have then given rise to specific learning processes implemented by the trainees in the MOOC-ecosystem/instrument.

To analyze the trainees' posts in the CMBs, in the light of the double learning process, it is important to note which verbs are used by the trainees.

- For *instrumentation/self-organization* the verbs are in the future tense (I will do it, I will re-propose, I will test it, I will use it, etc.) or there are verbs or adjectives to express the trainees' own judgment (I have noticed, I really appreciated, nice idea, etc.).

- For *instrumentalization/sharing* instead the verbs refer to one's own self (I reflect, I know, I thought, etc.) when one is creating new connection stimulated by the MOOC-ecosystem; while the verbs are in the present tense when one shares one's didactical praxeologies (I do this, I use that, etc.).

All the interventions are written in a normal type. If you find bold or underlined word, those will be "signs" inserted by me to accomplish the analysis. It is also important to note that all discussions take place asynchronously, at any time of day.

Each MOOC also includes three questionnaires (initial, intermediate, final) to assess the degree of satisfaction/appreciation for the educational offerings and how much/how the online course affects their professional development. We will then conclude the analysis by focusing on a question posed in the final questionnaires of the two MOOCs, relating to the sense of belonging to the online community felt by the trainees.

18.5 EXAMPLES FROM MOOC GEOMETRIA

18.5.1 Difference between Perpendicular and Vertical

Module 1 of *MOOC Geometria* is focused on the concept of distance between a point and a line (connected to other concepts, including perpendicular and height). The proposed activities are directed to avoid or overcome student misconceptions related to these concepts. One activity is "The mainmast" (for full version see: https://goo.gl/4F4rJW), for lower secondary school students (grades 6–8), and is developed from a concrete situation: the teacher gives each student a white circular sheet with a sketch of a boat on a sea wave (Figure 18.1).

School students are asked to draw, on a round sheet (to avoid references), the mainmast of the boat: they have to concentrate on the perpendicular. The activity continues with observations of the various drawings and with discussions of the various solutions. This activity should ground the discussion around the two concepts: (i) *vertical* (physical concept linked to the gravitational field); (ii) *perpendicular* (geometric concept linked to the right angle).

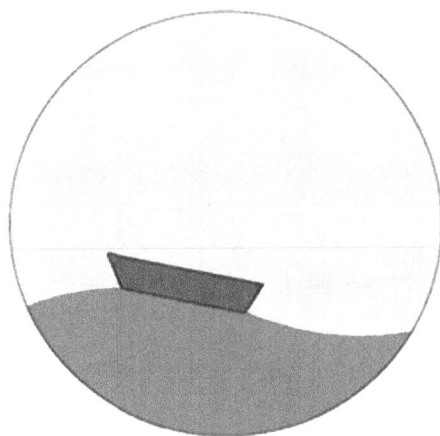

FIGURE 18.1   Boat on round sheet.

The CMB embedded into Module 1 was the forum. In the forum, the trainers have inserted an assignment to stimulate discussion among the trainees: *"Share your ideas and/ or teaching experiences related to the topics of the mainmast activity."* The forum collected 24 discussions, each containing from 0 to 62 response replicas, for 207 posts in total. The forum also keeps track of the date and time each post was published. Let's consider the following discussion:

A.P.—27/10/15; 6:50 p.m.—*The idea is to play with the heights of the triangles and I half minded to propose it to my pupils :) This is a draft of text.*

3 male friends Antonio, Bruno and Carlo are at the top of the triangle in the figure (Figure 18.2). 3 female friends Antonella, Barbara and Carlotta are also at the top of the triangle in the picture. Friends via whatsapp agree to find themselves in the orthocentre of the triangle while the friends will meet in the centroid of the triangle. Draw the meeting points of the two groups.

PS: I used the map of Latina, my city.

Didactic note: I deliberately chose an obtuse triangle and the position of the triangle is not that stereotyped by the boys.

P.R.—27/10/15; 11:16 p.m.—*This activity is **beautiful: I will propose** it next week (obviously using a map of a city closer to my boys, like Turin) to see how they have internalized the concepts of orthocentre and centroid, since they have just discovered heights and medians [...]*

M.L.—28/10/15; 10:40 a.m.—***I really like*** *the proposal and I hypothesize a variant of the text: in a treasure hunt the competitor Alberto of the team is in A, Bruno in B and so Caterina in C. The next clue will be given only when all three competitors will meet in the orthocentre of the triangle and communicate the position to the director ... etc ... it could also be said that there is a tolerance of a certain amount of meters for the possible*

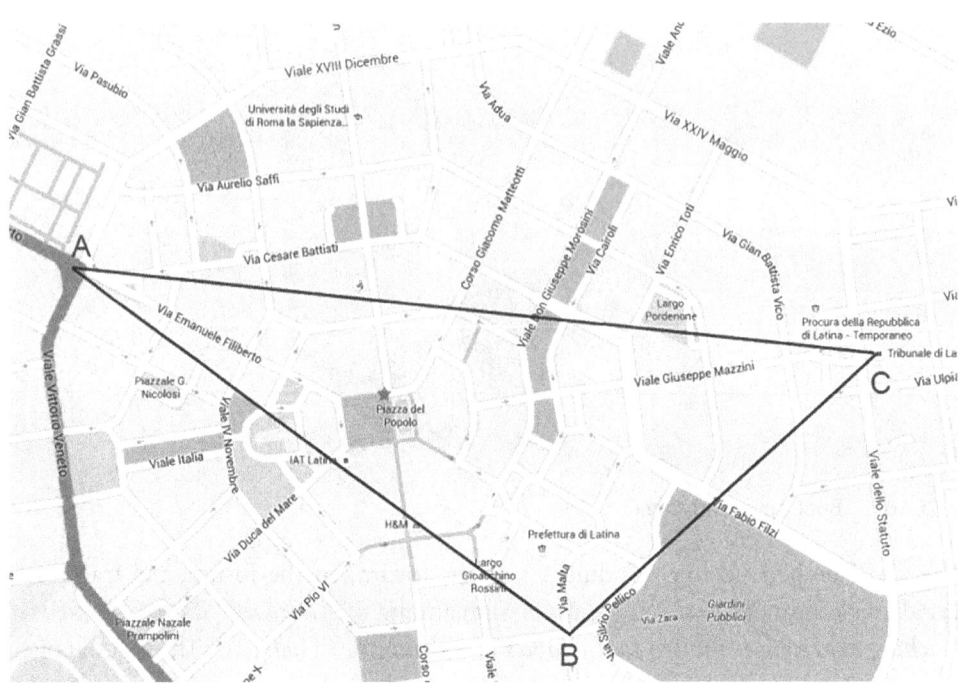

FIGURE 18.2   Latina's map created by A.P.

presence of buildings on the geometrically found point. Other points of discussion could arise on the comparison between the mutual positions of center of gravity and ortho-center. What do all of you think?

A.P.—29/10/15; 4:45 p.m.—***I really like*** *the use of tolerance! [...] Thanks for the idea :)*

A.P. opens the discussion with a proposal that is addressed directly to his colleagues in the course. Such is his involvement that, driven by the stimuli he has received by reading the materials and implementing a process of instrumentation/self-organization, he produces his own resource and shares it with others. A.P. is inspired, as suggested in the MOOC proposals, by a concrete situation: he took from Google Maps a map of his city, Latina, and joining three points, he drew a triangle (Figure 18.2). He would like to have some ideas on how to use this material didactically. P.R. congratulated A.P. and, in self-organizing herself, makes visible the fact that she has added a new node to her network of knowledge: "*I will propose it*," but at a specific time ("*next week*")—to link it to mathematical concepts that she has already treated with her classroom—rather than at an indefinite future time. M.L. positively evaluated A.P.'s idea and, in making it her own, hypothesized a variant, to stimulate reflection and argumentation in the students. A.P. replied to M.L., showing appreciation for the suggestion he received and indicating that A.P. had updated his net-work of knowledge.

Let's point out that this production of A.P. is completely spontaneous and also surprises the trainers: MOOC had not asked for the production or sharing of materials. The case of A.P. during the MOOC does not remain an isolated one: other trainees also spontaneously begin to share with the MOOC-ecosystem their own materials, asking for opinions and suggestions on them.

A.P. used the ideas he received from the MOOC to develop a new and original product. As such, the other trainees were provided with an additional source of learning: what they experienced in the MOOC and what they observed another trainee doing thanks to the MOOC. So, the trainees, other than A.P., are led to (i) trigger a process of instrumentation/self-organization and (ii) embed that product in their network. The process in which this mediation is inserted is that of the double learning process, and in this sense the object does not mediate knowledge, but enters into a genesis of usage schemes produced at that time.

18.5.2 Difference between Arc and Angle

Module 2 of *MOOC Geometria* opens with an activity called "The clock" (for full version see: https://goo.gl/HWGY5g) that is for lower secondary school (grade 6) and is focused on the concept of angle. The angle is one of the first entities of geometry shown to the students. However, the difficulty of reaching an adequate mastery of this concept is known. In fact, the students face an epistemological obstacle: understanding that the width of the angle does not change with the length of the sides. Starting from a problematic situation linked to the clock and enlarging it to the construction of a large clock (on the ground, maybe in the gym—Figures 18.3, 18.4, 18.5), we want the school students to obtain small angles (e.g., of a degree) on circumferences of different radii. In this way, they should associate the angle with the space between the two half-lines and not simply the arc used to indicate it. This experience has the purpose of helping students to avoid the widespread misunderstanding, that the angle is identified with the arc or with a finite region of the plane.

The CMB embedded into Module 2 was the forum that collected 31 discussions, each containing from 0 to 21 response replicas, for 152 posts in total. In the following, some

FIGURE 18.3   Division of the arc corresponding to the 90° angle in three equal parts to obtain the hour 1 corresponding to the 30° angle.

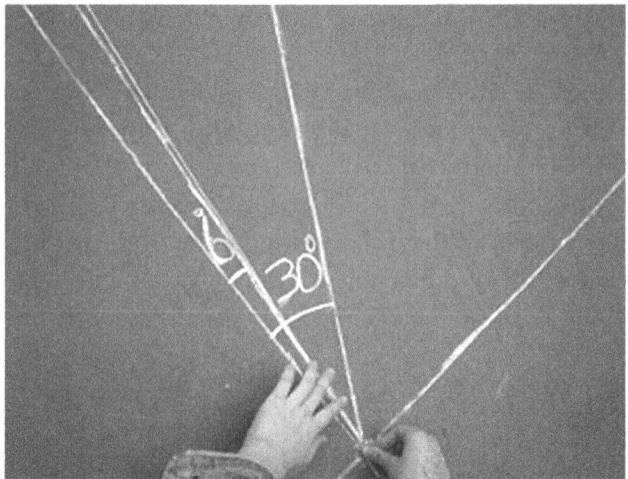

FIGURE 18.4   6° angle, corresponding to 1 minute, fifth part of the 30° angle corresponding to the hour 1.

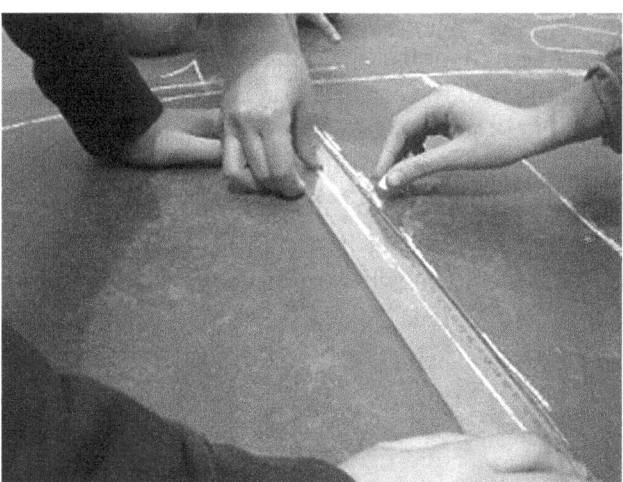

FIGURE 18.5   Division into six equal parts of the arc corresponding to the 6° angle to obtain the 1° angle.

trainees' first comments on this activity are given. They show how the first phase of the double learning process, the instrumentation/self-organization, takes place.

E.S.—2/11/15, 5:33 p.m.—"***Very interesting** and **very well documented**.*"

M.A.—3/11/15, 8:15 a.m.—"*I find the angle-clock association **very interesting**. The construction of the latter on the ground, **I believe**, allows us to ascertain directly the distinction between angle and arc.*"

F.G.—3/11/15, 11:27 p.m.—"*Tomorrow **I will try** it with my first classroom. The activity is **very well structured** and **I really like** the stimulus questions proposed [...].*"

All those who have intervened in the forum have certainly experimented with an instrumentation phase, exactly when they clicked the link that directed them to the activity of the 1° angle. The self-organization phase follows spontaneously: when the trainees read the information and consider it a benefit for their teaching practices, they are automatically expanding their network of knowledge. On the one hand, each one formulates her own judgment, putting in place an instrumentation phase. On the other hand, someone also begins to respond to comments from other trainees, putting in place an instrumentalization phase, i.e., one begins to build new connections independently using the MOOC-instrument, expanding the ecosystem network at the same time.

Let us consider a forum discussion called "Radian" (from the first trainee who started writing), which received 21 replies. Let us see the first post and some other answers.

> N.C.—2/11/15, 09:49 a.m.—"*The proposed activities have made me think about* (a) *how the conceptual articulation 'Angle vs. arc' is delicate. When the guys study trigonometry at high secondary school* (b)*, they know the Radian that [...] allows you to no longer distinguish between (width of) angle and (length of) arc. I would like to know your thoughts* (c)*, especially those who teach at lower secondary school.*"

In (a) there is an evident phase of instrumentalization: the trainee is creating new connections between his network of knowledge and that of the ecosystem. N.C. is a higher secondary school teacher. He was stimulated by the activities that he saw in this module, and he is connecting this thinking to his classroom (b). In particular, he invites other people to share their thoughts about this topic (c).

> D.L.—5/11/15, 6:19 p.m.—"*Hi, I teach in a linguistic high school [...]. When* **I introduce** *the radian, after having defined what it is,* **I make** *everyone draw a circle in which they must identify the width of a radian. Sometimes [the students] engineer themselves with ribbons or wires to bring the true length of the radius back to the circumference. At this point, [the students] have to cut out the circumferences, which will be all or almost of different sizes. Overlapping them, [the students] understand that the radian does not depend on the radius and even less on the arc. Regarding the graph of the goniometric functions [...]* **I have created** *a Geogebra file that leaves the projection segment trace while a P point on the circumference describes the angle. If someone is interested I enclose it [...].*"

D.L., in responding to N.C., follows the stages of the double learning process: he reads the comment of the colleague (instrumentation); he considers it interesting and inserts it into his network (self-organization); he connects the experience of the colleague with his own (instrumentalization), and he decides to share his didactical praxeology with N.C. in particular and more generally with the rest of the ecosystem (sharing).

Shortly after D.L. another trainee responds,

> E.G.—7/11/15, 4.16 pm—"*I think the file you are talking about is such as this ...* **I find** *enlightening the students build it!*"

E.G. immediately attaches a GeoGebra file.

Therefore, D.L. and E.G. spontaneously discuss their didactical practices and have no qualms about sharing their materials. This practice does not generally happen in face-to-face educational courses, and it was very surprising for the trainers because it happened in a completely natural way. In this way, we see how a trainee's didactical praxeology becomes something that can potentially become part of the meta-didactical (and not didactic)* praxeologies of another trainee.

18.6 EXAMPLES FROM MOOC NUMERI

18.6.1 MERLO Methodology

Module 2 of *MOOC Numeri* is dedicated to the MERLO methodology (Arzarello, Robutti & Carante, 2015). MERLO is a tool for teaching and learning, reusable in different circumstances or contexts, especially in mathematics. It is based on the commonality of meaning of the different representations of a concept, which school students are called upon to recognize. A MERLO card is composed of a delivery and five boxes, within which you have to insert the target statement (TS), one or more items sharing meaning with the TS, and the remaining boxes outside the boundary of meaning with the TS. The items that are designed, in addition to the common meaning, can have the same representation used in the TS, i.e., they can both be an image, or a graph, or statements in natural and/or symbolic language.

The module is divided into two parts. In the first part of the module, the trainees become familiar with the MERLO methodology. Two videos are presented: one that illustrates the theoretical foundations that frame the project and another that goes into the details of the design of the cards. Subsequently, the trainees try to solve the cards prepared by the trainers. In the second part of the module, the trainees have to try their hand at designing a MERLO card on a topic of their choice, linked to the Numbers core. Their designs are reviewed by the trainer who curates the module, who leaves feedback on their productions. Let's show some examples of the productions made by the trainees, which were then made available to all of them, on a repository outside the MOOC. We will not dwell on the detailed explanation of the cards, but rather on the quality of the result which the trainees have reached.

S.L.C. has developed a MERLO card on direct proportionality (Figure 18.6). She chose as TS an expression in natural language and connected it with two items that share the same meaning with TS, but are expressed by images (B, E). Then she chose an item (C) that does not share the same meaning, but has similarity in representation, and an item (D) that does not share the same meaning with TS and does not have similarity in representation with TS. The trainer of the module suggests to S.L.C. to modify the item D, because the graph has some traits of direct proportionality and this could cause confusion in the resolution of the card. S.L.C. accepts the suggestion and reloads a new card with D replaced as in Figure 18.7.

---

* Note that I say "meta" because if it came into the teaching practice, it means that the trainee who reads must incorporate and put into practice what he has read. We cannot have proof or certainty of this, unless he explicitly declares it on some CMB. In general, it is not said that any trainee has the opportunity or the interest to immediately implement a teaching practice proper to another trainee.

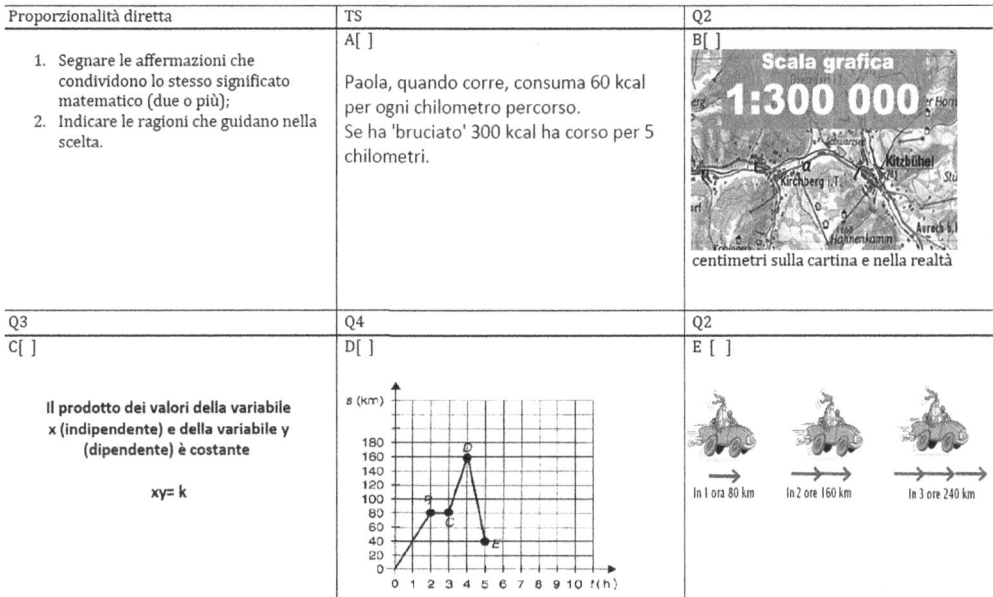

Proporzionalità diretta	TS	Q2
1. Segnare le affermazioni che condividono lo stesso significato matematico (due o più); 2. Indicare le ragioni che guidano nella scelta.	A[ ] Paola, quando corre, consuma 60 kcal per ogni chilometro percorso. Se ha 'bruciato' 300 kcal ha corso per 5 chilometri.	B[ ] Scala grafica **1:300 000** centimetri sulla cartina e nella realtà
Q3	Q4	Q2
C[ ] Il prodotto dei valori della variabile x (indipendente) e della variabile y (dipendente) è costante xy= k	D[ ]	E [ ]

FIGURE 18.6   S.L.C.'s MERLO card.

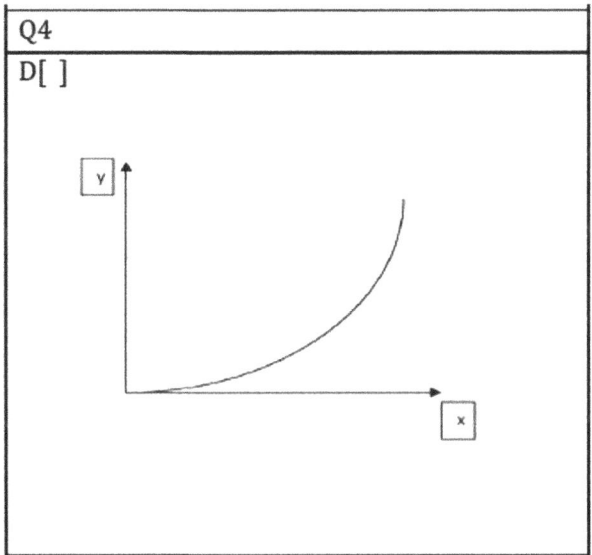

FIGURE 18.7   New D item chosen by S.L.C.

E.F. has created a MERLO board on the height of the triangles (Figure 18.8). She does not explain what the TS is for her and in what relation the other items are with it. The trainer points out to her, in particular: "*The distractor in B of the MERLO card is not very 'distracting', in the sense that it is trivial. Modify with something that makes you think a little more.*" E.F. does not modify the loaded card.

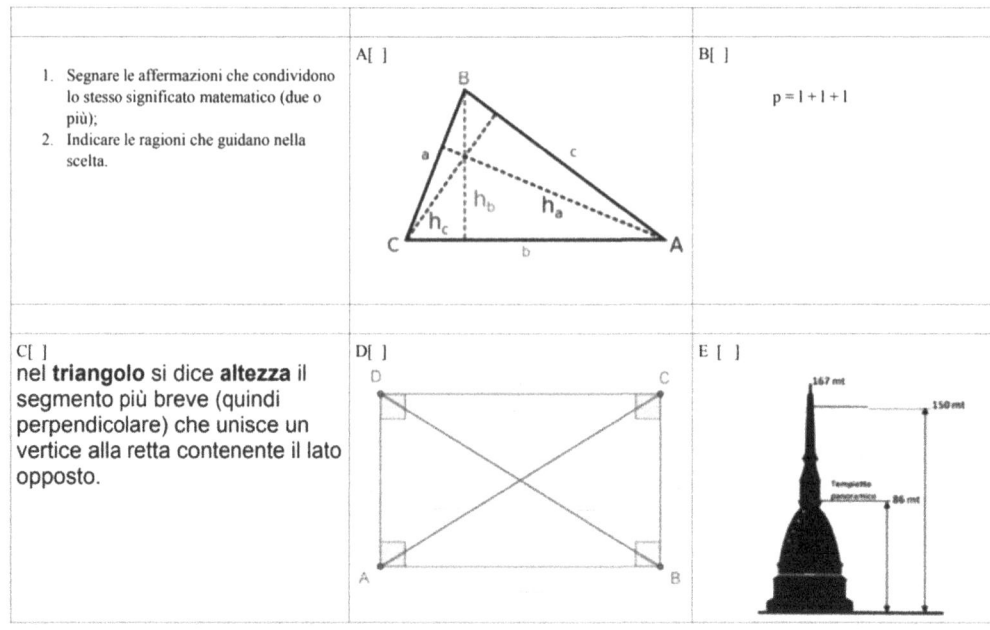

FIGURE 18.8   E.F.'s MERLO card.

Analyzing these protocols, we note that the design of a MERLO card is a work that requires both a lot of knowledge of the discipline and the ability to predict the reasoning that may arise in students who solve the card. To insert a distractor is not simple, nor is it simple to identify items that share the meaning with the TS and do not have a similarity in the representation with the TS.

Let's consider some of the trainee interventions released in the forum, starting from the second part of the module.

A.A.—21/11/2016, ore 10:39 p.m.—*"In my opinion, MERLO certainly stimulates discussion and argument and favors formative assessment, but [...] I think that design a MERLO card is difficult [...]"*

A.P.—19/11/2016, ore 4:46 p.m.—*"[...] Surely [MERLO] implies for the teacher a greater effort in the predisposition than a standard verification, but such difficulty could be lessened with the diffusion of the method and the familiarization with it."*

G.B.—22/11/2016, ore 3:20 p.m.—*"It could be stimulating to have these cards built by the students themselves, asking them to identify objects that have a common element (but without indicating it) and propose the card to their classmates as a game, challenging their classmates to discover the conceptual node that acts as a link."*

Faced with a totally new topic, the trainees implement the instrumentation/self-organization process. In fact, there are alternating comments that denote reflections on the practices carried out during the training, but also propositions to integrate this new methodology in one's own practices, while admitting that it will be neither immediate nor easy. It is certainly necessary to "maintain" the practice: it is not enough to know and understand, but it is necessary to practice exercises to refine the design skills of new cards, obtaining results that will then be visible in the long term.

18.6.2 Arithmetic and Algebra

Module 5 of *MOOC Numeri* opens with an activity called "Arithmetic helps algebra and algebra helps arithmetic" (for full version see: https://goo.gl/83Q7Bf) that is for higher secondary school (grades 9–10).

Games of mathematical "magic" and challenges of mental calculation abilities are at the heart of this activity for dealing with the conceptual nodes *natural language* and *algebraic language*. The activity refers to the introduction of the rules of algebra and the difficulties encountered when the student must translate a problem algebraically ("put it into a formula"). Concretely, the activity tries to give meaning to algebraic calculation, to ensure that school students do not interpret the algebraic formulas as pure sequences of signs. Several problems are proposed in which the language of algebra overcomes that of arithmetic and becomes a tool for expressing relationships and generalities. The following is one of the proposed problems: "*think of a number.*"

The teacher, addressing the entire class, proposes to each student to execute instructions in the notebook; the teacher does not know which number each student chose initially.

- Think of an integer.
- Add 12 to it.
- Multiply the result by 5.
- Subtract 4 times the original number.
- Add 40 to the result.

The teacher asks some students for the final result; then she subtracts 100 from this result and "guesses" the starting number. The teacher then justifies her "foresight" with the symbolic calculation. In particular, the teacher observes that the rules of calculation are none other than the application of the rules of arithmetic; in particular, she emphasizes the role of distributive property that allows us to "distribute" a product on a sum but also to "collect" a common factor, depending on how we interpret the equivalence:

$$a \cdot (x + y) = a \cdot x + a \cdot y$$

Finally, the teacher explains to the class how this calculation rule has a simple geometric interpretation. If we consider two rectangles, the first of sides a and x, and the second of sides a and y, these can be arranged to form a single rectangle of sides a and $(x + y)$. And the sum of the areas of the first two rectangles is equal to the area of the third one: $a \cdot (x + y) = a \cdot x + a \cdot y$

The CMB embedded into Module 5 was a forum that collected 35 discussions, each containing from zero to seven response replicas, for 61 posts in total. In the following, we show some trainees' comments on the forum dedicated to comment on this activity. Again, the phases of the double learning process are shown.

A.B.—5/12/16, 10:33 p.m.—*"I teach in a lower secondary school and, although it is clear that this is an activity [Arithmetic helps algebra and algebra help arithmetic] to be performed in a higher secondary school, I find the first two [stages]* **really stimulating.** *I* **like** *the problematic introduction and the enrichment with the geometric appearance, which* **I already use** *for the literal calculation in the third."*

G.P.—6/12/16; 11:47 p.m.—*"Indeed, the proposed activities are a bridge between lower and higher secondary school. However the stage 1 [think of a number] can also be proposed in the grade 6 when, by treating the four operations and their properties, the mental calculation is dealt with. [...] Even if you 'lose' a lesson maybe you give someone the chance to have an extra tool or it could be a way to enhance excellence."*

S.B.—11/12/15; 6:32 p.m.—*"Yes,* **I think** *it is essential not to wait for the third year [grade 8] to do algebra and literal calculation. Already from the first one [grade 6] (and from the primary) they are exposed to the letters, formulas of the perimeters and areas or simply when the properties of the operations are resumed and generalized. [...]* **I advise** *you to sow already in the first [grade 6] so then in the second [grade 7] they do not memorize all the inverse formulas of the areas, but get them ... :-)*

P.S. **beautiful** *games 'think of a number' [...]."*

The chosen trainees teach in a lower secondary school. The activity "Arithmetic helps algebra and algebra help arithmetic" is presented as an activity for grades 9–10. However, as was hoped, the trainees were not only positively affected, "*I like ...; beautiful games ...* ", but they also think they can adapt it for their classes.

There is therefore an instrumentation/self-organization process on the part of these trainees, which is leading them to reflect on a possible use of the activity in their own classes. In particular, G.P. shows a more organized network of knowledge than A.B., because he has already come to think about what to propose to which classes. He has already made estimations: "*Even if you 'lose' a lesson maybe you give someone the chance to have an extra tool or it could be a way to enhance excellence.*" He then put in place an instrumentalization phase when he completed his reflections and a sharing phase in the moment in which he shared them with the rest of the MOOC-ecosystem.

S.B. shares an own thought, that is, "do not wait the grade 8 to start to use letters or to talk about algebra." He underlines that the students, even at primary school, are exposed to them. Therefore, he urges his colleagues not to be reticent, and he does so with a smile!

18.7 RESULTS

As mentioned, the trainers had made the methodological choice to intervene as little as possible on the CMBs to encourage interaction of the trainees-only community. Although we have shown only examples of posts made on the forum, it is characteristic of how the trainees interact with each other within the online environment. In the examples of *MOOC Geometria*, we have seen how participants spontaneously share their own teaching practices and materials. In particular, the shared materials are both materials that the trainees already possessed before attending the MOOC (think of the GeoGebra file that E.G. shares) and materials that are created from the inspiration of the ideas that MOOC offers (think of the image that A.P. shares). In the examples of *MOOC Numeri*, we have seen how trainees engaged in the production of new materials in response to the acquisition of a new methodology (MERLO) and how this seems to affect their teaching practices, generating reflections. In addition, we have also seen how trainees confront each other by exchanging ideas, reflections, and suggestions on how to make algebra accessible as early as grade 6.

Therefore, without ever having had the opportunity to meet in person, in a spontaneous way, they exchange ideas, make their mathematics teaching experience available, and share their own materials, exactly as a *community of practice* (Wenger, 1998). Note that while the trainers did invite the trainees to exchange teaching experiences, the exchange of their materials is an entirely spontaneous act.

This community of trainees is different from those that usually characterize the traditional face-to-face educational courses. It was born spontaneously since participation in the MOOC took place voluntarily. Moreover, the trainees freely express themselves: there is no institutional component that wants to restrict them. The discussions proceed in a very free and spontaneous way, even with the use of emoticons (as seen previously in the comment by A.P. in *MOOC Geometria* and by S.B. in *MOOC Numeri*). This is a remote, voluntary, free, and collaborative community, not subject to institutional pressures.

Even the trainees are aware that they have formed a community; in fact, in an open question of the final questionnaire of both *MOOC Geometria* and *MOOC Numeri*, we asked: "*As a participant of MOOC, to what extent do you feel you are part of a community?*" As noted in Table 18.2, 50% of the *MOOC Numeri* members were also *MOOC Geometria* trainees. Among the trainees who claim to feel quite and very much part of an online community, we distinguish the answers given as follows: 66% of the *MOOC Geometria* trainees; 67% of the trainees who were only enrolled in MOOC Numbers; and 78% of the trainees who were students of both MOOCs. Below are some of their testimonies:

Geo: "*Being part of something that brings together people from all over Italy has a certain effect.*"

Num: "*I had the feeling of being part of a big family.*"

Geo&Num: "*It is the second MOOC I participate in, this idea of remote sharing has, in both cases, really made me feel part of a community of teachers who are willing to improve their teaching practices.*"

Geo&Num: "*I feel part of a community that learns at any time of day or night. It made me happy and made me smile.*"

This sense of community is certainly a possible partial explanation for the high completion rates of our MOOCs, although we do not think it is the only reason for this. As shown in the introduction, the participant's role is hotly contested across almost all literature and debate about MOOCs. Indeed, the key dilemmas in MOOCs center on what participation actually means, how it should be measured, and consequently, what metrics of success and quality are appropriate for these courses (Yang et al., 2013; Bayne & Ross, 2014; Onah, Sinclair & Boyatt, 2014). Part of this complexity seems to arise because there are simply so many people, doing so many different sorts of things, in any given MOOC. This presents a challenge for researchers, educators, and institutions accustomed to using "completion" as a fairly stable measure of the success and quality of an educational offering. As reported in the literature, it is probably not the right way to judge the quality of a MOOC or of participants' experiences, because this statistic is not taking sufficient account of those who may be engaging but "do not adhere to traditional expectations, centered around regular assessment and culminating in a certificate of completion" (Kizilcec, Piech & Schneider, 2013, p. 9). I will not go into these matters. However, I will only make some considerations about the completion rates of *MOOCs Geometria* and *Numeri*. Although the literature reports that they are not the most appropriate yardstick for assessing the success or quality of a MOOC, the completion rates of *MOOCs Geometria* and *Numeri* were anyway higher than those that generally relate to the completion of MOOCs for teacher education: 36% for *MOOC Geometria* and 42% for *MOOC Numeri* (Table 18.2).

The other trainers and I believe there are four reasons why we recorded this educational success:

1. The educational opportunity that each of our MOOCs offered was valid both in terms of time and content.

2. The trainees have constituted more than just an online community; in fact, we have talked about communities of practice.

3. The vigilant presence of the trainers has certainly "reassured" the trainees who did not feel abandoned in an online environment.

4. Last, but not least, teacher education is a right and a duty for Italian teachers. What better chance if you do not take advantage of a MOOC: online learning space (accessible wherever and whenever) and free!

18.8 DISCUSSION AND CONCLUSION

In this chapter, I have analyzed how the *Math MOOC UniTo* project for teacher education has actually had positive repercussions on the professional development of the teachers who took part in it. It was possible to observe how the interactions among the trainees have changed their meta-didactical praxeologies (remember footnote 4), generating the establishment of a community of practice among themselves. This complex process has been described by integrating and coordinating different theoretical models. In fact, I have considered the MDT, a typical model of face-to-face training programs, and I have enriched it with the Connectivism and the Instrumental Approach, allowing the definition the MOOC-MDT. The model of the double learning process, from the ecosystem to the individual and vice versa, was thus constructed. It effectively illustrates the dynamic framework that distinguishes a MOOC from a traditional course, and which explains how the initial inert course takes life in the ecosystem thanks to the contribution, mostly unpredictable, with which individuals interact with the MOOC environment.

The double learning process that is generated can be schematically summarized in two basic steps:

1. The inert material, i.e., the MOOC-artifact: the educational and methodological proposal offered by the trainers, who work according to their professional praxeologies.

2. The ecosystem network, namely the MOOC-ecosystem/instrument: it develops according to the intertwined dynamics of instrumentation/self-organization and instrumentalization/sharing.

The starting situation offers a variety of occasions where:

a) A teacher, e.g., A, stimulated by some component of the starting situation, produces something: for example, she makes reflections on CMBs, sharing her own ideas or any experiments conducted in class; possibly also shares her own materials.

b) Another teacher, e.g., B, benefits from the observations/shares of A, integrates them into her network and in turn intervenes in the MOOC, exposing and sharing her ideas.

The ecosystem enriched by the As influences the Bs, which in turn influence the ecosystem, and so on, in a process that feeds itself. Thanks to the CMBs present in the MOOC platform, the trainees begin a series of peer-to-peer communication processes (in fact the trainers generally abstain from intervening), which develop according to the methods and tools typical of social networks and produce an aggregation of interest groups that gradually grow and generate a real community of practice.

The meaning of these processes, still to be deepened in the research, can in the meantime be further clarified by two observations on the dynamics of the MOOCs described above. In the first place, collaborative participation and changing praxeologies are all practices that evolve on objects, that is, starting from the inert material that becomes an ecosystem. This process takes place in a "chaotic" way, in the sense that it is unpredictable and uncontrollable. This characteristic distinguishes in particular the MOOCs from traditional courses, in which everything happens in a generally predictable way. However, this "chaos" (in the sense of Siemens, 2005) guarantees the active and massive involvement of the participants (as described in the examples) that spontaneously intervene in a dialogue between peers.

A second observation concerns teachers' products (such as those produced by A in the example above). The teacher A's product is a meaning that she attributes to the interpretation that the trainers give to that object and then she uses it in the classroom. For the teacher B the situation is different: A's experience does not generally fall within her praxeologies. B can instead be moved by a certain tension (in the sense of Goos, 2013) to improve (for any reason) and, also inspired by the experience of A (which was accomplished freely and not "imposed" by the trainers), has an extra source of learning (what she sees in the MOOC; what she sees another trainee is doing thanks to the MOOC). Prompted by this,

B triggers an instrumentalization phase and acquires that object, considering utilization schemes that are interesting for her. On the one hand, this object mediates a praxeology between *A* and *B*. On the other hand, the process in which this mediation is inserted is that of the double learning process, and in this sense the object does not mediate knowledge, but enters into a genesis of utilization schemes produced at that time.

In conclusion, the chapter illustrates that MOOCs for mathematics teacher education can have high completion rates, if methodologies and strategies similar to those described here are followed:

- Reducing to the minimum trainers' interventions in CMBs, but being vigilant behind the scenes.

- Inserting stimulus questions in the CMBs to trigger the discussions.

- Proposing activities that are close to real contexts, but that offer new teaching methods to deal with students' misconceptions or difficulties.

Moreover, these MOOCs have offered several benefits to teachers who attended them. In fact, they have allowed fruitful exchanges, an evolution of their praxeologies, and have generated communities of practice even at a distance. Therefore, for the positive influence and impact they have on online mathematics professional learning, it is worth making more use of them.

Finally, yet importantly, the MOOCs for mathematics teacher education offer a promising ground for research as it seems necessary to develop a new theoretical framework compared to traditional ones. My contribution, here shown, goes in this direction, and hopefully much other research will be necessary to define a complete one.

18.9 LIMITATION OF THE STUDY AND FUTURE RESEARCH

The results of this study should be interpreted with caution. First of all, the MOOCs presented here are MOOCs for teachers and not for students. So, the conclusions we have reached cannot necessarily be extended to all MOOCs in general. We also point out that, although the number of participants in our courses is massive, it does not reach the thousands of users, as often happens in American MOOCs (in Avineri et al., about 1,700 teachers are enrolled in their MOOCs). Therefore, it is not said that the methodology adopted for the monitoring and also the involvement of the participants can be the same when the number of members is very large.

Our analysis shows that a real involvement of trainees in collaborative work needs to be triggered and supported by suitable tools added to the platform. The availability in the platform of tools consonant with the social networks used in everyday life increases the triggering of what Manlove et al. (2007) call co-regulated learning, in the sense that the trainees themselves regulate their tasks and collaboration. Our analysis leaves open the question of which devices are the best for improving active collaboration among the trainees: possibly further research and concrete experiments will be able to give a more definitive contribution to this crucial issue. What is interesting here is that our analysis centered on collaboration processes through the adaptation of the meta-didactical lens has made it possible to grasp this important problem in a clear way. This suggests that the method of research we have undertaken is promising and fruitful for further results along this stream.

ACKNOWLEDGMENT

I thank very much the entire MOOCs teams, especially Prof. F. Arzarello, Prof. O. Robutti, V. Alberti, S. Labasin, and A. Coviello, for their help and their crucial contribution to the design and management of the project. Without them neither the MOOCs nor this chapter would exist.

BIBLIOGRAPHY

Abeer, W., & Miri, B. (2014). Students' preferences and views about learning in a MOOC. *Procedia-Social and Behavioral Sciences, 152*, 318–323.

Aldon, G., Arzarello, F., Panero, M., Robutti,O., Taranto, E., & Trgalová, J. (2017). MOOC for mathematics teacher training: Design principles and assessment. In G. Aldon, & J. Trgalová (Eds.), *Proceedings of the 13th International Conference on Technology in Mathematics Teaching* (pp. 200–207). Lyon, France.

Aldon, G., Arzarello, F., Panero, M., Robutti, O., Taranto, E., & Trgalová, J. (2018). MOOC for mathematics teacher education to foster professional development: Design principles and assessment. In G. Aldon, & J. Trgalová (Eds.), *Technology in Mathematics Teaching - Selected Papers of the 13th ICTMT conference*, Switzerland: Springer International Publishing AG.

Anichini, G., Arzarello, F., Ciarrapico, L., Robutti, O., & Statale, L. S. (2004). *Matematica 2003. La matematica per il cittadino*. Lucca, Italy: Matteoni stampatore.

Arzarello, F., Robutti, O., & Carante, P. (2015). MERLO: A new tool and a new challenge in mathematics teaching and learning. In K. Beswick, T. Muir, & J. Wells (Eds.), *Proceedings of the 39th Conference of the International Group for the Psychology of Mathematics Education* (Vol. 2, pp. 57–64). Hobart, Australia: PME.

Arzarello, F., Robutti, O., Sabena, C., Cusi, A., Garuti, R., Malara, N., & Martignone, F. (2014). Meta-didactical transposition: A theoretical model for teacher education programmes. In N. Sinclair, A. Clark-Wilson, & O. Robutti (Eds.), *The Mathematics Teacher in the Digital Era* (pp. 347–372). Springer Netherlands.

Avineri, T., Lee, H. S., Lovett, J. N., Gibson, T., & Tran, D. (2017). Design and impact of MOOCs for mathematics teachers. In *Proceedings of the 13th International Congress on Mathematical Education*. (ICME13, July 24 – 31, 2016). Hamburg, Germany.

Bartolini Bussi, M. G., Boni, M., & Ferri, F. (1995). *Interazione sociale e conoscenza a scuola: la discussione matematica*. Modena: Centro Documentazione Educativa.

Bayne, S., & Ross, J. (2014). *The Pedagogy of the Massive Open Online Course: The UK View*. York, UK: The Higher Education Academy.

Borba, M. C., Askar, P., Engelbrecht, J., Gadanidis, G., Llinares, S., & Aguilar, M. S. (2017). Digital technology in mathematics education: Research over the last decade. In *Proceedings of the 13th International Congress on Mathematical Education* (pp. 221–233). Cham: Springer.

Campbell, M. P. (2009). Mathematics teachers and professional learning communities: Understanding professional development in collaborative settings. In S. L. Swars, D. W. Stinson, & S. Lemons-Smith (Eds.), *Proceedings of the 31st Annual Meeting of the North American Chapter of the International Group for the Psychology of Mathematics Education* (pp. 956–964). Atlanta, Georgia.

Clark-Wilson, A., Aldon, G., Cusi, A., Goos, M., Haspekian, M., Robutti, O., et al. (2014). The challenges of teaching mathematics with digital technologies—the evolving role of the teacher. In P. Liljedahl, C. Nichol, S. Oesterle, & D. Allan (Eds.), *Proceedings of the Joint Meeting of PME 38 and PMENA 36* (Vol. 1, pp. 87–116). Vancouver: University of British Columbia.

Daza, V., Makriyannis, N., & Rovira Riera, C. (2013). MOOC attack: closing the gap between pre-university and university mathematics. *Open Learning: The Journal of Open, Distance and e-Learning, 28*(3), 227-238.

Feng, W., Tang, J., & Liu, T. X. (2019). Understanding dropouts in MOOCs. Retrieved from http://keg.cs.tsinghua.edu.cn/jietang/publications/AAAI19-Feng-dropout-moocs.pdf [Accessed February 2019]

Goos, M. (2005). A sociocultural analysis of the development of preservice and beginning teachers' pedagogical identities as users of technology. *Journal of Mathematics Teacher Education, 8*(1), 35–59. doi:10.1007/s10857-005-0457-0.

Goos, M. (2013). Sociocultural perspectives in research on and with mathematics teachers: A zone theory approach. *ZDM, 45*(4), 521–533.

Joubert, M. (2013). Using digital technologies in mathematics teaching: Developing an understanding of the landscape using three "grand challenge" themes. *Educational Studies in Mathematics, 82*, 341–359. doi:10.1007/s10649-012-9430-x.

Kizilcec, R., Piech, C., & Schneider, E. (2013). Deconstructing disengagement: Analyzing learner subpopulations in massive open online courses. In *LAK '13*. Leuven, Belgium.

Manlove, S., Lazonder, A. W., & de Jong, T. (2007). Software scaffolds to promote regulation during scientific inquiry learning. *Metacognition and Learning, 2*, 141–155.

Onah, D. F., Sinclair, J., & Boyatt, R. (2014). Dropout rates of massive open online courses: Behavioural patterns. *EDULEARN14 Proceedings*, 5825–5834.

Panero, M., Aldon, G., Trgalová, J., & Trouche, L. (2017). Analysing MOOCs in terms of their potential for teacher collaboration: The French experience. In T. Dooley, & G. Gueudet (Eds.), *Proceedings of the Tenth Congress of European Society for Research in Mathematics Education (CERME10, February 1 – 5, 2017)* (pp. 2446–2453). Dublin, Ireland: DCU Institute of Education and ERME.

Pappano, L. (2012). The year of the MOOC. *The New York Times, 2*(12), 26–32.

Robutti, O., Cusi, A., Clark-Wilson, A., Jaworski, B., Chapman, O., Esteley, C., & Joubert, M. (2016). ICME international survey on teachers working and learning through collaboration: June 2016. *ZDM, 48*(5), 651–690.

Siemens, G. (2005). Connectivism: A learning theory for the digital age. *International Journal of Instructional Technology and Distance Learning, 2*(1), 3–10. Retrieved from http://www.itdl.org/Journal/Jan_05/article01.htm [Accessed 20 August 2018].

Taranto, E. (2018). *MOOC's Zone Theory: Creating a MOOC Environment for Professional Learning in Mathematics Teaching Education*. PhD Dissertation. Turin University.

Taranto, E., Arzarello, F., & Robutti, O. (2017a). MOOC: Repository di strategie e metodologie didattiche in matematica. In Borgato, M. T., Pancaldi, S. (A cura di), *Annali online della Didattica e della Formazione Docente* (Vol. 14, pp. 257–279), ISSN: 2038-1034.

Taranto, E., Arzarello, F., Robutti, O., Alberti, V., Labasin, S., & Gaido, S. (2017b). Analysing MOOCs in terms of their potential for teacher collaboration: The Italian experience. In T. Dooley, & G. Gueudet (Eds.), *Proceedings of the Tenth Congress of European Society for Research in Mathematics Education (CERME10, February 1 – 5, 2017)* (pp. 2446–2453). Dublin, Ireland: DCU Institute of Education and ERME.

Vérillon, P., & Rabardel, P. (1995). Cognition and artifacts: A contribution to the study of though in relation to instrumented activity. *European Journal of Psychology of Education, 10*(1), 77–101.

Wenger, E. (1998). *Communities of Practice: Learning, Meaning, and Identity*. Cambridge University Press.

Yang, D., Sinha, T., Adamson, D., & Rosé, C. P. (2013). Turn on, tune in, drop out: Anticipating student dropouts in massive open online courses. In *Proceedings of the 2013 NIPS Data-driven Education Workshop* (Vol. 11, p. 14).

CHAPTER 19

Online Mathematics "Self-Help Kiosks" to Support Pre-Service Teachers

Helen Forgasz, Jennifer Hall, and Simone Zmood

CONTENTS

19.1 INTRODUCTION

There has been increasing attention paid to the need to improve the numeracy capabilities of adults and children around the world. The terminology used for numeracy varies internationally. Other terms commonly used include mathematical literacy and quantitative literacy; for discussions of various definitions adopted, see Vacher (2014) and Forgasz, Leder, and Hall (2017).

There are two international testing regimes in which numeracy performance levels are measured. Every three years, 15-year-old students from many nations complete the Organisation for Economic Co-operation and Development's (OECD) Programme for International Student Assessment (PISA). One of the tests that the students complete is the test of mathematical literacy—see OECD (2018a) for details. Adults around the world complete the OECD's Survey of Adult Skills (known as Programme for the International Assessment of Adult Competencies [PIAAC])—see OECD (2018b) for details.

In PISA, mathematical literacy is defined as:

> formulating, employing and interpreting mathematics in a variety of contexts. It includes reasoning mathematically and using mathematical concepts, procedures, facts and tools to describe, explain and predict phenomena. It assesses the capacity of individuals to recognise the role that mathematics plays in the world and to make the well-founded judgements and decisions needed to be constructive, engaged and reflective citizens.
>
> *(OECD, 2017, p. 1)*

In PIAAC, numeracy is considered a skill parallel to literacy. Numeracy is defined as:

> the ability to use, apply, interpret, and communicate mathematical information and ideas. It is an essential skill in an age when individuals encounter an increasing amount and wide range of quantitative and mathematical information in their daily lives.
>
> *(OECD, n.d., p. 1)*

While many countries participate in PISA and PIACC, only in some countries has numeracy been embraced as a component of school curricula. We focus here on four English-speaking countries: the U.K., U.S.A., Canada, and Australia.

19.1.1 Numeracy and Schooling

The U.K. was one of the first countries to recognize the importance of numeracy in education—Crowther (1959) defined numeracy as the mirror image of literacy, and Cockcroft (1982) contended that it was "the responsibility of teachers of mathematics and other subjects to equip children with the skills of numeracy" (p. ix). Although no clear definition of numeracy is provided in the U.K. curriculum, the importance of numeracy development as a responsibility of all teachers is discussed in the U.K. curriculum framework for Key Stages 1 and 2 (Grades 1–6) in Section 5.2 as follows:

> Teachers should develop pupils' numeracy and mathematical reasoning in all subjects so that they understand and appreciate the importance of mathematics.
>
> *(Department of Education, 2014c)*

Yet, no distinction is drawn between numeracy and mathematics in the current U.K. mathematics curriculum (Department of Education, 2014d). With the aim of numeracy for all in the U.K., an online resource has been established (National Numeracy, 2019a). Numeracy is defined as "the ability to use mathematics in everyday life" (National Numeracy, 2019b). The resource is jointly funded by the U.K. Department of Education, corporate sponsors, and charitable organizations.

In the U.S., the Common Core State Standards in Mathematics (CCSSM) provide a mathematics curriculum for kindergarten to Grade 12 students in 41 U.S. states and the District of Columbia (Common Core State Standards Initiative [CCSSI], 2018b). In the CCSSM (CCSSI, 2018a), the term "mathematical literacy" does not appear, while "numeracy" and "quantitative literacy" only appear in the reference list. Madison (2015) examined the CCSSM for links to quantitative literacy (QL), particularly six competencies used when applying mathematics in context: interpretation, representation, calculation, analysis/synthesis, assumption, and communication. Madison suggested that the CCSSM are supportive of calculation and somewhat supportive of representation and analysis/synthesis, with little evidence shown for support of the other competencies. Nevertheless, Madison claimed that "there are standards at every grade level that are very supportive of QL" (p. 7). His argument was based on mathematical content rather than other aspects of numeracy.

In Canada, education is the responsibility of individual provinces and territories. Since Ontario is the province with the highest proportion of the country's population (approximately 40%; Statistics Canada, 2016), we discuss numeracy within that context. Making connections among mathematical topics, with other subject areas, and with students' everyday lives is a key focus of the Ontario mathematics curriculum at all grade levels (Ontario Ministry of Education [OME], 2005a, 2005b, 2007); this stance is clearly connected to conceptions of numeracy. The term "numeracy" is only found once in the Ontario mathematics curriculum, in the Grades 11 and 12 document (OME, 2007), but "mathematical literacy" is used in both the Grades 9 and 10, and the Grades 11 and 12 documents (OME, 2005b, 2007) when referring to skills needed for the workforce. In a related publication, *Leading Math Success: Mathematical Literacy Grades 7–12*, the Expert Panel on Student Success in Ontario (2004) suggested that mathematical literacy involves both mathematical skills and "understanding the value of mathematics and having the inclination and the confidence to use it" (p. 24). It was also argued that mathematical literacy is comprised of various dimensions (e.g., financial literacy, spatial literacy) and is the responsibility of all teachers.

In Australia, although the state and territory governments oversee education, it was agreed by all state and territory education ministers that each state/territory curriculum

be consistent with the Australian Curriculum (AC). The AC for Foundation to Grade 10*
(Australian Curriculum, Assessment and Reporting Authority [ACARA], n.d.-a) has eight
content domains (e.g., English, mathematics), seven general capabilities (e.g., numeracy, ethical understanding), and three cross-curricular priorities (e.g., Aboriginal and Torres Strait
Islander histories and cultures). Teachers of all grade levels and across all content domains are
expected to develop students' competence in all the general capabilities and expose students
to the cross-curricular priorities. In the AC, the numeracy general capability is defined as:

> The knowledge, skills, behaviours and dispositions that students need to use math-
> ematics in a wide range of situations. It involves students recognising and under-
> standing the role of mathematics in the world and having the dispositions and
> capacities to use mathematical knowledge and skills purposefully.
>
> *(ACARA, n.d.-c)*

ACARA (n.d.-c) has argued that:

> When teachers identify numeracy demands across the curriculum, students have
> opportunities to transfer their mathematical knowledge and skills to contexts out-
> side the mathematics classroom. These opportunities help students recognise the
> interconnected nature of mathematical knowledge, other learning areas and the
> wider world, and encourage them to use their mathematical skills broadly.

In summary, the working definitions of numeracy/mathematical literacy are fairly similar
in the U.K., Canada, and Australia, and are consistent with the OECD (n.d., 2017) definitions. In the U.S., less attention is paid to numeracy, at least in the CCSSM. In Australia,
there is greater emphasis on the development of students' numeracy capabilities in cross-
curricular contexts.

While numeracy and related concepts are mentioned in the U.S. and Canadian curriculum documents examined, there is a paucity of information provided regarding how
pre-service teachers in these countries will develop the needed numeracy skills for their
profession. In contrast, in the U.K. and Australia, there is a substantial focus on pre-service teachers' numeracy skills. Thus, we focus on numeracy and teacher education in the
U.K. and Australia, as both countries have similar mandated requirements for pre-service
teachers regarding their personal numeracy capabilities.

19.1.2 Numeracy and Teacher Education

In the U.K., "all current and prospective trainee teachers must pass the skills tests in numeracy and literacy before they can be recommended for the award of qualified teacher status"
(Department of Education, 2018b). To ensure future teachers' numeracy and literacy competence, prospective teachers are tested on core skills that are needed "to fulfil their professional role in schools, rather than the subject knowledge needed for teaching" (Department

---

* Foundation to Grade 10 is the equivalent of kindergarten to Grade 10 in North America.

of Education, 2018b). The providers of pre-service teacher education are required to check that prospective students meet entry requirements for the skills tests before commencing their studies. For numeracy skills, there are two tests: an aurally administered mental arithmetic test and a written test focusing on arithmetic and data interpretation (Department of Education, 2018a). Based on what is written on the website, the support materials provided (Department of Education, 2014a), a glossary of numeracy terms (Department of Education, 2014b), and the sample written tests (downloadable from Department of Education, 2018a), the focus is on basic arithmetic and statistics skills and the interpretation of data as might be found in the context of a teacher's professional world.

Similar to the requirements in the U.K., it is now mandated in Australia that, prior to graduation from a teacher education program, all pre-service teachers must pass tests of personal numeracy competence as well as of personal literacy skills (Australian Government Department of Education and Training, 2017), the Literacy and Numeracy Test for Initial Teacher Education students (LANTITE). In LANTITE, numeracy is defined as the capacity for pre-service teachers "to apply mathematics to solve appropriate real-world problems" (Australian Council for Educational Research [ACER], 2017) and is focused on three numeracy processes: "identifying mathematical information and meaning in activities and texts," "using and applying mathematical knowledge and problem solving processes," and "interpreting, evaluating, communicating and representing mathematics" (ACER, 2017). Sample literacy and numeracy LANTITE questions are provided on the ACER website (n.d.). The contextual settings of the items and mathematical processes required to solve them are consistent with the definition of numeracy and the three mathematical processes described by ACER (2017).

Prior to the introduction of LANTITE, the Australian Institute of Teaching and School Leadership (AITSL) developed professional standards for the teaching profession at three levels: graduate teachers, accomplished teachers, and leading teachers (AITSL, 2015). For accreditation as a tertiary teacher education program, it is necessary to demonstrate that all the professional standards for graduate teachers are being met. There are two standards directly related to numeracy skills—one related to teaching, the other to the work of a teacher:

> 2.5: Literacy and numeracy strategies. Know and understand literacy and numeracy teaching strategies and their application in teaching areas.
>
> *(AITSL, 2015, p. 21)*

> 5.4: Interpret student data. Demonstrate the capacity to interpret student assessment data to evaluate student learning and modify teaching practice.
>
> *(AITSL, 2015, p. 24)*

For accreditation as a provider of teacher education, changes were needed in the teacher education programs offered at Monash University in line with the AITSL professional standards at the graduate teacher level. In this chapter, we provide an example of online resources, the Self-Help Kiosks, which were developed to support pre-service teachers to prepare for the numeracy demands of their profession.

19.2 SELF-HELP KIOSKS

In this section, we begin by providing information about the context in which the Self-Help Kiosks (SHKs) were developed—the Master of Teaching program at Monash University. Then, we describe the rationale behind the development of the SHKs and provide information about their format and structure. We conclude by providing a detailed example of one SHK.

19.2.1 The Monash University Master of Teaching Program

For the two-year graduate level Master of Teaching (MTeach) program to be accredited, as well as to prepare pre-service teachers to meet the expectations of the Australian Curriculum, a new subject was devised, entitled Numeracy for Learners and Teachers (NLT). Underpinning the content of NLT was the Model of 21st Century Numeracy described by Goos, Geiger, and Dole (2014). NLT is a compulsory subject for all MTeach students, except those preparing to be Early Years teachers (birth to age eight), and was first taught in 2015.

All elementary teachers in Australia are generalists who teach mathematics. Researchers (e.g., Ernest, 1988; Norton, 2017) have found that elementary pre-service teachers, particularly those with weak mathematics backgrounds, are often anxious about mathematics, lack confidence with mathematics, and have low achievement in mathematics. Secondary pre-service teachers at Monash University focus on 2 of the 24 teaching specialisms offered (See Monash University, 2018 for details). Therefore, secondary teachers have discipline-specific expertise, and not all secondary pre-service teachers will have strong mathematics backgrounds.

To support the MTeach pre-service teachers studying NLT who may have had weaker mathematics backgrounds, we devised a series of online modules, named Self-Help Kiosks (SHKs), aligned with the numeracy expectations of the Australian Curriculum. We did not teach mathematics content in NLT, but students were encouraged to access the SHKs to review pertinent mathematics content. They were able to answer quizzes to check their understanding of the content presented.

19.2.2 Rationale

The rationale for creating the SHKs was to provide pre-service teachers with easy access to a collection of resources in order to enable revision of the mathematical content and skills that underpin numeracy. The SHKs were initially designed to support pre-service teachers enrolled in the compulsory MTeach subject, NLT, and were made available in Semester 1, 2015, the first year that the subject was taught. Feedback on the SHKs from pre-service teachers and those teaching the subject was very positive, and the SHKs were subsequently made available to other pre-service teachers enrolled in mathematics or numeracy units in the MTeach and BEd programs in Semester 1, 2016. This was achieved by copying the SHKs from the NLT Moodle site to the other units' Moodle sites. This approach proved unsatisfactory, as changes made in the SHKs in the NLT Moodle site for 2016 did not copy over to the Moodle sites of the other subjects. Thus, in 2017, a separate Moodle site was created in order to address this issue and to make the SHKs available to all pre-service teachers.

19.2.3 Structure

The Faculty of Education provided funds to support the development of the SHKs. Budget constraints, however, dictated the scope of what could be included in the first version. A research assistant was engaged to identify the key mathematical concepts, and realistic real-world applications of them, associated with each of the following four mathematics topics identified by the NLT teaching staff:

1. Proportional reasoning (fractions, decimals, percentages, ratios, and rates)

2. Basic algebra

3. Collecting and analyzing data

4. Representing and interpreting data

Since Monash University uses the Moodle online platform to support student learning and to communicate with students, Moodle was the logical environment in which to build the SHKs. For each key mathematical idea, the most appropriate resources were used to build the SHKs. Rather than re-inventing teaching materials, the research assistant was required to locate existing exemplary relevant online teaching resources. The NLT teaching staff reviewed the SHKs and provided feedback and suggestions for modifications before the SHKs were made available to the pre-service teachers enrolled in NLT.

At the end of each semester, all Monash University students complete university-mandated online evaluations of each subject they have studied. As well as a number of scorable items, students are asked to comment on what was best about the subject, how the subject might be improved, and anything else. In 2015, in response to the general open-ended question regarding positive aspects of the subject, several students commented on the SHKs. For example:

> I was initially a bit nervous about undertaking a unit on numeracy but the resources such as the self-help kiosks, the readings and the lectures made everything really clear. Inclusion of the online kiosks for extra help was very thoughtful and encouraging. I would like the self-help kiosks and resources to be available on Moodle beyond the course duration. There's so much to work through and it'd be a great resource to use once out on the job.

In response to the positive feedback received from the subject evaluations about the existing SHKs, extra funding was sought and approved to develop four more:

5. Space and location

6. Chance and probability

Welcome to the
Mathematics Self-Help Kiosks!

You are invited to review the materials in each section to brush up on your mathematical skills. While these kiosks were originally created to support students in EDF5017: Numeracy for Learners and Teachers, they are also a valuable resource for other units and for your preparations for the ACER Literacy and Numeracy Test for Initial Teacher Education Students. At the end of each section, there is a self-check quiz that allows you to check your understanding.

FIGURE 19.1   Screen capture of the SHK homepage (welcome message).

7. From VCE (Victorian Certificate of Education) study scores to ATAR (Australian Tertiary Admission Rank)*

8. Videos on the interpretation of national testing data and PowerPoint files (Phillip Holmes-Smith, a visiting expert, conducted workshops in 2015. The videos of the workshops and his PowerPoint presentations were made available as one SHK.)

As noted earlier, in 2017, the eight SHKs were mounted on their own Moodle site and made available to all pre-service teachers. The content of each SHK is reviewed regularly to ensure its currency, but the eight-SHK structure has remained consistent since 2017. The design of the Moodle homepage for the SHKs is shown in Figure 19.1, and the tile design (to access each SHK) is shown in Figure 19.2.

Each of the first six SHKs focusing on mathematical topics was comprised of the following sections:

a. Introduction

b. Key mathematical topics

c. Issues and misconceptions

d. Examples for teaching and learning

e. Self-check quiz

The purpose of the introduction was to provide a context for how the specific mathematical topic relates to numeracy by connecting it to citizenship, work, or

---

* Each state and territory in Australia oversees its own school completion certification. Monash University is located in the state of Victoria, where Grade 11 and 12 students complete the two-year Victorian Certificate of Education (VCE). For each subject studied at the Grade 12 level, students receive a study score. Each study score is standardized and scaled, and all study scores are then combined to produce an Australian Tertiary Admission Rank (ATAR). The ATAR is used for university selection. It was considered important that pre-service teachers should understand the process and underlying mathematics used to determine ATARs.

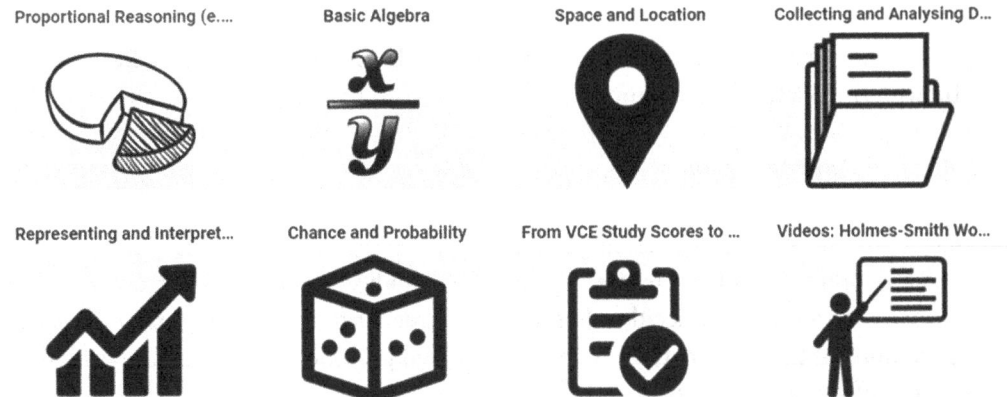

FIGURE 19.2   Screen capture of the SHK homepage (tiles to access individual SHKs).

personal/social life, the three dimensions of numeracy described by Goos et al. (2014). The introduction began with a quotation or video chosen to stimulate interest and motivate students to explore the topic further. For example, the introduction to the Basic Algebra SHK featured a link to Terry Moore's TED talk: *Why is 'x' the Unknown?* (TED, 2012).

The key mathematical concepts and skills for each of these six SHKs were provided as website links and/or video links to cater for pre-service teachers who preferred to read or those who preferred to watch and listen. We limited the mathematical demands to Grade 9 or 10 mathematics (in line with the *Australian Curriculum: Mathematics*—see ACARA, n.d.-b) and the contextual demands to settings that were relevant to teachers as professionals and to common everyday life experiences. The resources provided also depended on the most appropriate way to share the specific information. For example, a video was included showing step-by-step how to create graphs and charts in Excel. A brief overview was also provided of the issues and possible misconceptions that school students might have related to the mathematical topic; these also served to alert the pre-service teachers as to what to watch for in their own understanding. We also felt that it was important to provide examples of each mathematical topic that school students would encounter, as well as examples of pertinent teaching strategies/practices. Our aim was to raise the pre-service teachers' awareness that numeracy pervades their everyday experiences and to provide them with examples that they might use or adapt in their own teaching. Last, a short self-check quiz of 5 to 15 questions was available to enable the pre-service teachers to check their understanding of the topic and to identify areas needing further work.

The last two SHKs ("From VCE study scores to ATAR" and "Videos: Holmes-Smith workshops (Numeracy for PSTs)") focused on very context-specific numeracy demands for Australian teachers, that is, how Grade 12 results are converted into scores used for university entrance, and how to interpret national and international testing results. In some respects, the contents of these SHKs built on the mathematical skills included in the other six SHKS.

19.2.4 Details of One SHK: Representing and Interpreting Data

In this section, we describe more fully what was included in one of the SHKs, Representing and Interpreting Data, as an example.

19.2.4.1 Part A: Introduction

The introduction to the Representing and Interpreting Data SHK was utilized to remind pre-service teachers that statistics and data displays can be used and misused in communication. Developing skills for interpreting statistical analyses and data representations will help pre-service teachers to decide whether they agree with claims being made by others' interpretations of the information. A short, animated graphic was included to provide an illustration of the evolution of data representations. A link to Hans Rosling's video *200 Countries, 200 Years, 4 Minutes – The Joy of Statistics* (BBC, 2010) was also provided, as an example of a powerful form of modern communication of statistics.

19.2.4.2 Part B: Key Mathematical Topics

As a starting point for pre-service teachers to revise what they should have learned in school and/or to refresh their understanding, websites were provided where key ideas were summarized or explained. A few website links were included to provide an overview of statistics and to clarify the difference between descriptive statistics and inferential statistics. (Details on data collection and statistical calculations had been included in the prior topic, Collecting and Analysing Data.)

It was felt that videos would be the easiest way to demonstrate how to represent data, as well as how to read and interpret graphs, charts, and tables. Appropriate videos were located from publicly available sources such as university websites and the Khan Academy website. The video links for the three dimensions of Representing and Interpreting Data are shown in Figure 19.3. Understanding how graphs can be used to mislead people was deemed to illustrate the critical orientation of numeracy, and an appropriate video link was included.

19.2.4.3 Part C: Issues and Misconceptions

The list of potential issues and misconceptions related to statistical data and inferences drawn from them is lengthy. To avoid overwhelming the pre-service teachers, whilst also trying to encourage a critical orientation towards the consumption of "facts," only a short list was provided, which included ideas such as:

- Care must be taken when making inferences about the population from a small sample.
- Correlation doesn't always imply causation.

19.2.4.4 Part D: Examples for Teaching and Learning

As examples of school student learning, links were provided to potential sources of data that Australian school students might commonly encounter, such as sport websites,

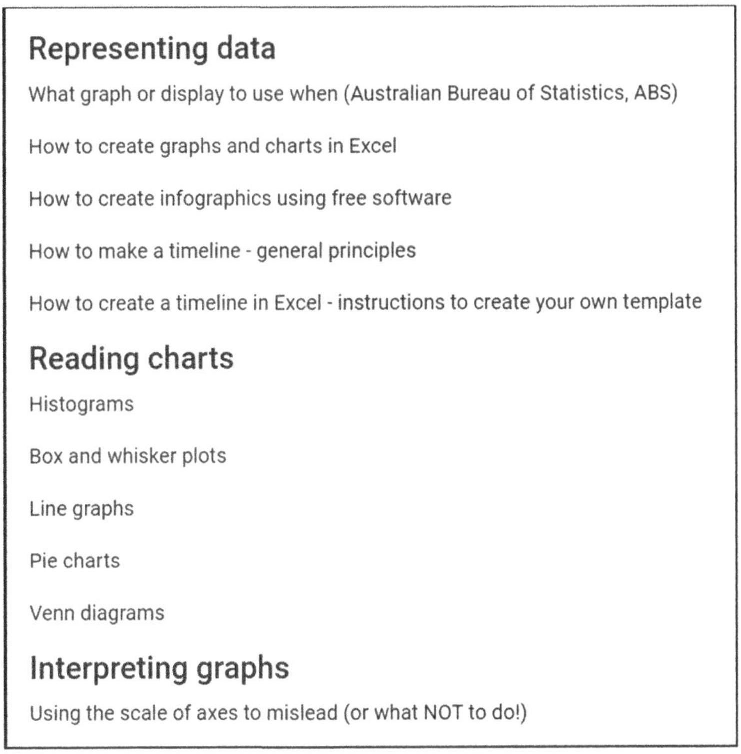

> ## Representing data
>
> What graph or display to use when (Australian Bureau of Statistics, ABS)
>
> How to create graphs and charts in Excel
>
> How to create infographics using free software
>
> How to make a timeline - general principles
>
> How to create a timeline in Excel - instructions to create your own template
>
> ## Reading charts
>
> Histograms
>
> Box and whisker plots
>
> Line graphs
>
> Pie charts
>
> Venn diagrams
>
> ## Interpreting graphs
>
> Using the scale of axes to mislead (or what NOT to do!)

FIGURE 19.3   Screen capture of the list of videos for the key aspects of the Representing and Interpreting Data SHK.

newspaper websites, and opinion polls. A link to the Australian Bureau of Statistics (ABS) website was also included. The ABS is the national statistics agency of the Australian government, and ABS data are often referenced in the media. We considered it important for pre-service teachers to learn where to locate reliable source data to enable informed critique.

Examples related to the work of teachers were also provided. Website links were chosen to address important issues such as what teachers and parents know and believe about technology use in the classroom, as well as how teachers can interpret the results and reports from international and national testing regimes.

19.2.4.5 Part E: Self-Check Quiz

The resources in each SHK provided an overview and starting point for pre-service teachers who felt that they needed to refresh their mathematics knowledge and skills. The self-check quiz gave them an opportunity to assess their understanding of the concepts covered, and to identify any areas for further learning. In Figures 19.4 and 19.5, screen captures of self-check quiz questions are provided as examples.

For future development and/or refinements of the SHKs, we felt it important to find out which SHKs had been used, and what the pre-service teachers who had used them had gained.

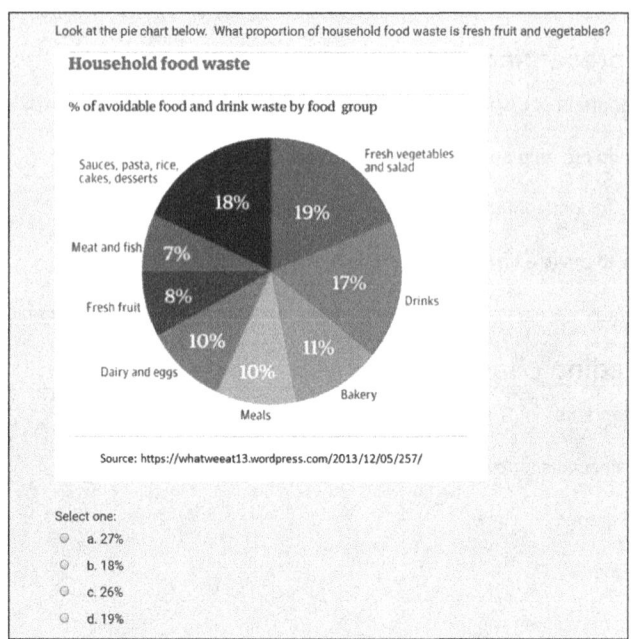

FIGURE 19.4   Screen capture of Question #2 from the Representing and Interpreting Data self-check quiz.

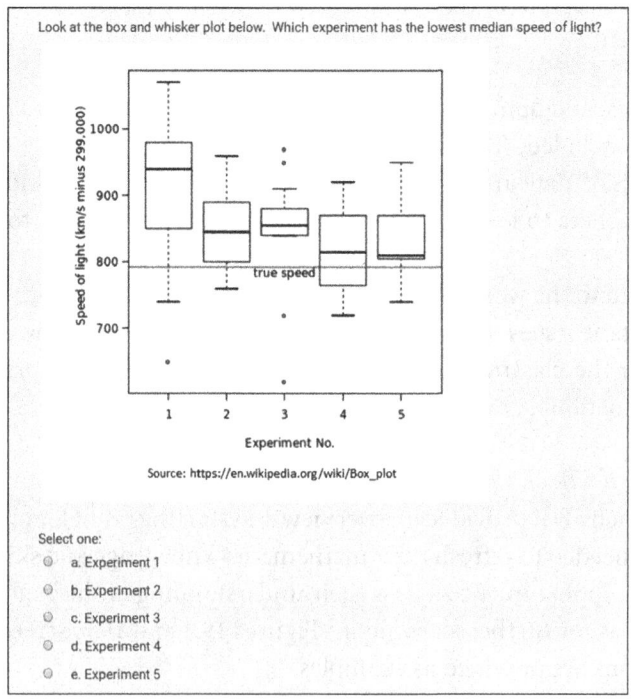

FIGURE 19.5   Screen capture of Question #4 from the Representing and Interpreting Data self-check quiz.

19.3 STUDENTS' REACTIONS TO AND USE OF THE SHKS

We gathered data on students' reactions to and use of the SHKs from a variety of sources. We begin by discussing SHK-related findings from questionnaires and interviews about the LANTITE that were conducted as part of a broader study within the Faculty of Education. Then, we discuss findings from an SHK-focused questionnaire, conducted in 2018, of students who had enrolled in the SHKs Moodle site. Finally, we discuss insights gained from our analysis of Moodle usage data from 2017 onwards, when the SHKs became a separate Moodle site.

19.3.1 LANTITE Questionnaire and Interviews

Monash University is a prestigious Australian university, and students accepted into the teacher education programs offered by the Faculty of Education are generally very high-achieving. The majority of students passed the LANTITE on their first attempt. However, there are some students who struggle with literacy (mainly international students) and some with weak mathematics backgrounds. A Moodle site was developed to assist all pre-service education students to prepare for the literacy and the numeracy components of the LANTITE. A link to the SHKs was included on this site.

A link to an anonymous online questionnaire about the LANTITE was included at the top of the LANTITE preparation Moodle site. Students were asked a variety of questions about their experiences preparing for and completing the LANTITE test. They were asked about the resources they had used to prepare for the numeracy component, and one of the listed resources was the SHKs. Of the 120 students who responded in 2017 to questions about their preparation for the numeracy component of LANTITE test, 80 provided feedback on their usage of the SHK resources. The majority (61%) indicated that they had not used the SHKs in their test preparation, while 28% indicated that they had used the SHKs a little or a moderate amount, and 11% used the SHKs a lot. At the end of the questionnaire, students were invited to provide their contact details if they were willing to participate in a follow-up interview, and 12 interviews were subsequently conducted. In the interviews, students were asked about their experiences in preparing for the numeracy component of the LANTITE test. Only one student mentioned the SHKs:

> If I came across something that I was like, "Oh, I can't quite remember the best method" […] I would just go back and look at the video like, "Oh, that's how you do it." […] and when I got it, and I saw all those YouTube videos and like the maths helpdesk or kiosk, whatever it was called, and the actual practice tests, I was like, "Oh, this is great."

Because the interview protocol did not include a question about specific resources used by the students in preparing for the numeracy component of the LANTITE test, it is likely that more of the interviewees did actually access the SHKs.

19.3.2 SHKs Questionnaire

In 2018, a questionnaire link was provided on the SHKs Moodle site, and students were made aware of it through an announcement on the site, as well as announcements about the SHKs on the Moodle sites of individual mathematics/numeracy subjects studied by pre-service teachers. The questionnaire was comprised of demographic questions (e.g., gender, age, program of study), as well as open-ended and closed questions about the students' experiences with the SHKs, such as how they learned about the site and which aspects they used and found useful.

The SHKs questionnaire was completed by 21 participants, most of whom (71.4%) were preparing to be Primary or Primary/Secondary teachers. Interestingly, the majority (*n* = 14) of participants reported that they had studied university-level mathematics. The most common way that the participants learned about the SHKs was from a lecturer (*n* = 9), followed by noticing the SHKs icon on another Moodle site (*n* = 7) and being informed about them through LANTITE resources/workshops (*n* = 5). The participants reported learning about the SHKs as a consequence of being enrolled in one (or more) of five different subjects, either in class or via the subject's Moodle site; NLT was the modal response (*n* = 4).

Eight of the 17 respondents who discussed why they enrolled in the SHKs Moodle site stated that they did so to help with their coursework, while four respondents said that they enrolled as a way to prepare for the numeracy component of the LANTITE test. The SHKs that were reported to have been used the most often were Basic Algebra, Proportional Reasoning, and Representing and Interpreting Data, topics that all feature prominently in the pre-service teacher education mathematics/numeracy subjects, as well as on the numeracy component of the LANTITE test. The participants reported that all aspects of the SHKs (e.g., videos, self-check quizzes) were helpful to their learning.

19.3.3 Moodle Usage Data

Since February of 2017, 181 students have enrolled in the SHKs Moodle site. Of these students, 126 completed a brief demographic questionnaire to be able to access all of the materials on the SHK's Moodle site. Approximately two-thirds of these respondents were MTeach students (69.0%), with the rest (31.0%) being Bachelor of Education (BEd) students. Other than a very small proportion of those studying to be teachers of students in the Early Years (birth to age eight; 1.6%), the respondents were fairly evenly spread across the other streams: 25.4% Early Years/Primary (birth to Grade 6), 20.6% Primary (Foundation to Grade 6), 32.5% Primary/Secondary (Foundation to Grade 12), and 19.8% Secondary (Grades 7–12).

As described earlier, there were eight SHKs, each of which contained a range of resources. Questionnaire responses indicated that the self-check quizzes were the most commonly accessed resources. This popularity may be because they are interactive resources; another explanation might be that the students wanted to check their understanding first, and then if they struggled with the quizzes, they would engage with the other resources for the particular SHK topic. The numbers of students attempting and completing the self-check quizzes by SHK topic are shown in Table 19.1.

As can be seen in Table 19.1, for each SHK, the vast majority of students who attempted the self-check quiz completed it. The self-check quizzes for the first two SHKs were accessed by far more students than the self-check quizzes for any of the other four. The difference may be due to students' perceived greater need to review these two topics, as they often are

TABLE 19.1   Users of SHK Self-Check Quizzes, by Topic

SHK Mathematical Topic	Number Attempting Quiz	Percentage Completing Quiz
Proportional reasoning	126	71.4%
Basic algebra	108	86.1%
Space and location	75	80.0%
Collecting and analyzing data	53	83.0%
Representing and interpreting data	45	100.0%
Chance and probability	57	84.2%

challenging for students, or it may simply be that students worked through the SHKs in order and ran out of time or energy to complete the others.

After the self-check quizzes, the website links were the most commonly used resources in each mathematical SHK. Again, the first two SHKs—Proportional Reasoning and Basic Algebra—were used by far more students than the other four mathematical SHKs, with 25 and 20 students, respectively, accessing the website links for these SHKs (compared to between 4 and 11 students for the others).

The files from the Phillip Holmes-Smith workshops were also widely accessed, possibly because there is an assignment in NLT that involves analyzing data from the Australian National Assessment Program – Literacy and Numeracy (NAPLAN; see National Assessment Program, 2016). For instance, between 128 and 135 students accessed the Part A and Part B videos and PowerPoint files from the workshops. In fact, all of the resources in this section were accessed by at least 50 students, indicating a demand by students for resources about analyzing and understanding large-scale assessment data in the Australian context.

19.4 CONCLUSIONS

As shown by the findings from the various data sources, the pre-service teachers at Monash University found the SHKs to be a useful resource to support their learning in mathematics/numeracy subjects, as well as in their preparation for the numeracy component of the LANTITE test. In particular, the self-check quizzes were used by many of the students enrolled in the SHKs Moodle site. This suggests that students are keen to test their knowledge and understanding, possibly as a strategy to guide their preparation for the LANTITE test and their classes, as well as for reviewing class material. Another likely reason for the popularity of the self-check quizzes was their interactive nature; researchers (e.g., Phillips, 2005) have suggested that online learners prefer active forms of learning such as quizzes to passive forms of learning such as reading information. Elementary pre-service teachers may have a particular need and/or desire to revise the fundamental mathematics topics encompassed by the SHKs. Interestingly, some pre-service teachers with secondary mathematics specialisms indicated that they had drawn on resources in the SHKs as support materials for students without strong foundations in elementary-level mathematics whom they had encountered during teaching placements in schools.

The SHKs are unique mathematics support resources in that they are specifically targeted to pre-service teachers, rather than to students enrolled in mathematics or statistics subjects/degrees, or the general public. Since the SHKs go beyond simply providing mathematical content to review, they have the potential to support pre-service teachers to review mathematics for their own understanding and build confidence to teach mathematics in future, as well as to foster the numeracy development of their future students across any curriculum discipline that they may be required to teach. Furthermore, building mathematical understanding by using the SHK materials can help the Monash pre-service teachers to succeed on the numeracy component of the LANTITE test, and to demonstrate their capacity to meet the Australian Professional Standards for Teachers, thus meeting the expectations of their profession. By using self-directed learning resources like the SHKs, pre-service teachers, particularly those with limited mathematical background or anxiety over their mathematical skills, can develop a richer understanding of, and appreciation for, mathematics and its relevance and place in all curricular disciplines. These pre-service teachers can also become more numerate members of society and the workforce, as well as potential mentors and advocates for others in their profession.

Resources have been provided in Australian universities to students in mathematics and statistics (e.g., MacGillivray, 2008). See University of Melbourne (n.d.) for a specific example of resources for first-year mathematics students. There are also excellent publicly available online mathematics learning resources (e.g., Khan Academy and Wootube). Instructors at other universities concerned about the background mathematics content knowledge of their students, whether pre-service teachers or students in other fields, may like to consider creating tailored online mathematics learning resources similar to the SHKs that would be available to their students.

ACKNOWLEDGMENTS

We wish to acknowledge the support provided by Brendan Boniface, e-Education Support Administrator at Monash University, in the design of the Self-Help Kiosks Moodle site, and in the retrieval of the Moodle site usage data.

REFERENCES

Australian Council for Educational Research. (n.d.). *Literacy and numeracy test for initial teacher education students: Sample questions.* Retrieved from https://teacheredtest.acer.edu.au/files/Literacy_and_Numeracy_Test_for_Initial_Teacher_Education_students_-_Sample_Questions.pdf.

Australian Council for Educational Research. (2017). *Test content.* Retrieved from https://teacheredtest.acer.edu.au/about/test-content.

Australian Curriculum, Assessment and Reporting Authority. (n.d.-a). *Australian curriculum.* Retrieved from https://www.australiancurriculum.edu.au/.

Australian Curriculum, Assessment and Reporting Authority. (n.d.-b). *Mathematics.* Retrieved from https://www.australiancurriculum.edu.au/f-10-curriculum/mathematics/.

Australian Curriculum, Assessment and Reporting Authority. (n.d.-c). *Numeracy.* Retrieved from https://www.australiancurriculum.edu.au/f-10-curriculum/general-capabilities/numeracy.

Australian Government Department of Education and Training. (2017). *Literacy and numeracy test for initial teacher education students.* Retrieved from https://www.education.gov.au/literacy-and-numeracy-test-initial-teacher-education-students.

Australian Institute for Teaching and School Leadership. (2015). *Accreditation of initial teacher education programs in Australia: Standards and procedures.* Retrieved from https://www.ait sl.edu.au/docs/default-source/general/accreditation-of-ite-programs-in-australia.pdf?sfvrs n=3013e33c_2.

BBC. (2010, November 26). *Hans Rosling's 200 countries, 200 years, 4 minutes - The Joy of Stats* [Video file]. Retrieved from https://www.youtube.com/watch?v=jbkSRLYSojo.

Cockcroft, W. H. (1982). *Mathematics counts: Report of the committee of inquiry into the teaching of mathematics in schools.* London, England: Her Majesty's Stationery Office. Retrieved from http://www.educationengland.org.uk/documents/cockcroft/cockcroft1982.html.

Common Core State Standards Initiative. (2018a). *Common core state standards for mathematics.* Retrieved from http://www.corestandards.org/wp-content/uploads/Math_Standards1.pdf.

Common Core State Standards Initiative. (2018b). *Standards in your state.* Retrieved from http://www.corestandards.org/standards-in-your-state/.

Crowther, G. (1959). *The Crowther report. 15 to 18. A report of the Central Advisory Council for Education (England).* London, England: Her Majesty's Stationery Office. Retrieved from http://www.educationengland.org.uk/documents/crowther/.

Department of Education. (2014a). *Areas of numeracy covered by the professional skills test.* Retrieved from http://sta.education.gov.uk/system/resources/W1siZiIsIjIwMTQvMTEvMjkvMTdfND JfMzZfMzY0X0FyZWFzX29mX251bWVyYWN5LnBkZiJdXQ/Areas%20of%20numerac y.pdf.

Department of Education. (2014b). *Glossary of numeracy terms.* Retrieved from http://sta.educ ation.gov.uk/system/resources/W1siZiIsIjIwMTQvMTEvMjkvMTdfNDJfMzZfMzEzX051b WVyYWN5X2dsb3NzYXJ5LnBkZiJdXQ/Numeracy%20glossary.pdf.

Department of Education. (2014c). *Statutory guidance: National curriculum in England: Framework for Key Stages 1 to 4.* Retrieved from https://www.gov.uk/government/publications/nationa l-curriculum-in-england-framework-for-key-stages-1-to-4/the-national-curriculum-in-en gland-framework-for-key-stages-1-to-4.

Department of Education. (2014d). *Statutory guidance: National curriculum in England: Mathematics programmes of study.* Retrieved from https://www.gov.uk/government/publi cations/national-curriculum-in-england-mathematics-programmes-of-study/national-curri culum-in-england-mathematics-programmes-of-study.

Department of Education. (2018a). *The numeracy professional skills tests.* Retrieved from http://sta .education.gov.uk/professional-skills-tests/numeracy-skills-tests.

Department of Education. (2018b). *Professional skills tests.* Retrieved from http://sta.education.gov. uk/.

Ernest, P. (1988). The attitudes and practices of student teachers of primary school mathematics. In A. Borbás (Ed.), *Proceedings of the 12th international conference on the psychology of mathematics education* (Vol. 1, pp. 288–295). Veszprém, Hungary: OOK Printing House.

Expert Panel on Student Success in Ontario. (2004). *Leading math success: Mathematical literacy Grades 7–12.* Toronto, Canada: Queen's Printer for Ontario.

Forgasz, H. J., Leder, G., & Hall, J. (2017). Numeracy across the curriculum in Australian schools: Teacher education students' and practicing teachers' views and understandings of numeracy. *Numeracy, 10*(2), 1–20. doi:10.5038/1936-4660.10.2.2.

Goos, M., Geiger, V., & Dole, S. (2014). Transforming professional practice in numeracy teaching. In Y. Li, E. Silver, & S. Li (Eds.), *Transforming mathematics instruction: Multiple approaches and practices* (pp. 81–102). New York, NY: Springer.

MacGillivray, H. (2008). *Learning support in mathematics and statistics in Australian universities: A guide for the university sector.* Retrieved from http://www.mathcentre.ac.uk/resources/up loaded/guide--altc-learning-support-in-maths-and-stats.pdf.

Madison, B. L. (2015). Quantitative literacy and the common core state standards in mathematics. *Numeracy, 8*(1), 1–13. doi:10.5038/1936-4660.8.1.11.

Monash University. (2018). *Secondary teaching specialist areas and prerequisites*. Retrieved from https://www.monash.edu/education/future-students/teacher-education/master-of-teaching/secondary-specialisms.

National Assessment Program. (2016). *NAPLAN*. Retrieved from https://www.nap.edu.au/naplan.

National Numeracy. (2019a). *National numeracy for everyone, for life*. Retrieved from https://www.nationalnumeracy.org.uk/.

National Numeracy. (2019b). *What is numeracy?* Retrieved from https://www.nationalnumeracy.org.uk/what-numeracy.

Norton, S. J. (2017). Primary mathematics trainee teacher confidence and its relationship to mathematical knowledge. *Australian Journal of Teacher Education, 42*(2), 47–61. doi:10.14221/ajte.2017v42n2.4.

Ontario Ministry of Education. (2005a). *The Ontario curriculum Grades 1–8: Mathematics (Revised)*. Toronto, Canada: Queen's Printer for Ontario.

Ontario Ministry of Education. (2005b). *The Ontario curriculum Grades 9 and 10: Mathematics (Revised)*. Toronto, Canada: Queen's Printer for Ontario.

Ontario Ministry of Education. (2007). *The Ontario curriculum Grades 11 and 12: Mathematics (Revised)*. Toronto, Canada: Queen's Printer for Ontario.

Organisation for Economic Co-operation and Development. (n.d.). *PIAAC numeracy – Sample items*. Retrieved from http://www.oecd.org/skills/piaac/Numeracy%20Sample%20Items.pdf.

Organisation for Economic Co-operation and Development. (2017). *PISA for development brief*. Retrieved from https://www.oecd.org/pisa/aboutpisa/9-How-PISA-D-measures-math-literacy.pdf.

Organisation for Economic Co-operation and Development. (2018a). *Programme for international student assessment*. Retrieved from http://www.oecd.org/pisa/.

Organisation for Economic Co-operation and Development. (2018b). *Survey of Adult Skills (PIAAC)*. Retrieved from http://www.oecd.org/skills/piaac/.

Phillips, J. M. (2005). Strategies for active learning in online continuing education. *The Journal of Continuing Education in Nursing, 36*(2), 77–83. Retrieved from https://www.healio.com/nursing/journals/jcen.

Statistics Canada. (2016). *Population by year, by province and territory*. Retrieved from http://www.statcan.gc.ca/tables-tableaux/sum-som/l01/cst01/demo02a-eng.htm.

TED. (2012, June 6). *Why is 'x' the unknown? Terry Moore* [Video file]. Retrieved from https://www.youtube.com/watch?v=YX_OxBfsvbk&feature=youtu.be.

The University of Melbourne. (n.d.). *Reading packs*. Retrieved from https://ms.unimelb.edu.au/study/mslc/oasis/reading-packs.

Vacher, H. L. (2014). Looking at the multiple meanings of numeracy, quantitative literacy, and quantitative reasoning. *Numeracy, 7*(2), 1–14. doi:10.5038/1936-4660.7.2.1.

PART 5

Commentary

CHAPTER 20

Online Mathematics Education, the Good, the Bad, and the General Overview

Sarah Ferguson

CONTENTS

20.1 INTRODUCTION

In current education systems, types of classrooms vary widely. Traditional classrooms are giving way to web-facilitated, hybrid, and online courses as learning through technology is becoming a mainstream learning modality. Figure 20.1 shows Allen and Seaman's (2016) course classifications based on the percentage of content delivered in through online means.

Allen and Seaman (2016) share the Integrated Postsecondary Education Data System (IPEDS) definition of distance education, stating that distance education is:

> Education that uses one or more technologies to deliver instruction to students who are separated from the instructor and to support regular and substantive interaction between the students and the instructor synchronously or asynchronously.

(p. 41)

Proportion of Content Delivered Online	Type of Course	Typical Description
0%	Traditional	Course where no online technology used — content is delivered in writing or orally.
I to 29%	Web Facilitated	Course that uses web-based technology to facilitate what is essentially a face-to-face course. May use a learning management system (LMS) or web pages to post the syllabus and assignments.
30 to 79%	Blended/Hybrid	Course that blends online and face-to-face delivery. Substantial proportion of the content is delivered online, typically uses online discussions, and typically has a reduced number of face-to-face meetings.
80+%	Online	A course where most or all of the content is delivered online. Typically have no face-to-face meetings.

FIGURE 20.1   Course classifications.

Allen and Seaman (2016) have been tracking the growth and acceptance of online learning modalities at the higher education level since 2003. In their 2016 report on online learning in higher education in the United States, Allen and Seaman reported that 5.8 million students embarked on online courses during the fall 2014 semester. Of these 5.8 million students, 2.85 million were purely online students, meaning they were taking all of their coursework in an online format, while 2.97 million were taking some, but not all, of their courses online (Allen & Seaman, 2016). Additionally, Allen and Seaman (2016) reported that, in the fall 2014 semester, 69.2% of degree-granting institution offered distance education courses.

Online learning presents unique opportunities, experiences, and challenges for teachers and students. When referencing online education, it is important to remember that merely being online does not constitute an education experience. Experienced online teachers are finding different types of support and assistance are needed as students navigate various issues and problems while learning in an online environment (Burden, 2008; Taylor & Galligan, 2006). Courses that are taught through an online platform provide different opportunities for students and are often organized differently than traditional face-to-face courses (Moore & Kearsley, 2012). In an online environment, blending connectivity and personal learning freedom becomes a focal point as content delivery and access to information concerns are built into many learning system designs (Garrison, 2011).

Online learning requires a change in learning traditions for students as well. Successful online students are often self-disciplined, acutely focused on their coursework, self-starters, and comfortable with online interaction. In online courses, students must learn to use technology to communicate and fully express their ideas, often using only written text. Students who possess these characteristics are more likely to complete their online course endeavors (Smith & Ferguson, 2005). From the teacher perspective, online courses necessitate specific teaching practices; online teachers must be willing to revise their teaching techniques to adapt to online course properties and needs. While Allen and Seaman (2010, 2011) report that faculty and administration support of online learning is growing, a reluctance to revise teaching practices remains present among faculty members (Beaudoin, 2002).

As seen throughout the chapters of this book, teaching mathematics online presents a unique set of attributes. From course design (as seen in Chapters 3, 4, 16, 12, and 18) to upholding course integrity (as seen in Chapter 5), planning an online course requires focusing on attributes that often are not components of face-to-face course preparations. To explore the challenges that online course instructors face, a survey of practitioners was sent to online mathematics educators through a list serv, social media, and by word of mouth. Forty-six responses were received from online mathematics educators, and responses were tallied to provide substantiation for the themes explored in this chapter. All survey participants were experienced online mathematics educators in middle school through post-secondary settings. Experience levels ranged from less than one year of online teaching experience up to 15–20 years of online teaching experience with a mean between four and six years of online mathematics education experience.

In the study of practitioners, participants were asked how they teach online mathematics courses differently than face-to-face mathematics courses. One participant commented "I would argue that there is a need for a 'different' way to teach online math other than just recreating the F2F classroom in an online setting" while others commented that they utilize different testing techniques, supply video lectures for students to watch, schedule times to talk one on one with students, rely more heavily on written communication, implement stronger structure and more detailed syllabi to ensure students are able to progress through their coursework independently and uninhibited, and redesign activities around available online engagement tools. Throughout the remainder of this chapter, additional struggles, benefits, and teaching modifications will be explored.

Throughout the chapters of this book, some unique online teaching techniques are explored. In Chapter 17, Sedaghatjou, Kaur, and Kagizmanli explore interactive embedded diagrams and pre-service teachers' use of 2D and 3D modeling tools to enhance mathematical understanding. Chapter 16 explores the use of Open Educational Resources (OER) to counteract the perceived challenges of teaching mathematics in an online environment. In Chapter 16, Mainali discusses the spectrum of OER available while also exploring their effectiveness and utilization styles.

In Chapter 14, VanDieren, Moore-Russo, and Seeburger discuss the use of CalcPlot3D as a dynamic visualization activity that can be implemented into calculus courses from both the teacher and student perspectives. Continuing on the theme of student content mastery, in Chapter 15, Shahbazi discusses the unique learning modules the University of Toronto has implemented to assist students with improving both their foundational and advanced mathematics skills. Chapter 15 details the learning modules and the use of the Math In Action journal to disseminate students' work. Like the modules discussed in Chapter 15, Chapter 19 focuses on "Self-Help Kiosks" that were developed by Monash University in Australia to support pre-service teachers in their quest to review, enhance, and reinforce their mathematics skills. To promote communication within the online classroom. Chapters 8 and 12 detail the use of discussions in online mathematics courses. Chapter 8 explores Odysseys2Sense (O2S) as an online discussion forum with unique collaborative tools while Chapter 12 argues the positives and negatives regarding discussions and explores strategies for successfully implementing discussions in mathematics courses.

20.2 BENEFITS OF TEACHING MATHEMATICS ONLINE

Technology changes the role of teachers, the role of students, and the manner through which information is passed in educational settings (Kelly, 2003). Kelly (2003) states, "A characteristic of the information age is that knowledge is more widely held, openly shared, and easily accessed" (p. 1038). As online education opportunities become increasingly prevalent, desired, and promoted, the question is raised, "Do the benefits of online mathematics education opportunities outweigh the difficulties?" Although there are struggles associated with teaching mathematics online, Chapters 9, 12, and 19 explore some of the benefits of online mathematics instruction. In the survey of practitioners, participants were asked "what benefits do you see of online mathematics courses?" Comments provided by the survey participants were reviewed and the key works were extrapolated. Figure 20.2 shows a word cluster depicting the responses.

Some participants listed multiple benefits in their survey response; of the 62 benefits listed, 48 mentioned pacing, accommodating student needs, or flexibility, suggesting the largest benefit on teaching mathematics online, from the instructor perspective, is the ability for students to progress through content at a pace aligned to their learning goals with structured flexibility to accommodate each student's unique learning needs. This result is in keeping with Allen and Seaman's (2010) argument that online instruction breaks down the barriers of time and place to allow students to work on their scholastic endeavors when they are able, while receiving the educational opportunities they desire, structured around other personal or professional needs.

One survey participant commented that the most beneficial aspect of online mathematics courses is that "students are able to view multiple [procedural solution] ways in a lesson through videos and then worked examples," while another participant commented "students can re-watch instruction videos," enabling a customized learning experience that ensures the pacing of instruction is appropriate for student learning. Other participants

FIGURE 20.2 Benefits word cloud.

commented "students are able to re-read the math concepts multiple times" and "students can engage with their course work from anywhere under a multitude of diverse learning circumstances." One participant, who teaches online mathematics courses at the high school level, praised the ability online mathematics education gives students in remote areas, or who attend smaller schools, to have access to mathematics courses their school does not offer. A second online high school mathematics teacher commented that students can use mathematics course to "recover credit for the classes they failed but need to graduate, or students can take AP courses which are not offered at their school," signifying the boundless limits of information acquisition learning online enables. These participant comments are substantiated by Braude and Merrill's (2013) research in which they suggested pausing online lectures, reviewing and replaying online material, and alleviation of peer pressure to determine pacing, are valuable components of online education opportunities.

While pacing and locational flexibility are praised as benefits of online learning, 21st-century skills and self-promoted learning were also cited as benefits of online mathematics courses by the study of practitioners' participants. One participant commented that taking online mathematics courses "teaches students to be self-sufficient and encourages [students to become] lifelong learners because they [learn they] can teach themselves through reading and research." Another participant further commented: "Students learn how to be independent and experience self-discovery of content and time-management."

Unlike a traditional classroom setting in which teachers move forward at a dictated lesson pace, online learning promotes individualized learning, content mastery, and provides students an opportunity to focus their attention on the content they need to more deeply examine and master (Kennedy, Ellis, & Oien, 2007). Encouraging student-led pacing can enhance student learning but can also be potentially detrimental to students who are not self-motivated to structure their time in their online course environment (Wadswroth, Husman, Duggan, & Pennington, 2007). In a qualitative study reviewing student learning strategies and motivation in an online developmental mathematics course, Wadsworth et al. (2007) found that time management was predictive of students' course performance. Wadsworth et al. (2007) reviewed three question surveys from a set of 89 developmental mathematics students and concluded that success in an online developmental mathematics course is partially dependent on "the learning strategies and self-efficacy of the students" (p. 12).

20.3 DIFFICULTIES OF TEACHING MATHEMATICS ONLINE

Online learning has many benefits, but learning mathematics online also causes some difficulties for teachers and students. Survey of practitioners' participants were asked to identify areas where teaching mathematics online posed difficulty. Initially, participants were provided a list of characteristics from which to select, as shown in Figure 20.3, and then, further into the survey, participants were asked to dictate what they considered to be the struggles they, as teachers, faced when teaching mathematics online. Two different styles of questions were asked to provide opportunities for cross referencing what practitioners identified as issues while also looking at what practitioners suggested as issues when unprompted from a list.

Which of the following do you see as a difficulty relative to teaching mathematics online? (Select all that apply.)

More Details

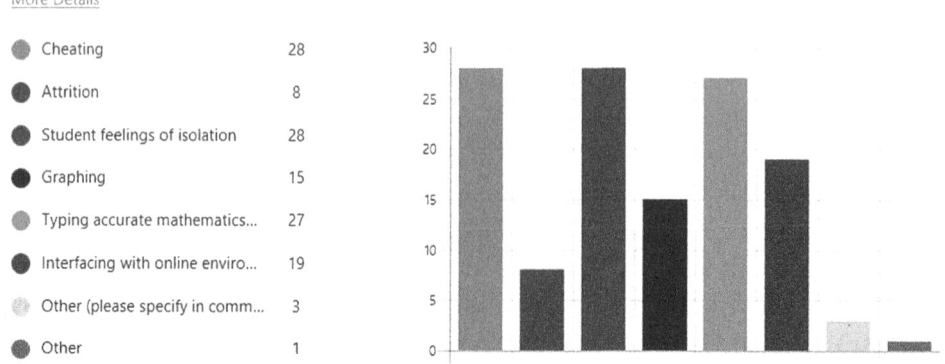

- ● Cheating — 28
- ● Attrition — 8
- ● Student feelings of isolation — 28
- ● Graphing — 15
- ● Typing accurate mathematics... — 27
- ● Interfacing with online enviro... — 19
- ● Other (please specify in comm... — 3
- ● Other — 1

FIGURE 20.3   Areas of difficulty.

As shown in Figure 20.3, cheating and perceived student feelings of isolation were the most frequently reported difficulties, with 28 votes each, and typing accurate mathematics received 27 votes. Online learning critics warn that education models centered upon technology place less emphasis on real-time decision making, stifle real-time oral discourse, necessitate new forms of student-monitoring practices, and foster a digital divide amongst students (Anderson, 2008). When the survey of practitioners' participants were asked about their least favorite aspect of teaching mathematics online, 23 participants noted the inability to or difficulty with building relationships with students. Participants expressed discouragement in their inability to get to know their students personally, to be able to provide specific and individual assistance when needed throughout the course content, feelings of depersonalization of the course experience, the inability to "look directly into students faces" and build a rapport with the students, and the inability to make personal connections to motivate and encourage students throughout their scholastic endeavors, not just through one select course. In addition to frustrations related to lacking personal engagement with students, three participants discussed items related to cheating as their least favorite aspects. One participant commented "you never really know who is doing the work" while another explained their dislike for "dealing with security and dishonesty issues" and another reflected on the connectedness between students' lack of interest and concern only for grades, rather than understanding, and subsequently encountered cheating by the students.

20.3.1 Cheating

As seen in Chapter 2 and Figure 20.3, cheating in an online mathematics course is an issue to be considered. Cheating is sometimes accidental and sometimes purposeful. Krause and Putnam (2016) report that students used online calculator applications to assist with correcting calculation errors, but also to avoid problem computations. Whether the use of a calculator is permitted in a course or not is school- and instructor-dependent, but Dorko commented in Chapter 2 that "when circumventing a problem, students would type a question directly into the calculator before attempting it on their own." Krause and Putnam (2016) also found that students looked at instructional websites and online forums when doing online homework, and the study explored different web-searching practices that students implemented when seeking assistance from online resources. In Chapter 2, Dorko described a scenario where one student searched "how do you find a second derivative?" to gain informational assistance to enable self-reflection and question completion, while another student searched online by entering the exact question verbiage into the search engine and was able to retrieve a full pre-completed solution. With the full solution provided, the student was able to copy and paste the answer directly into their homework, claiming the work as their own. Dorko commented that students sometimes use online calculation or solution tools as learning aids to help overcome obstacles encountered with embarking upon specific solution techniques, but iterate that without knowing the students, instructors are not able to identify instances when online calculation tools are used to negate problem-solving endeavors or to assist in learning while overcoming problem-solving obstacles.

20.3.2 Isolation

In addition to cheating, feelings of isolation are perceived as a hinderance to learning in an online environment. In Chapter 9, Kidd described feelings of isolation as a common difficulty expressed by online students (Bambara, Harbour, Davies, & Athey, 2009). Like the survey of practitioners' participants who commented on "missing face-to-face" interactions with students or longing to look at their students' faces to help build rapport or discern unspoken needs, students yearn for interactions with their teachers and peers. Kidd also comments on the effort that is necessary, by teachers and students, to facilitate interaction. In an online environment, instructors strive to create engagement opportunities and take leadership roles in facilitating both teacher-to-student and student-to-student engagement initiatives. One participant in the survey of practitioners commented: "Many students are scared of mathematics. When they start an online program and have to learn the system and learn the mathematics, this is difficult for them." In addition to learning the learning management system used by the course, the mathematics, and the course expectations, students are also learning to engage with their teacher and their peers in a virtual setting.

Teaching students to engage with their instructor and their peers is an important aspect of teaching online. In Chapter 12, communication in an online mathematics course was compared to communication in a face-to-face mathematics course with the conclusion that both course modalities relied on strong, successful communication, but clearly there are differences in the communication formats. In an online course, instructors spend time learning how to model, facilitate, and promote beneficial virtual discussions. One survey of practitioners' participant commented "I have to create ways for students to interact which involves changing how activities are presented." This participant went on to explain that they seek to use online tools and engagement strategies but are sometimes forced to "revert to process driven direct instruction due to an inability to create an online engagement."

20.3.3 Typing Math Online

While opportunities exist for students to use technology to write directly into documents on their computer or hand write and then scan or photograph their work before submitting, these are not universally accepted practices, and writing mathematics accurately is an issue in online mathematics courses. In Chapter 2, Dorko explored the works of Ellis, Hanson, Nuñez, and Rasmussen, 2015; Hauk and Segalla, 2005; Gage, Pizer, and Roth, 2003; Heenehan and Khorami, 2016; Leong and Alexander, 2014; Roth, Ivanchenko, and Record, 2008; and Yushau and Khan, 2014, all of whom note students' negative feedback related to difficulties inputting answers into online homework programs. Inputting mathematical symbols or equations can be very difficult and is not an issue isolated to just homework applications. One survey of practitioners' participant commented: "I think math is a more difficult course to take online, compared to other subjects. Students often get hung up with typing math symbols and graphing online." This difficulty of typing mathematics was showcased in the vignette explained in Chapter 9 related to the "e symbol."

Smith and Ferguson (2005) found that "online environments are not well adapted to mathematics" (p. 331). In their research, Smith and Ferguson (2005) found that learning management systems do not support complex mathematics diagrams or notations. Students often try to replace mathematical symbols with keyboard-enabled characters, resulting in online instructors and students seemingly communicating in code (Smith & Ferguson, 2005). In addition to mathematics symbols and equation structures, online mathematics curriculums must accommodate graphing. As seen in Chapter 2, researchers have found that students become frustrated with online platforms that are cumbersome to use or do not have a margin for error (Leong & Alexander, 2014). In Leong and Alexander's (2014) example, students became particularly frustrated when attempting to graph coordinate points, with non-fluid online systems, and ensuing incorrect responses due to small variations in point placement. Graphing tends to be a problematic concept for many students and is a pivotal component of many mathematics courses. Involving strategic competence, conceptual understanding, and relational observations, graphing is a representation activity which requires students to make meaning from abstract concepts through the use of anticipatory thinking (Cavanaugh, Gillan, Bosnick, Hess, & Scott, 2008). Differentiating between procedural understanding errors and online navigation issues, as described by Leong and Alexander (2014), compounds student and instructor frustrations when typing mathematics or utilizing some online mathematics tools.

20.4 FAVORITE ASPECT OF TEACHING MATH ONLINE

In addition to sharing their thoughts regarding the benefit and difficulties of teaching math online, in a survey of practitioners, participants were asked to explain their favorite aspects of teaching mathematics online; some participants listed multiple favorite aspects. From their perspective as instructors of online mathematics courses, learning flexibility and meeting individual student needs are their favorite aspects. Figure 20.4 shows a pie chart depicting the frequency of responses by survey participants.

Favorite Aspects of Teaching Math Online

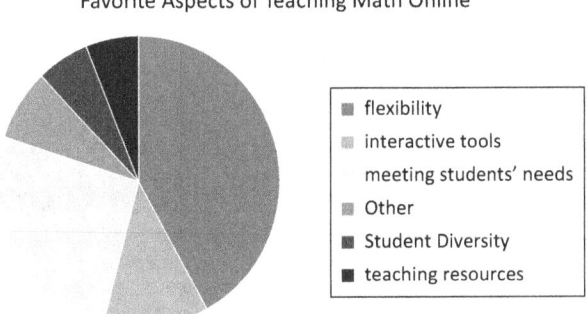

FIGURE 20.4   Practitioners' favorite aspects of teaching math online.

Flexibility was the most commented favorite attribute response and made up 42% of responses. The ability to meet unique student learning needs made up 26% of responses, and the availability of interactive online learning tools, such as discussion boards, online assessments, and multimedia capabilities, accounted for 12% of responses. Teachers appreciated the ability to work at times conducive to their personal schedules and the convenience of meeting students online. One survey participant commented "I can sit comfortably at my desk, eating a sandwich, drinking a soda, and wearing my workout clothing. I can also choose my hours." A second survey participant commented "I can work pretty much at my convenience" while multiple others wrote "flexibility of schedule" and "it allows me to be able to work from home." Additional survey participants mention that they have "more time to find resources to help students" and "time to focus on improving content." Participants commented on their enjoyment of meeting diverse student populations and interacting with students who they otherwise would not have in face-to-face settings. Participants commented on their enjoyment of "helping students succeed in mathematics who have struggled to do so otherwise" and that online teaching enables them to "do all my favorite parts of teaching-explaining mathematics and helping students." Related to meeting students' needs, other participants commented on their appreciation of being able to provide "immediate feedback on student learning" and the "the ability to give specific and detailed feedback to each student."

20.5 LEAST FAVORITE ASPECT OF TEACHING MATH ONLINE

While flexibility and meeting student needs were instructors' favorite aspects of teaching mathematics online, not having direct contact with students to build teacher-to-student interactions and build rapport and to actually see the students accounted for 51% of responses when survey participants were asked about their least favorite aspect of teaching math online. Figure 20.5 shows a word cloud of survey of practitioners' participant responses relative to their least favorite aspects of teaching math online.

In Chapter 9, Kidd discusses strategies to engage students you likely will not meet face-to-face. Kidd comments "the face-to-face interaction was [the students'] preferred method" and indicates that commonly students only take mathematics courses online because they are not able to always attend face-to-face class meetings. From the survey of practitioners'

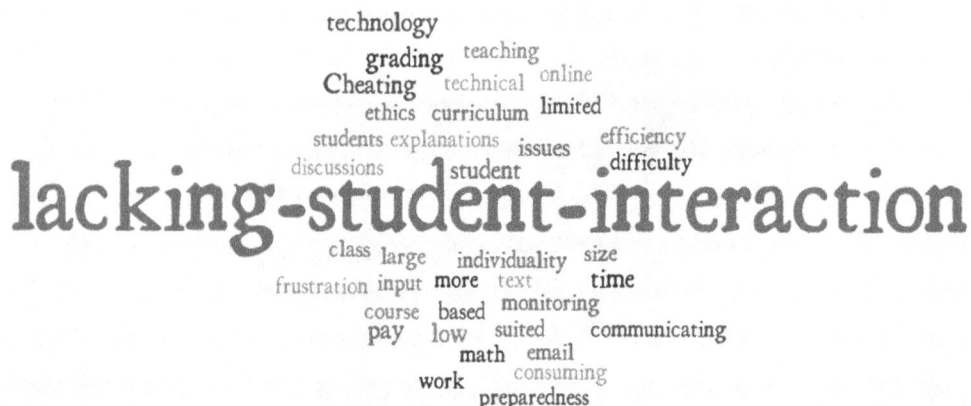

FIGURE 20.5   Word cluster of practitioners' least favorite aspects.

responses, it can be concluded that teachers also value face-to-face interaction with students. In their survey responses, practitioners commented "I don't like not being able to contact all my students," expressed dislike of "not seeing the actual students," and highlighted the difficulties online platforms present related to "building relationships and getting to know my students well." Survey participants also commented that they "miss looking directly into students faces, [and] the rapport that develops"; online teachers miss the "human interaction" aspect of teaching. Additional survey participants commented that it is more difficult them, as online instructors, to bond to "bond" with students and explain that "not knowing students personally" makes it hard to "help more specifically and individually when they don't initiate."

20.6 MATHEMATICS COURSES BEST AND POOREST SUITED FOR ONLINE STUDY

Recognizing the unique benefits and difficulties associated with teaching and learning mathematics online, the survey participants were asked to share their opinions related to which math courses they felt were best suited for online study. When looking at actual mathematics content and its adaptability to online study, survey participants had varying opinions. Developmental through advanced-level courses, such as middle school math, algebra 1, honors courses, pre-calculus, geometry, calculus, and statistics, were all listed as well-suited for online study. When asked which mathematics courses are not well-suited for online study, this same list of developmental through advanced-level courses appeared.

With no consensus regarding the best-suited and poorest-suited courses for online study, other factors, such as student characteristics, should be considered. Interestingly, comments often strayed from content themes to student attributes and course design attributes. For example, nine survey participants commented along the lines of the content is not the deciding factor but rather "it depends on how hard the student wants to work." Participants commented "I think any mathematics content can be presented [online]; success is more dependent on student study skills and computer literacy rather than content," "I think all math content is suitable for online study as long as the student is motivated," and "I am not sure it is content that matters but students motivation." One survey participant explained: "I think it depends on the student. If you have a motivated student, any course is doable. If you have a student who doesn't want to put in the effort required to learn independently, any course will be challenging." Following this same theme of student dependence, another survey participant commented: "I don't feel that one content is more suited for online studying than another. I feel it depends on the level of independence and maturity of the learner." Similar to participant responses, in a qualitative study of over 3,000 asynchronous online courses offered through the State University of New York (SUNY) system, Smith and Ferguson (2005) found that characteristics of successful online students are self-discipline, ability to focus on their coursework, self-starters, and comfort with online interaction. Smith and Ferguson argued that students who possess these characteristics are more likely to complete their online course endeavors

In addition to student characteristics, survey participants also highlighted course attributes as generating positive or negative course experiences. Participants commented "with properly created and maintained material, any mathematics content is suitable for online study," and "I feel that if a mathematics course has the right resources built in for the students and the students understand how to use them, any Algebra or Geometry based content can be studied online." Similarly, from the least suited responses, participants suggested: "I think highly technical mathematics may be difficult to study online, but a large piece of this puzzle relates to quality of instruction and how the course is put together."

When discussing mathematics courses least suited for online study, themes explored related to the difficulties of teaching mathematics online resurfaced. For example, one practitioner commented: "I think highly technical mathematics may be difficult to study online, but a large piece of this puzzle relates to quality of instruction and how the course is put together." Related to the graphing typing mathematics content themes explored previously, another practitioner commented: "Courses that involve a lot of graphing are hard for students to do the work online, also when showing a lot of work is required but must be typed." Another common theme related to mathematics courses least suited for online study centered around the lack of direct teacher-to-student interaction. Three practitioners commented that foundational level courses are difficult to implement online because of the scaffolding nature of mathematics content. One practitioner cautioned: "a great deal of important concepts are taught in [foundational] classes and students will struggle later if they don't get it." Similarly, another practitioner commented "I think foundational courses like Algebra 1 (and even MS math) can be difficult to do online. It's so, so important that students have a mastery in Algebra 1 in order to be successful in upper level math courses." A third comment along this theme was: "foundations need guidance and most students need someone present to help motivate."

The theme of teacher guidance did not end with foundational courses. In the practitioner's comment below, the need for positive teacher–student relationships in upper-level courses is iterated.

> Once you get into PreCalc and Calculus it becomes more challenging for students who do not have confidence their math skills as they often miss the basic concepts that layer in but without the daily teacher guidance they can get lost quickly.

Further responses highlighted the need for well-designed courses and quality content resources. Comments highlighting these needs include: "I feel that if a mathematics course does not have the right resources built in to help the students learn the mathematics content will not be suitable" and "I think highly technical mathematics may be difficult to study online, but a large piece of this puzzle relates to quality of instruction and how the course is put together."

20.7 CONCLUSION

Online learning presents a unique set of challenges for teachers and students. To utilize online learning to its fullest potential, it is necessary that both students and teachers take advantage of the unique opportunities available in an online classroom such as collaboration tools, internet-housed resources, and simulations. Merely being online does not constitute an education experience. Teachers are finding that students need different types of support and assistance with various issues or circumstances in an online environment (Burden, 2008; Taylor & Galligan, 2006). Courses that are taught through an online platform are organized differently than traditional face-to-face courses and provide different opportunities for students (Moore & Kearsley, 2012). In an online environment, content delivery and access to information concerns give way to blending connectivity and personal learning freedom (Garrison, 2011). From the students' perspective, online learning requires a change in learning traditions. In online courses, students must learn to use technology to communicate and fully express their ideas, often using only written text.

Throughout the chapters of this book, case studies, exploratory research, teaching techniques, benefits, and struggles related to teaching mathematics online have been explored. From MOOCs that are typically offered free of charge and cater to large groups of students seeking to work independently through course content (Chapters 6, 13 and 18) to focused assets which assist online mathematics learners (Chapters 12, 14, 15, 16, 17, and 19), online learning techniques are being implemented in K-12, collegiate, and graduate level environments to bring learning experience to individuals desiring to advance their cognitive mathematics understanding.

The way in which society functions and the manners in which people work, learn, live, and play are being increasingly impacted by technology. Smartphones, tablets, laptops, and the multitude of devices through which the internet can be accessed and communication promoted are enhancing the rate at which knowledge can be retrieved and transferred. Allen and Seaman (2014) report that over 7.1 million students are turning to online means for educational opportunities and Garrison (2011) comments "we are just beginning to discover and understand the extent to which these technologies will transform expectations for, and approaches to, learning" (p. 5). As students increasingly embark on online learning experiences, the benefits, struggles, and unique attributes of teaching and learning mathematics online will continue to evolve.

REFERENCES

Allen, I. E., & Seaman, J. (2010). *Class Differences: Online Education in the United States, 2010.* Sloan Consortium.

Allen, I. E., & Seaman, J. (2011). *Going the Distance: Online Education in the United States, 2011.* Newburyport, MA: Sloan Consortium.

Allen, I. E., & Seaman, J. (2014). Opening the Curriculum: Open Educational Resources in US Higher Education, 2014. *Babson Survey Research Group.*

Allen, I. E., & Seaman, J. (2016). *Online Report Card: Tracking Online Education in the United States.* Babson Survey Research Group.

Anderson, T. (2008). *The Theory and Practice of Online Learning.* Edmonton, AB: Athabasca University Press.

Bambara, C. S., Harbour, C. P., Davies, T. G., & Athey, S. (2009). The lived experiences of community college students enrolled in high-risk online courses. *Community College Review, 36*(3), 219–238.

Beaudoin, M. F. (2002). Distance education leadership: An essential role for the new century. *Journal of Leadership Studies, 8*(3), 131–144.

Burden, A. M. (2008). An overview of several popular web-enhanced instructional products: Part I. *The AMATYC Review, 29*(1), 69-80.

Braude, S., & Merrill, J. (2013). The chancellor's new robes: Online education. *Creative Education, 4*(7A2), 50.

Cavanaugh, C., Gillan, K. J., Bosnick, J., Hess, M., & Scott, H. (2008). Effectiveness of interactive online algebra learning tools. *Journal of Educational Computing Research, 38*(1), 67–95.

Ellis, J., Hanson, K., Nuñez, G., & Rasmussen, C. (2015). Beyond plug and chug: An analysis of Calculus I homework. *International Journal of Research in Undergraduate Mathematics Education, 1*(2), 268–287.

Gage, M., Pizer, A., & Roth, V. (1999). WeBWorK: An internet-based system for generating and delivering homework problems. http://archives.math.utk.edu/ICTCM/i/12/P003.html.

Garrison, D. R. (2011). *E-learning in the 21st Century: A Framework for Research and Practice.* New York, NY: Taylor & Francis.

Hauk, S., & Segalla, A. (2005). Student perceptions of the web-based homework program WeBWorK in moderate enrollment college algebra classes. *The Journal of Computers in Mathematics and Science Teaching, 24*(3), 229.

Heenehan, M. E., & Khorami, M. (2016). Students' reactions to the homework assessment system WeBWorK. *Mathematics and Computer Education, 50*(1), 42.

Kelly, B. (2003). The emergence of technology in mathematics education. *A History of School Mathematics, 2,* 1037–1084.

Kennedy, P., Ellis, W., & Oien, J. (2007). Mastery with meaning: Access to mathematics online. *Mathematics and Computer Education, 41*(2), 118–126.

Krause, A., & Putnam, R. (2016). Online calculus homework: The student experience. In T. Fukawa-Connelly, N. Infante, M. Wawro, & S. Brown (Eds.), *Proceedings of the 19th Annual Conference on Research in Undergraduate Mathematics Education,* (pp. 266–280). Pittsburgh, PA: West Virginia University.

Leong, K. E., & Alexander, N. (2014). College students attitude and mathematics achievement using web based homework. *Eurasia Journal of Mathematics, Science & Technology Education, 10*(6).

Moore, M. G., & Kearsley, G. (2012). *Distance Education: A Systems View of Online Learning.*

Roth, V., Ivanchenko, V., & Record, N.. (2008). Evaluating student response to WeBWorK, a web-based homework delivery and grading system. *Computers & Education, 50,* 1462–1482.

Smith, G. G., & Ferguson, D. (2005). Student attrition in mathematics e-learning. *Australasian Journal of Educational Technology, 21*(3), 323.

Taylor, J., & Galligan, L. (2006). Mathematics for maths anxious tertiary students: Integrating the cognitive and affective domains using interactive multimedia. *Literacy & Numeracy Studies, 15*(1), 23–43.

Wadsworth, L. M., Husman, J., & Duggan, M. A. (2007). Online mathematics achievement: Effects of learning strategies and self-efficacy. *Journal Of Developmental Education, 30*(3), 6–14.

Yushau, B., & Khan, M. A. (2014). Student perceptions of online homework in preparatory year pre-calculus courses. *International Journal of Mathematics Trends and Technology, 8*(1), 12–17.

Index

Printed in the United States
by Baker & Taylor Publisher Services